Deep Neural Networks for Multimodal Imaging and Biomedical Applications

Annamalai Suresh
Department of Computer Science and Engineering, Nehru Institute of Engineering and Technology, Coimbatore, India

R. Udendhran
Department of Computer Science and Engineering, Bharathidasan University, India

S. Vimal
Department of Information Technology, National Engineering College (Autonomous), Kovilpatti, India

A volume in the Advances in Bioinformatics and Biomedical Engineering (ABBE) Book Series

Published in the United States of America by
    IGI Global
    Medical Information Science Reference (an imprint of IGI Global)
    701 E. Chocolate Avenue
    Hershey PA, USA 17033
    Tel: 717-533-8845
    Fax: 717-533-8661
    E-mail: cust@igi-global.com
    Web site: http://www.igi-global.com

Library of Congress Cataloging-in-Publication Data

Names: Suresh, Annamalai, 1977- editor. | Udendran, R., 1992- editor. |
    Vimal, S., 1984- editor.
Title: Deep neural networks for multimodal imaging and biomedical
    applications / Annamalai Suresh, R. Udendran, and S. Vimal, editors.
Description: Hershey, PA : Medical Information Science Reference, [2020] |
    Includes bibliographical references and index. | Summary: "This book
    provides research exploring the theoretical and practical aspects of
    emerging data computing methods and imaging techniques within healthcare
    and biomedicine. The publication provides a complete set of information
    in a single module starting from developing deep neural networks to
    predicting disease by employing multi-modal imaging"-- Provided by
    publisher.
Identifiers: LCCN 2019057105 (print) | LCCN 2019057106 (ebook) | ISBN
    9781799835912 (hardcover) | ISBN 9781799835929 (ebook)
Subjects: MESH: Deep Learning | Multimodal Imaging--methods | Image
    Interpretation, Computer-Assisted--methods | Biomedical
    Technology--methods
Classification: LCC R855.3 (print) | LCC R855.3 (ebook) | NLM WN 26.5 |
    DDC 610.285--dc23
LC record available at https://lccn.loc.gov/2019057105
LC ebook record available at https://lccn.loc.gov/2019057106

This book is published in the IGI Global book series Advances in Bioinformatics and Biomedical Engineering (ABBE) (ISSN: 2327-7033; eISSN: 2327-7041)

British Cataloguing in Publication Data
A Cataloguing in Publication record for this book is available from the British Library.

For electronic access to this publication, please contact: eresources@igi-global.com.

Advances in Bioinformatics and Biomedical Engineering (ABBE) Book Series

Ahmad Taher Azar

Prince Sultan University, Riyadh, Kingdom of Saudi Arabi
and Benha University, Egypt

ISSN:2327-7033
EISSN:2327-7041

MISSION

The fields of biology and medicine are constantly changing as research evolves and novel engineering applications and methods of data analysis are developed. Continued research in the areas of bioinformatics and biomedical engineering is essential to continuing to advance the available knowledge and tools available to medical and healthcare professionals.

The **Advances in Bioinformatics and Biomedical Engineering (ABBE) Book Series** publishes research on all areas of bioinformatics and bioengineering including the development and testing of new computational methods, the management and analysis of biological data, and the implementation of novel engineering applications in all areas of medicine and biology. Through showcasing the latest in bioinformatics and biomedical engineering research, ABBE aims to be an essential resource for healthcare and medical professionals.

COVERAGE

- Data Mining
- Tissue Engineering
- Gene regulation
- Biomechanical Engineering
- Health Monitoring Systems
- Orthopedic Bioengineering
- Rehabilitation Engineering
- Finite Elements
- Protein Engineering
- DNA Sequencing

IGI Global is currently accepting manuscripts for publication within this series. To submit a proposal for a volume in this series, please contact our Acquisition Editors at Acquisitions@igi-global.com or visit: http://www.igi-global.com/publish/.

Titles in this Series

For a list of additional titles in this series, please visit:
http://www.igi-global.com/book-series/advances-bioinformatics-biomedical-engineering/73671

Biomedical and Clinical Engineering for Healthcare Avancement
N. Sriraam (Ramaiah Institute of Technology, India)
Medical Information Science Reference • © 2020 • 275pp • H/C (ISBN: 9781799803263) • US $285.00

Attractors and Higher Dimensions in Population and Molecular Biology Emerging Research and Opprtunities
Gennadiy Vladimirovich Zhizhin (Russian Academy of Natural Sciences, Russia)
Engineering Science Reference • © 2019 • 232pp • H/C (ISBN: 9781522596516) • US $165.00

Computational Models for Biomedical Reasoning and Problem Solving
Chung-Hao Chen (Old Dominion University, USA) and Sen-Ching Samson Cheung (University of Kentucky, USA)
Medical Information Science Reference • © 2019 • 353pp • H/C (ISBN: 9781522574675) • US $275.00

Medical Data Security for Bioengineers
Butta Singh (Guru Nanak Dev University, India) Barjinder Singh Saini (Dr. B. R. Ambedkar National Institute of Technology, India) Dilbag Singh (Dr. B. R. Ambedkar National Institute of Technology, India) and Anukul Pandey (Dumka Engineering College, India)
Medical Information Science Reference • © 2019 • 340pp • H/C (ISBN: 9781522579526) • US $365.00

Examining the Causal Relationship Between Genes, Epigenetics, and Human Health
Oscar J. Wambuguh (California State University – East Bay, USA)
Medical Information Science Reference • © 2019 • 603pp • H/C (ISBN: 9781522580669) • US $295.00

Expert System Techniques in Biomedical Science Practice
Prasant Kumar Pattnaik (KIIT University (Deemed), India) Aleena Swetapadma (KIIT University, India) and Jay Sarraf (KIIT University, India)
Medical Information Science Reference • © 2018 • 280pp • H/C (ISBN: 9781522551492) • US $205.00

Nature-Inspired Intelligent Techniques for Solving Biomedical Engineering Problems
Utku Kose (Suleyman Demirel University, Turkey) Gur Emre Guraksin (Afyon Kocatepe University, Turkey) and Omer Deperlioglu (Afyon Kocatepe University, Turkey)
Medical Information Science Reference • © 2018 • 381pp • H/C (ISBN: 9781522547693) • US $255.00

701 East Chocolate Avenue, Hershey, PA 17033, USA
Tel: 717-533-8845 x100 • Fax: 717-533-8661
E-Mail: cust@igi-global.com • www.igi-global.com

Table of Contents

Detailed Table of Contents

Chapter 1

Reyana A., Nehru Institute of Engineering and Technology, India
Krishnaprasath V. T., Nehru Institute of Engineering and Technology, India
Suresh A., Nehru Institute of Engineering and Technology, India

Critical health information is gathered, compiled, and analysed from image scanning, laboratory testing reports, health trackers, etc. Healthcare requires a support of real-time digital information for effective decision making. As industries bring better care to consumer health systems, providers achieve their goal in delivering better diagnosis and efficient treatment to consumers. This chapter contributes to the major health components—content, infrastructure, analytics, communication, and interfaces—forming the core of all health IT. The challenge is further exaggerated on the current issues related to health data and the contribution of IT systems in healthcare management. Further, the scenarios on intelligent care systems that are capable of early diagnosis are described. This impacts the future of global healthcare services by rapidly increasing the development and bridging the gap between the personal and relevant datasets creating a new era in the world of healthcare.

Chapter 2

V. R. S. Mani, National Engineering College, Kovilpatti, India

In this chapter, the author paints a comprehensive picture of different deep learning models used in different multi-modal image segmentation tasks. This chapter is an introduction for those new to the field, an overview for those working in the field, and a reference for those searching for literature on a specific application. Methods are classified according to the different types of multi-modal images and the corresponding types of convolution neural networks used in the segmentation task. The chapter starts with an introduction to CNN topology and describes various models like Hyper Dense Net, Organ Attention Net, UNet, VNet, Dilated Fully Convolutional Network, Transfer Learning, etc.

Sivakami A., Bharat Institute of Engineering and Technology, India
Balamurugan K. S., Bharat Institute of Engineering and Technology, India
Bagyalakshmi Shanmugam, Sri Ramakrishna Institute of Technology, India
Sudhagar Pitchaimuthu, Swansea University, UK

Biomedical image analysis is very relevant to public health and welfare. Deep learning is quickly growing and has shown enhanced performance in medical applications. It has also been widely extended in academia and industry. The utilization of various deep learning methods on medical imaging endeavours to create systems that can help in the identification of disease and the automation of interpreting biomedical images to help treatment planning. New advancements in machine learning are primarily about deep learning employed for identifying, classifying, and quantifying patterns in images in the medical field. Deep learning, a more precise convolutional neural network has given excellent performance over machine learning in solving visual problems. This chapter summarizes a review of different deep learning techniques used and how they are applied in medical image interpretation and future directions.

Karthika Paramasivam, Kalasalingam Academy of Research and Education, India
Prathap M., Wollo University, Ethiopia
Hussain Sharif, Wollo University, Ethiopia

Tensor flow is an interface for communicating AI calculations and a use for performing calculations like this. A calculation communicated using tensor flow can be done with virtually zero changes in a wide range of heterogeneous frameworks, ranging from cell phones, for example, telephones and tablets to massive scale-appropriate structures of many computers and a large number of computational gadgets, for example, GPU cards. The framework is adaptable and can be used to communicate a wide range of calculations, including the preparation and derivation of calculations for deep neural network models, and has been used to guide the analysis and send AI frameworks to more than twelve software engineering zones and different fields, including discourse recognition, sight of PCs, electronic technology, data recovery, everyday language handling, retrieval of spatial data, and discovery of device medication. This chapter demonstrates the tensor flow interface and the interface we worked with at Google.

Dhinesh Kumar R., Capgemini US, USA

Commonly, the sensor installed by these entities collects the citizen and infrastructure data by explicit approval. The performance based on the outline of the citizens could be done through AI depending on the data they offer that could promote advancements in the cities, or it could be utilized for the profit of the corporation. Certain barriers should be crossed in order to succeed in achieving the deployments of the cities. Commonly, the only target is to inform AI through the data gathered by the IoT in the real-time process. At certain times, the generation of data is the main part that is done by several elements, and this could be employed in a deliberate way. Certain challenges are also faced in the economic growth and its impact on biomedical industry. This chapter presents the impact of smart technology and economic growth for biomedical applications.

Ahmed Alenezi, Taibah University, Al-Ula Campus, Madhina, Saudi Arabia
M. S. Irfan Ahamed, Taibah University, Al-Ula Campus, Madhina, Saudi Arabia

Generally, the sensors employed in healthcare are used for real-time monitoring of patients, such devices are termed IoT-driven sensors. These type of sensors are deployed for serious patients because of the non-invasive monitoring, for instance physiological status of patients will be monitored by the IoT-driven sensors, which gathers physiological information regarding the patient through gateways and later analysed by the doctors and then stored in cloud, which enhances quality of healthcare and lessens the cost burden of the patient. The working principle of IoT in remote health monitoring systems is that it tracks the vital signs of the patient in real-time, and if the vital signs are abnormal, then it acts based on the problem in patient and notifies the doctor for further analysis. The IoT-driven sensor is attached to the patient and transmits the data regarding the vital signs from the patient's location by employing a telecom network with a transmitter to a hospital that has a remote monitoring system that reads the incoming data about the patient's vital signs.

Hmidi Alaeddine, Laboratory of Electronics and Microelectronics, Faculty of Sciences of
Monastir, Monastir University, Monastir, Tunisia
Malek Jihene, Higher Institute of Applied Sciences and Technology of Sousse, Sousse
University, Sousse, Tunisia

Deep Learning is a relatively modern area that is a very important key in various fields such as computer vision with a trend of rapid exponential growth so that data are increasing. Since the introduction of AlexNet, the evolution of image analysis, recognition, and classification have become increasingly rapid and capable of replacing conventional algorithms used in vision tasks. This study focuses on the evolution (depth, width, multiple paths) presented in deep CNN architectures that are trained on the ImageNET database. In addition, an analysis of different characteristics of existing topologies is detailed in order to extract the various strategies used to obtain better performance.

Reyana A., Nehru Institute of Engineering and Technology, India
Krishnaprasath V. T., Nehru Institute of Engineering and Technology, India
Preethi J., Anna University, India

The wide acceptance of applying computer technology in medical imaging system for manipulation, display, and analysis contribute better improvement in achieving diagnostic confidence and accuracy on predicting diseases. Therefore, the need for biomedical image analysis to diagnose a particular type of disease or disorder by combining diverse images of human organs is a major challenge in most of the biomedical application systems. This chapter contributes an overview on the nature of biomedical images in electronic form facilitating computer processing and analysis of data. This describes the different types of images in the context of information gathering, screening, diagnosis, monitor, therapy, and control and evaluation. The characterization and digitization of the image content is important in the analysis and design of image transmission.

In deep learning, the main techniques of neural networks, namely artificial neural network, convolutional neural network, recurrent neural network, and deep neural networks, are found to be very effective for medical data analyses. In this chapter, application of the techniques, viz., ANN, CNN, DNN, for detection of tumors in numerical and image data of brain tumor is presented. First, the case of ANN application is discussed for the prediction of the brain tumor for which the disease symptoms data in numerical form is the input. ANN modelling was implemented for classification of human ethnicity. Next the detection of the tumors from images is discussed for which CNN and DNN techniques are implemented. Other techniques discussed in this study are HSV color space, watershed segmentation and morphological operation, fuzzy entropy level set, which are used for segmenting tumor in brain tumor images. The FCN-8 and FCN-16 models are used to produce a semantic segmentation on the various images. In general terms, the techniques of deep learning detected the tumors by training image dataset.

Chapter 10

The recent growth of big data has ushered in a new era of deep learning algorithms in every sphere of technological advance, including medicine, as well as in medical imaging, particularly radiology. However, the recent achievements of deep learning, in particular biomedical applications, have, to some extent, masked decades-long developments in computational technology for medical image analysis. The methods of multi-modality medical imaging have been implemented in clinical as well as research studies. Due to the reason that multi-modal image analysis and deep learning algorithms have seen fast development and provide certain benefits to biomedical applications, this chapter presents the importance of deep learning-driven medical imaging applications, future advancements, and techniques to enhance biomedical applications by employing deep learning.

Artificial intelligence (AI) in medical imaging is one of the most innovative healthcare applications. The work is mainly concentrated on certain regions of the human body that include neuroradiology, cardiovascular, abdomen, lung/thorax, breast, musculoskeletal injuries, etc. A perspective skill could be obtained from the increased amount of data and a range of possible options could be obtained from the AI though they are difficult to detect with the human eye. Experts, who occupy as a spearhead in the field of medicine in the digital era, could gather the information of the AI into healthcare. But the field of radiology includes many considerations such as diagnostic communication, medical judgment, policymaking, quality assurance, considering patient desire and values, etc. Through AI, doctors could easily gain the multidisciplinary clinical platform with more efficiency and execute the value-added task.

Chapter 12

Electronic health records (EHR) have been adopted in many countries as they tend to play a major role in the healthcare systems. This is due to the fact that the high quality of data could be achieved at a very low cost. EHR is a platform where the data are stored digitally and the users could access and exchange data in a secured manner. The main objective of this chapter is to summarize recent development in wearable sensors integrated with the internet of things (IoT) system and their application to monitor patients with chronic disease and older people in their homes and community. The records are transmitted digitally through wireless communication devices through gateways and stored in the cloud computing environment.

Chapter 13

Artificial intelligence integrated with the internet of things network could be used in the healthcare sector to improve patient care. The data obtained from the patient with the help of certain medical healthcare devices that include fitness trackers, mobile healthcare applications, and several wireless sensor networks integrated into the body of the patients promoted digital data that could be stored in the form of digital records. AI integrated with IoT could be able to predict diseases, monitor heartbeat rate, recommend preventive maintenance, measure temperature and body mass, and promote drug administration by having a review with the patient's medical history and detecting health defects. This chapter explores IoT and artificial intelligence for smart healthcare solutions.

The global market for IoT medical devices is expected to hit a peak of 500 billion by the year 2025, which could signal a significant paradigm shift in healthcare technology. This is possible due to the on-premises data centers or the cloud. Cloud computing and the internet of things (IoT) are the two technologies that have an explicit impact on our day-to-day living. These two technologies combined together are referred to as CloudIoT, which deals with several sectors including healthcare, agriculture, surveillance systems, etc. Therefore, the emergence of edge computing was required, which could reduce the network latency by pushing the computation to the "edge of the network." Several concerns such as power consumption, real-time responses, and bandwidth consumption cost could also be addressed by edge computing. In the present situation, patient health data could be regularly monitored by certain wearable devices known as the smart telehealth systems that send an enormous amount of data through the wireless sensor network (WSN).

Chapter 15

Karthick Raghunath K. M., Malla Reddy Institute of Engineering and Technology, India
Anantha Raman G. R., Malla Reddy Institute of Engineering and Technology, India

As a decorative flower, marigolds have become one of the most attractive flowers, especially on the social and religious arena. Thus, this chapter reveals the potential positive resultants in the production of marigold through neuro-fuzzy-based smart irrigation technique in the static-clustered wireless sensor network. The entire system is sectionalized into clustering phase and operational phase. The clustering phase comprises three modules whereas the operational phase also includes three primary modules. The neuro-fuzzy term refers to a system that characterizes the structure of a fuzzy controller where the fuzzy sets and rules are adjusted using neural networks iteratively tuning techniques with input and output system data. The neuro-fuzzy system includes two distinct way of behavior. The vital concern of the system is to prevent unnecessary or unwarranted irrigation. Finally, on the utilization of multimodal image analysis and neuro-fuzzy methods, it is observed that the system reduces the overall utilization of water (~32-34%).

Chapter 16

Sebastin S., National Engineering College, Thoothukudi, India
Murali Ram Kumar S. M., National Engineering College, Thoothukudi, India

The recent researches show that cement mortar containing eggshell as a partial replacement of sand contained radiation absorption property. In these cement mortar samples, 5%, 10%, 15%, and 20% by mass of sand was replaced by crushed eggshells, and there was increase in radiation absorption. The result also showed that using eggshell as a partial replacement for sand leads to decrease in compressive strength of the cement mortar. Waste of any kind in the environment when its concentration is in excess can become a critical factor for humans, animals, and vegetation. The utilization of the waste is a priority today in order to achieve sustainable development. So, we have planned to increase that compressive strength by means of using seashell as a partial replacement for cement. The main objective of the project is to maintain radiation absorption by means of using eggshell along with seashell as a partial replacement for cement to increase the compressive strength.

Preface

Deep learning is being touted as the one-stop solution for all kinds of IT requirements, given that deep learning is on development path that includes a number of innovations and disruptions being brought in this exotic and enigmatic discipline. Deep learning has progressed rapidly in recent years. Techniques of deep learning has played important role in medical field like medical image processing, computer-aided diagnosis, image interpretation, image fusion, image registration, image segmentation, image-guided therapy, image retrieval and analysis. Deep learning has been steadily growing across industry and academic domains. The main reason is that deeper and greater understanding of the deep learning by worldwide businesses, government organizations. Academic institutions and research abs through industry collaborations bring forth a variety of advancements in the deep learning arena. Deep learning methods are highly effective when the numbers of available samples are large during a training stage. For example, in Image Net Large Scale Visual Recognition Challenge (ILSVRC), offered more than 1 million annotated images. The most important limitations as well as issues of the deep learning phenomenon are being meticulously addressed in order to enhance greater deployment of deep learning networks and make it strategically effective technology. After the advent of deep learning, the next generation of artificial intelligence are being fully accomplished through this ground breaker technological paradigm. This edited book is produced and prepared with the sole aim of providing deeper insights into deep learning. IT practitioners and academic professors have together contributed the book chapters. The book provides novel theories of the deep learning as well as a unified theory that captures the abstract principles that they share.

Chapter 1 contributes on the major health components like content, infrastructure, analytics, communication and interfaces forming core of all health IT. The challenge is further exaggerated on the current issues related to health data and the contribution of IT systems in healthcare management. Further the scenarios on the intelligent care systems that are capable on early diagnosis are described. This impact the future of global healthcare services rapidly increasing the development of bridging gap between the personal and relevant datasets creating a new era in the world of healthcare.

Chapter 2 paints a comprehensive picture of different deep learning models used in different multi modal image segmentation task. This paper is an introduction for those new to the field, an overview for those working in the field and a reference for those searching for literature on a specific application. Methods are classified according to the different types of multi modal images and the corresponding type of Convolution Neural Network used in the segmentation task. The chapter starts with an introduction to CNN topology and describes various models like Hyper Dense Net, Organ Attention Net, UNet, VNet, Dilated Fully Convolutional Network, Transfer Learning etc.

Chapter 3 summarizes up-to-date review of different deep learning techniques used and how they are applied in medical image interpretation and future directions and utilization of various deep learning methods on medical imaging endeavours to create systems that can help in the identification of disease and the automation of interpreting biomedical images to help treatment planning. New advancements in Machine learning, primarily about deep learning employed for identifying, classifying and quantifying patterns in images in the medical field

Chapter 4 contributes on Tensorflow that is adaptable and can be used to communicate a wide range of calculations, including the preparation and derivation of calculations for deep neural network models, and has been used to guide the analysis and send AI frameworks to more than twelve software engineering zones and different fields, including discourse recognition. This chapter demonstrates the Tensor Flow interface and its interface.

Chapter 5 presents certain challenges are also faced in the economic growth and its impact on biomedical industry. This chapter presents the impact of smart technology and economic growth for biomedical applications.

Chapter 6 focuses on the working principle of IoT in remote health monitoring systems is that it tracks the vital signs of the patient in real-time and if the vital signs are abnormal then it acts based on the problem in patient and notifies the doctor for further analysis. The IoT-driven sensor is attached to the patient which transmit the data regarding the vital signs from the patient's location and employing a telecom network with a transmitter to a hospital which consists of remote monitoring system that reads the incoming data about the patient's vital signs.

Chapter 7 focuses on the evolution (depth, width, multiple paths...) presented in deep CNN architectures that are trained on the Image NET database. In addition, an analysis of different characteristics of existing topologies is detailed in order to extract the various strategies used to obtain better performances.

Chapter 8 presents the overview on the nature of biomedical images in electronic form facilitating computer processing and analysis of data. This describes the different types of images in the context of information gathering, screening, diagnosis, monitor, therapy and control and evaluation. The characterization and digitization of the image content is important in the analysis and design of image transmission

Chapter 9 presents the application of the techniques viz., ANN, CNN, DNN for detection of tumors in numerical and image data of brain tumor is presented. First, the case of ANN application is discussed for the prediction of the brain tumor for which the disease symptoms data in numerical form is the input. ANN modelling was implemented for classification of Human Ethnicity. Next the detection of the tumors from images is discussed for which CNN and DNN techniques are implemented. Other techniques discussed in this study are HSV color space, Watershed Segmentation and Morphological operation, Fuzzy Entropy Level set, that are used for segmenting tumor in brain tumor images. The FCN-8 and FCN-16 models are used to produce a semantic segmentation on the various images. In general terms, the techniques of deep learning detected the tumors by training image dataset.

Chapter 10 presents the recent achievements of deep learning in particular biomedical applications has, to some extent, masked decades-long developments in computational technology development for medical image analysis. The methods of Multi-modality medical imaging have been implemented in clinical as well as research studies. Due to the reason that multi-modal image analysis and deep learning algorithms have seen fast development and provide certain benefits to biomedical applications. This chapter presents the importance of deep learning driven medical imaging applications, its future advancements and techniques to enhance biomedical applications by employing deep learning.

Chapter 11 discusses the work mainly concentrated on certain regions of the human body that includes Neuroradiology, cardiovascular, abdomen, lung/thorax, breast, Musculoskeletal Injuries, etc. A perspective skill could be obtained from the increased amount of data and a range of possible options could be obtained from the AI though they are difficult to detect with the human eye. Experts, who occupy as a spearhead in the field of medicine in the digital era, could gather the information of the AI into healthcare. But the field of radiology includes many considerations such as diagnostic communication, medical judgment, policy-making, quality assurance, considering patient's desire and values, etc. Through AI, doctors could easily gain the multidisciplinary clinical platform with more efficiency and execute the value-added task, thus making more visibility to the patients.

Chapter 12 presents the recent development in wearable sensors integrated with the Internet of Things (IoT) system and their application to monitor patients with chronic disease and older people in their homes and community. The records are transmitted digitally through wireless communication devices through gateways and get stored in the cloud computing environment. Electronic Health Record (EHR) has been adopted in many countries as they tend to play a major role in the healthcare systems. This is due to the fact that the high quality of data could not be achieved at a very low cost. EHR had taken away in reaching out to be an effective healthcare information system, which promises positive health outcomes. EHR is a platform, where the data are stored digitally and the users could access and exchange data in a secured manner.

Chapter 13 discusses AI integrated with IoT that can predict diseases, monitor heartbeat rate, recommend preventive maintenance, temperature, body mass and promote drug administration by having a review with the patient's medical history.

Chapter 14 presents cloud computing and the Internet of Thing (IoT) that have an explicit impact on our day-to-day living. These two technologies combined together are referred to as CloudIoT, which deals with several sectors including healthcare, agriculture, surveillance systems, etc. Therefore, emerge of edge computing was required that could reduce the network latency by pushing the computation to the "edge of the network". Several concerns such as power consumption, real-time responses, and bandwidth consumption cost could also be addressed by edge computing. In the present situation, patient's health data could be regularly monitored by certain wearable devices known as the smart telehealth systems that send an enormous amount of data through the Wireless Sensor Network (WSN).

Chapter 15 reveals the potential positive resultants in the production of marigold through neuro-fuzzy based smart irrigation technique in the static-clustered Wireless Sensor Network. The entire system is sectionalized into clustering phase and operational phase. The clustering phase comprises three modules whereas the operational phase also includes three primary modules.

Finally, Chapter 16 presents the need for construction of sustainable healthcare building for effective healthcare management.

Chapter 1
Emerging Information Technology Scenarios and Their Contributions in Reinventing Future Healthcare Systems:
Applications and Practical Systems for Healthcare

Reyana A.
Nehru Institute of Engineering and Technology, India

Krishnaprasath V. T.
Nehru Institute of Engineering and Technology, India

Suresh A.
Nehru Institute of Engineering and Technology, India

ABSTRACT

Critical health information is gathered, compiled, and analysed from image scanning, laboratory testing reports, health trackers, etc. Healthcare requires a support of real-time digital information for effective decision making. As industries bring better care to consumer health systems, providers achieve their goal in delivering better diagnosis and efficient treatment to consumers. This chapter contributes to the major health components—content, infrastructure, analytics, communication, and interfaces—forming the core of all health IT. The challenge is further exaggerated on the current issues related to health data and the contribution of IT systems in healthcare management. Further, the scenarios on intelligent care systems that are capable of early diagnosis are described. This impacts the future of global healthcare services by rapidly increasing the development and bridging the gap between the personal and relevant datasets creating a new era in the world of healthcare.

DOI: 10.4018/978-1-7998-3591-2.ch001

INTRODUCTION TO MAJOR HEALTH COMPONENTS

Examination and evaluation of parameters that affects patient's well being regarding intellectual, physical, social, emotional, environmental, spiritual, cognitive and nutritional wellness in patients life is the health history. This is required to identify health care deviations for reporting wellness improvement occurred or needed. The health aspect's is the person's wellness leading to a healthy life. Many international organizations like World Health Organization work towards educating people embracing healthier life changes. This chapter contributes on the major health components like content, infrastructure, analytics, communication and interfaces forming core of all health IT. The challenge is further exaggerated on the current issues related to health data and the contribution of IT systems in healthcare management. Further the scenarios on the intelligent care systems that are capable on early diagnosis are described. This impact the future of global healthcare services rapidly, increasing the development of bridging gap between the personal and relevant datasets creating new era in healthcare stated by the author Friedman (2012). The new perspective in health IT in improving healthcare services, clinical decision making, collaborative work using artificial intelligence, groupware, etc is the current perspective in the context of healthcare services. Thus quality display of health data, error free decision on clinical diagnosis, quality image compression, digitization and other techniques yield the essential knowledge required to guide clinical services enhancing future healthcare services. To achieve healthiest lifestyle these components are needed to be taken into consideration. Health and wellness go hand in hand; to live a long healthy life it is necessary for one to understand these components. The major components of health history (figure 1) are summarized below:

Figure 1. Wellness Components

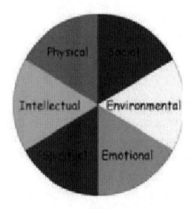

Intellectual/Mental: the inner strength to maintain proper discipline and face reality in fighting against illness.

Physical: based on our daily actions one can judge the functioning of our human body. Exercise, diet plan, weight maintenance, drug and alcohol avoidance keep us free from diseases. This is probably an important aspect, being active adding years to life. Also, physical health reduces anxiety and depression. Activities like running, swimming, walking, dancing, playing shall strengthen your muscles. Some people exercise regularly which increases the serotonin and nor epinephrine levels active for depression relief.

Regular exercise improves skin tone; strengthen bones, reducing chronic diseases. Thus physical wellness indicates flexibility and strength. Taking care on own health, attention to minor illness, knowledge on common medication practices are also concerned with physical wellness.

Social: temperament with friends, relatives and dear ones. A study conducted has proven that social isolation leads to obesity. To maintain social wellness it is necessary to handle issues without violence.

Emotional: the positivity like supportive and encouraging attitude towards others as depicted in figure 2. Emotional health possesses the ability to recognize reality and go along with the daily life demands. Treatments for depression and mental disorders like extreme anxiety, schizophrenia are available today.

Figure 2. Positivity for Wellness

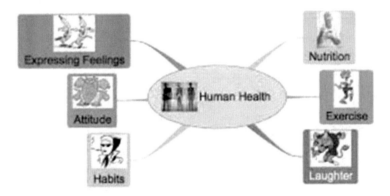

Environmental: the air, water, food and atmosphere around us speaks on the air purity, hygiene and nutritional diet practiced. To maintain good health good working environment and healthy living influence one's life.

Spiritual: moral and ethical values, dignity and harmonious temperament towards surrounding people helps in dealing tough situations in life like death, financial hardships, etc.,

Cognitive: brains ability to carry out functions like judgments, memory power, learning new language improves cognitive health. Playing puzzles, watching TV, reading crossword, social interactions are some of the activities that can be undergone.

Nutritional: maintaining balanced diet, awareness on illness due to improper diet plans.

In addition, working to help communities adapt to new means of nutritional wellness in their community can bestow the valued asset of lifelong lessons that allow communities to grow and age in a healthy manner. Knowledge on these is necessary to improve the health status of families and individuals against financial consequences, ill-health, etc.

Current Perspective and Issues Related to Health Data

Every patient's expect the health care professionals to use and protect their clinical records in best way. Today researchers aligning to the incentives for data sharing, share quickly letting other patients to know their side effects. Health care systems at times, make decisions helping companies with health data showing commercial interest. These raise questions on whether there exists a control on clinical trial data described by Ward & Barker (2013) and Wang & Krishnan (2014). These pay way for researchers to intensify ethical awareness among public people.

Though discussions on data sharing takes place among clinical traits who collect, analyze and reuse them in novel ways. Patients and their representative's feel privacy loss in health care. In traditional ways doctors protect health information of patient's. Nowadays doctors and other health care workers share health data among institutions in different countries. Complexity in sharing patient data in clinical trials includes ownership shift and control patients everywhere. Current technology seeks stay in touch to renewed consent forms developed and tested.

Threats to Patient Safety

There are challenges identified that can be harmful to patients due to medication errors and patient mortality. Also challenges related to missing values, interpretation problem, dependence problem, sampling size and handling methodologies are faced on the emergence of big data by Logan, K.(2016). When analyzing big data techniques at times it provides meaningless correlations. Another major issue is the protection of individual data. Protecting privacy has become an ethical challenge in big data for use in Public Health as stated by the authors Vayena, Salathé, Madoff, & Brownstein (2015).

Interoperability and Data Quality

Maintaining data quality is questionable with big data in healthcare. Need for capturing patient's points during interaction in health care system limits the number of information exchanges. Another issue is that heterogeneous and diverse data, among various characteristics of patients cannot be controlled. Data analytics tools can transform health care in various ways as describe by the author Vest & Gamm (2010). For example visit of medical practitioner replaced by remote consultation monitoring the health status. And patient's care can be given from expert's judgment around the world. Support of skilled interpreters required to customize person's genetic information in order to diagnose and treat genetic based illness. Policy makers have many new tools bringing insights in treatment and spending trends. Developments in data analytics remove barriers for substantial healthcare in data conventions, decision making and health care delivery to different actors in health industry. These policies and barriers emphasize and prioritize health reforms to encourage the use of data analytics.

Security, Visualization and Data Integrity Concerns

Electronic Health Records integrated into clinical process is shown in figure 3. To understand this we need to take a look to collect, analyze, store and present clinical data to different to different members.

Capture: Healthcare providers require accurate, complete and correctly formatted data for multiple users without conflict. Incomplete understanding on EHR contributes to quality issues throughout its lifecycle.

Cleaning: it is vital to cleanse dirty data, ensuring consistent and relevant datasets, those not corrupted, expensive and sophisticated tools required, to ensure accuracy and integrity of data in a data warehouse.

Storage: the volume of healthcare data managed on data centers, promise security control access. Cloud storage the popular storage medium drops costs as reliability grows. Disaster recovery, easier expansion, etc., are the important concerns before choosing partners on security and compliance issues.

Security: clinical data are subjected to high profile vulnerabilities like breaches and hackings. The technical safeguards are listed in security rules of HIPAA, authentication protocols and transmission security; auditing and integrity procedures are stored in every organization to protect health information. The procedures like anti-virus software, firewalls, encryption of sensitive data, etc. protect high valued data assets from malicious parties.

Stewardship: data reused and reexamined for quality measurements, therefore it is mandatory to know information like why, when and how the researchers created the data and whether the purpose accomplished. Organizations dealing with healthcare develop meaningful metadata from assigned steward. These include the standard formats and definitions including creation, deletion and updation tasks.

Querying: query reporting, analytics, interoperability problems prevent access on an organizations repository information is dealt in querying. Structured query language is used for relational and large databases. User trust only on completeness, accuracy and standardization of data and the target audience generate report that is accessible, clear and concise. The integrity and accuracy of data has critical impacts. The data must be examined before presenting it to his or her inferences.

Visualization: it's similar to color coding with red, yellow, green to understand stop, caution and go. Apart from this the other practices like charts, figures, labeling to reduce confusion. Low quality graphics, text overlapping, etc frustrate viewers leading to misinterpretation. Examples include bar charts, pie charts, scatter plots etc.

Updating: clinical records require frequent updation maintaining the dataset quality

Sharing: data sharing with external partners for health management and value-based care.

Figure 3. Healthcare Data

Contribution of IT Systems in Healthcare

Patients are coordinated through hospitals that include large departments and units relying itself on Hospital Information System that assist in diagnosis, management and improved services as depicted in figure 4. The quality of an information system includes user satisfaction, information and system quality. Rapid growth in information technology influences business. Due to increase in population, there is high demand in handlings patients using hospital information system. The system and operational activities enhance cost control, administration, and patient care reducing revenue.

Information Systems should manage and improve the quality of care. All levels of management should take into account the key actions required to maximize the system benefits. Healthcare has greater importance in developing countries enabling better quality and value to patients by Wise (2003).

In the healthcare context, the economic evaluation method has proven that the emergence of health information system improves time and reduces cost. Many organizations supersede the financial factors and performance using computer based technologies.

Health Information Technology

Health information technology provides technical infrastructure to record, analyze and share health data. The tools like app, smart devices, paper provide better care to achieve patient's health equity stated by Mitchell (1999). However using healthcare systems for information analysis shall support practitioners and health government ministry. The policies should better and reduce the spread of diseases, provide quality health care, reduction in medical errors, increase patients safety, etc.

Figure 4. Healthcare Information System

Health IT delivers accurate clinical records allowing doctors to have better understanding on patient's health history. This shall empower doctors to treat ailments and prevent fatal medications. The interrogation between practitioners and patient on health history shall lead to inaccurate drug name, ailments. Patients suffering from diseases benefit from EHR improving aspects in patient care like communication, safety, equity, timeliness and efficiency.

Developers all over the world provide technical support by contributing code to application implemented in local clinics. Health information technologies include both software and hardware to build health information systems. Health informatics is the combination of computer science, information science and healthcare. The tools included to this are clinical guidelines, medical terminologies, communication information systems those applied in the areas of dentistry, clinical care, nursing, public health, pharmacy and bio-medical research.

Information Technology assists medical colleges by implementing Computer Assisted Learning, Human Simulations and Virtual Reality. The Advanced Life Support simulators interpret ECG for intervention in drugs, defibrillation, injecting without applying it on real patients. The sophisticated simulation technologies availability for complex situations is very supportive to healthcare sector. The advancements in electronic medical records convert all medical transcriptions into single database reducing paper cost, availability of patient's medical history, supportive insurance, etc.,

The ability to care patient's record providing information service to diagnose and treat patient is improved instantly in all clinics with the usage of EMR. Health information technology support secure exchange of health information between organizations, consumers, providers and payers.

The benefits are listed below:

- Improve in effectiveness and quality of healthcare
- Increase in efficiency and productivity of healthcare
- Preventing medical emergencies
- Accuracy and correctness in medical procedures
- Reduce in treatment cost
- Effective administration
- Reduction in paper work and idle work time
- Extension to real-time communication
- Finally providing qualitied and affordable care to all

The various technological innovations are listed below:

- Clinical Decision Support (CDS)
- Electronic Medical Records (EMR)
- Computerized Physician Order Entry (CPOE)
- Bar-Coding at Medication Dispensing (BarD)
- Robot for Medication Dispensing (ROBOT)
- Automated Dispensing Machines (ADM)
- Electronic Medication Administration Records (eMAR), etc

ELECTRONIC HEALTH RECORD (EHR)

EHR reduces errors in preventive care, prescription of drugs and test results. Treatment process as electronic record provides clinical guidelines. Today clinical records are available in healthcare websites due to the adoption of electronic health records in clinics. This has reduced the training costs, complexity and the adoption to HER continue to raise since 2005. However recent surveys say the patient safety features has become the main concern in HIT when adopted physicians. Health information workflows improve the needs of end users, e.g. user-friendly interface, simplicity, speed of systems, etc. Standardized bar code system prevents much of the drug errors. The key opportunities are:

- Medical treatment and quality care
- Health monitoring, disease diagnosis
- Optimization in clinical performance
- Pharmaceutical research and development

International health systems understand the complexities and finding better opportunities to health information technology. It gives policy developers a chance to compare and contrast established indicators leading to adverse policies.

Data from millions of patients that is collected across various healthcare institutions could be framed as Electronic Health Record (EHR). It comprises of data in various forms that includes laboratory results, diagnosis reports, medical images, demographic information about patient and clinical notes. Transformation in the healthcare could be done with the help of deep learning techniques since these techniques outperform than the conventional machine learning methods. In addition, a huge and complex datasets gets generated day by day and thus it enables the need for the deployment of deep learning models. It is observed that the clinical decision in certain fields that vary from analysis of medical research issues and to suggest as well as prioritize the treatments in order to detect abnormalities and in addition, the identification of genomic markers in tissue samples is done. Deep learning is applied in EHR system to derive patient representation to provide enhanced clinical predictions, and augments clinical decision systems. Moreover, it is applied in order to detect the disease in the earlier stages, determining the clinical risk, forecasting the future need of regular checkups and in addition prediction of hospitalization in the near future if required. According to statistical report, the deep learning software market is estimate to reach 180 million U.S dollars in size by 2020.

The availability of huge amount of clinical information, particularly EHR has stimulated the growth of Deep Learning techniques which assist in rapid analysis of patient's data. Nowadays, EHR is being incorporated with deep learning techniques and tools into EHR systems which provide deep insights for health outcomes. The market for EHR deep learning tools is predicted to exceed $34 billion by the mid of 2020s, motivated massively by an emerging desire to automate tasks and provide deeper insights into clinical issues. The global health IT market is anticipated to be worth a surprising $223.16 billion by 2023, driven in part by deep learning. EHR is a thriving research domain for deep learning researchers since it helps to achieve higher accuracies in medical conditions. The process to extract clinical information from clinical notes using deep learning includes concept extraction, relation extraction, temporal event extraction and abbreviation expansion. In EHR representation learning process, concept and patient representations are included to obtain detailed analysis and precise predictive tasks. Outcome prediction in deep EHR is categorized as static and temporal prediction to predict patient outcomes. The clinical deep learning is more interpretable under the category of maximum activation, constraints, qualitative clustering and mimic learning in EHR analysis

Emerging Technologies in Healthcare

Health technologies include devices, procedures, vaccines and medicines to streamline the healthcare quality. The emerging technologies are Artificial Intelligence, Blockchain, Voice search, Chatbots, and Virtual reality is the most promising technologies in healthcare by the year 2019.

Artificial Intelligence

Usage of artificial intelligence in healthcare expected to increase at a rate of 40% by 2022. The artificial intelligence engine prevents risk and medical scenarios in three ways. The patient's get medication timing help from automate reminders that alert medical staff based on the personalized dosages.

Blockchain

Blockchain keeps digital record creating ledger of transactions claiming the way technology change to operate in digital healthcare marketing. Blockchain collect information stored on servers entertaining direct marketing without a media platform. The Blockchain is simple to restructure the change the data collections, ownership and digital advertisements.

Chatbots in Healthcare

Chatbot in healthcare offers improvement in patient's pathways, management of medication in emergency situations. Personal experiences proves vital role in chatbot adding touchpoints to people.

Case Scenario: Intelligent Care and Early Diagnosis Systems

Future envisage on the early detection of diseases and continuous monitoring of health. In the National Academy of Medicine report 2015, diagnostic errors, inaccurate diagnosis exists in all settings to an unacceptable rate bringing harmful effects to patient's community. Therefore perfect diagnostic sources from molecular contents saliva, sweat, feces and urine excreted every day is essential information. Researchers around the world provide clues and guidelines to maintain good health. Cardiologists with the vision of fixing pacemakers and other devices in heart require support to monitor their patient's on any crisis. Wearable devices deliver plenty of information and health-care systems track both individuals health, helps researchers to study on the effectiveness of treatments and take preventive measures. The future is to intercept diseases early and prevent them. Doing something on disease like aggressive cancer is worth monitoring and preventing it.

CONCLUSION

The key challenge in today's world concerned with healthcare is patients are digitally empowered. Many healthcare organizations adopt new technologies for collection, transfer, security and privacy concerns of clinical data. Internet of Things and predictive analysis shall shape the future of healthcare industries increasing the survivability and expectancy. The modernization of healthcare includes ageing population, rise in treatment cost and increase in victims suffering from chronic diseases requiring long term care. The digital era in intelligence centric patient transformation shall cut down the cost and improve the patients care. Most of the IT investors collaborate digitization sharing patients care episodes in bringing insights to meet business demands in healthcare industry. Telemedicine lack in health services and technology advances delivering quality healthcare.

REFERENCES

Frakt, A. B., & Pizer, S. D. (2016). The promise and perils of big data in health care. *The American Journal of Managed Care*, 22(2), 98. PMID:26885669

Friedman, U. (2012). Big Data: A Short History. *Foreign Policy*.

King, S. B. III. (2016). Big Trials or Big Data. *JACC: Cardiovascular Interventions*, 9(8), 869–870. doi:10.1016/j.jcin.2016.03.003 PMID:27101918

Logan, K. (2016). What happens when big data blunders? *Communications of the ACM*, 59(6), 15–16. doi:10.1145/2911975

Markle Foundation. (2003). *Connecting for Health—A Public–Private Collaborative: Key Themes and Guiding Principles*. Markle Foundation.

Maviglia, S. M., Zielstorff, R. D., Paterno, M., Teich, J. M., Bates, D. W., & Kuperman, D. J. (2003). Automating complex guidelines for chronic disease: Lessons learned. *Journal of the American Medical Informatics Association*, 10(2), 154–165. doi:10.1197/jamia.M1181 PMID:12595405

Metzger, J., & Fortin, J. (2003). *California Health Care Foundation. Computerized Physician Order Entry in Community Hospitals*. http://www.fcg.com/healthcare/ps-cpoe-research-and-publications.asp

Mitchell, T. (1999). Machine learning and data mining. *Communications of the ACM*, 42(11), 31–36. doi:10.1145/319382.319388

Neff, G. (2013). Why big data won't cure us. *Big Data*, 1(3), 117–123. doi:10.1089/big.2013.0029 PMID:25161827

Sukumar, S. R., Natarajan, R., & Ferrell, R. K. (2015). Quality of Big Data in health care. *International Journal of Health Care Quality Assurance*, 28(6), 621–634. doi:10.1108/IJHCQA-07-2014-0080 PMID:26156435

Vayena, E., Salathé, M., Madoff, L. C., & Brownstein, J. S. (2015). Ethical challenges of big data in public health. *PLoS Computational Biology*, 11(2), e1003904. doi:10.1371/journal.pcbi.1003904 PMID:25664461

Vest, J. R., & Gamm, L. D. (2010). Health information exchange: Persistent challenges and new strategies. *Journal of the American Medical Informatics Association, 17*(3), 288–294. doi:10.1136/jamia.2010.003673 PMID:20442146

Wang, W., & Krishnan, E. (2014). Big data and clinicians: A review on the state of the science. *JMIR Medical Informatics, 2*(1), e1. doi:10.2196/medinform.2913 PMID:25600256

Ward, J. S., & Barker, A. (2013). *Undefined by data, a survey of big data definitions.* arXiv preprint arXiv:1309.5821

Zhang, Z. (2014). Big data and clinical research: Perspective from a clinician. *Journal of Thoracic Disease, 6*(12), 1659–1664. PMID:25589956

Chapter 2
Deep Learning Models for Semantic Multi-Modal Medical Image Segmentation

V. R. S. Mani

ⓘ https://orcid.org/0000-0002-3234-0486

National Engineering College, Kovilpatti, India

ABSTRACT

In this chapter, the author paints a comprehensive picture of different deep learning models used in different multi-modal image segmentation tasks. This chapter is an introduction for those new to the field, an overview for those working in the field, and a reference for those searching for literature on a specific application. Methods are classified according to the different types of multi-modal images and the corresponding types of convolution neural networks used in the segmentation task. The chapter starts with an introduction to CNN topology and describes various models like Hyper Dense Net, Organ Attention Net, UNet, VNet, Dilated Fully Convolutional Network, Transfer Learning, etc.

INTRODUCTION

A deep neural network has extremely large number of layers of neurons, forming a hierarchical feature representation. The number of layers now rises to over 2,000. With this extremely large modeling capacity, a deep network can effectively remember all possible transformations after good training with an extremely large training data set and make smart predictions, they identify from new data sets which were untrained earlier. Thus, deep learning is used in many computer vision and medical imaging tasks. Convolutional Neural Network (CNN) models like AlexNet, VGGNet, RESnet or GoogLeNet have shown accurate results in various competitive bench marks. Different ideas were explored regarding CNN design, like use of different activation functions, loss functions, and optimization and regularization functions. Approximately there are seven different categories of CNN architectures which were used widely in different applications. All these architectures are based on depth, width, spatial exploitation, multi path, attention, channel boosting and feature map exploitation. In CNN, the topology is divided

DOI: 10.4018/978-1-7998-3591-2.ch002

into different learning stages like convolutional layer, non-linear processing units, and sub sampling layers. Multiple transformations are done in each layer using a group of convolutional kernels. Locally correlated features are extracted in the convolutional operation.

Semantic segmentation explains how each pixel of an image is labelled with a class label, (such as water, sand, grass, sky, ocean, or car). Semantic segmentation applications include: Autonomous driving, Robotic Vision, Remote Sensing, Medical Imaging, Video Processing, etc., Due to the tremendous advancements in multimodal image acquisition devices, the amount of imaging data to be analyzed is increasing day by day. Medical imaging data is heterogeneous and multidimensional and also sensitive to human error and varies across different subjects. The segmentation and interpretation of medical imaging data is therefore a time consuming process. Semantic segmentation relates each pixel of an image to a class label. Though closely related to semantic segmentation, medical image segmentation includes specific challenges that need to be addressed, such as the scarcity of labeled data, the high class imbalance found in the ground truth and the high memory demand of three-dimensional images. In this chapter, CNN-based method with three-dimensional filters applied to different structural and functional images are to be discussed in depth.

MULTI MODAL BRAIN MRI IMAGE SEGMENTATION:

Brain MR image segmentation is a key function in radio therapy and image guided surgery. Quantitative analysis of brain abnormality requires an accurate segmentation of MR images of brain which is a complex and hard challenge. The heterogeneous nature of the lesions which include its large scale variability in phrases of length, shape and area make segmentation assignment extremely difficult. Manual segmentation the use of a human expert is the high-quality approach that is time eating, tedious, high priced and impractical in larger research and introduces inter observer variability. A couple of image sequences with varying contrast want to be taken into consideration for identifying whether a selected location is a lesion or not. And additionally the extent of expert knowledge and enjoy impacts the segmentation accuracy. In mind imaging studies special modality pix are mixed to enhance the short comings of character imaging techniques.T1 weighted photos produce correct evaluation among gray remember (GM) and white matter (WM) tissues. T2 weighted and Proton Density (PD) photographs enables in visualizing the lesions and different abnormalities Fluid Attenuated Inversion recovery (aptitude) photographs improve the photograph assessment of white remember lesions on account of a couple of sclerosis [2]. To enhance the accuracy in brain photo segmentation, fusing distinct modality photograph is important. Fusing multi modality photograph is important within the case of toddler brains which has poor assessment.

Recently unique Deep studying techniques based totally on 2d CNNs (Pereira et al., 2015) are utilized in mind image segmentation duties. In second CNN primarily based strategies the 3-d brain segmentation is finished by using processing individual 2nd slices independently which is a non optimal method. CNNs can research both the capabilities and classifiers simultaneously from statistics. Currently densely related networks are utilized in scientific image segmentation (). Most of the present Multi modal CNN segmentation techniques comply with an early fusion strategy in which MRI T1, T2 and fractional anisotropy (FA) pix are certainly merged at the input of the network. These methods count on that the relationship among specific modalities is sincerely linear. Some different methods study complimentary facts from the other modalities. But the courting among different modalities are tons extra complicated. In Dense Nets direct connections from any layer to all the next layers in a feed-forward manner makes schooling easier and more correct. Because of direct connections among all layers there may be a glide of information and gradients all through the whole community. Additionally dense connections have a regularizing effect, which in turn reduces over-becoming on obligations with smaller schooling sets.

A fully linked 3D Hyper Dense structure became utilized by Dolz et al., (2019) which extends the dense connectivity of multimodal segmentation. Each MRI modality has a direction and dense connection going on among layers across the equal course and between layers across distinct nearby paths, which is different from the existing CNNs. Hyper Dense networks have the freedom to research complicated versions inside and across special modalities. Here in Hyper Dense networks outputs belonging to layers in special streams, each related to a unique picture modality are connected. Hyper dense community yields significant improvement over the existing state of the artwork segmentation techniques over bench mark data units. One challenge of the deep studying architecture is the vanishing or exploding gradients, which save you convergence during schooling.

A section of the Hyper Dense community from Dolz et al., (2019) is proven in figure: 1. Here each grey block is a convolution layer. The red arrows imply convolution operation and black arrows are the direct connections between feature maps from exceptional layers across the identical direction and unique nearby paths. The input of every convolution block (maps earlier than the crimson arrow) is the concatenation of the outputs (maps after the crimson arrow) of all the previous layers from both paths. Dense connections taking place in among the flow of man or woman modalities and across streams of different modalities give the community total freedom to explore complicated combos between functions of various modalities, within and in-between all degrees of abstraction.

Figure 1. Section of Hyper Dense Network from Dolz et al., (2019)

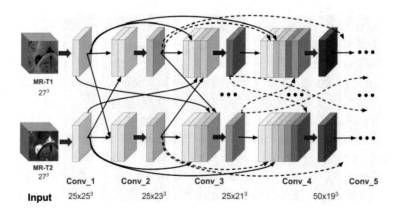

SEGMENTATION OF ABDOMINAL ORGANS IN CT IMAGES USING ORGAN ATTENTION NETWORK

In automatic computer aided prognosis and surgical procedure sturdy and accurate segmentation of exceptional abdominal organs in CT images is crucial, which is quiet tough due to the weak boundary among spatially adjoining systems in the abdominal area, various sizes of various organs, and complexity of the heritage. Morphological and topological complexity of the stomach shape includes anatomically linked structures like Gastro intestinal stucture (stomach, duodenum, small bowel and colon) and vascular structures. There are a lot of relative size variations among extraordinary organs of the body. Variation in relative length of the organ causes problems whilst applying deep networks to multi-organ segmentation, due to the fact that lower layers generally lack semantic information whilst segmenting small systems. To address these challenges an Organ attention Network with Opposite Connection (OAN-OC) is proposed by Wang et al., (2019). It is a two stage Convolution Neural network wherein features from the primary level are mixed with the original image in the 2nd stage to reduce the complexity of the heritage and to decorate organ boundaries via increasing the discriminative statistics approximately the organ of hobby. Impact of complicated historical past is decreased by focusing interest to the organs of interest. Reverse Connections are added to the primary level and lower layers get greater semantic data, thereby allowing them to adapt to the sizes of various organs.

Figure 2. Organ Attention Network with Opposite Connection

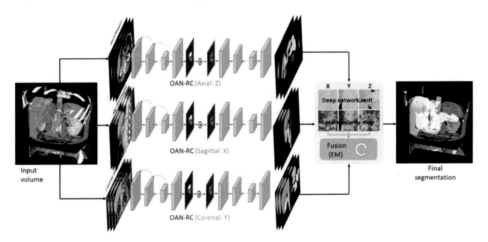

OAN-OC is carried out to each sectional slice, which is an extreme form of anisotropic nearby patches but the complete semantic (i.e. Quantity) information from one viewing route is also blanketed. This yields segmentation statistics from separate units of multi-sectional images (axial, coronal, and sagittal). The three resources of facts are statistically fused the use of nearby isotropic 3-d patches based on route - dependent local structural similarity. The fundamental fusion technique uses Expectation Maximization algorithm. Right here structural similarity is used and a path-based neighborhood assets is computed at each voxel. This models the structural similarity from the 2Dimages to the original 3-D shape (inside the 3-D volume) through neighborhood weights. This structural statistical fusion improves the overall performance through combining the records from the three unique perspectives in a principled manner and additionally imposing local shape.

UNET

In case of lack of training data, U-Net is a good choice. U-Nets have an ability to learn in environments of low to medium quantities of training data, and the amount of training data available is considered low. Also, a U-Net can be adapted to recent research, since its architecture is quite similar to the PSP Net or the One Hundred Layers Tiramisu, which are recent improvements such as for when dealing with more data. It is called a U-Net because it makes the shape of a U.

A U-Net is like a convolution auto encoder, but it also has skip-like connections with the feature maps located before the bottleneck (compressed embedding) layer, in such a way that in the decoder part some information comes from previous layers, bypassing the compressive bottleneck. See the Fig. 3 below.

Thus, in the decoder, data is not only recovered from a compression, but is also concatenated with the information's state before it was passed into the compression bottleneck so as to augment context for the next decoding layers to come. That way, the neural networks still learns to generalize in the compressed latent representation (located at the bottom of the "U" shape in the figure), but also recovers its latent generalizations to a spatial representation with the proper per-pixel semantic alignment in the right part of the U of the U-Net.

It is called a U-Net because it makes the shape of a U. To optimally train a U-Net, one needs to considerably augment the dataset. The 3rd place winners used the 4 possible 90 degrees rotations, as well as using a mirrored version of those rotation, which can increase the training data 8-fold: this data transformation belongs to the D4 Dihedral group. We proceeded with an extra random 45 degrees rotation, augmenting the data 16-fold, which represents the D8 Dihedral group. Thus every time an image is fed to the network, it is randomly rotated and mirrored before proceeding to train on that image. In the official U-Net paper, some elastic transforms are also used in the preprocessing. I think it would be interesting to use such data transformations, as depicted on a Kaggle kernel from another similar competition. In our pipeline, we also normalize every channel of the inputs and then use per-channel randomized linear transformations, for every patches.

Especially in the case of satellite imaging, it is possible to use the CCCI, NWDI, SAVI and EVI indexes. Those are extra channels (bands) computed from the naturally available bands. Such channels are used to extract more information in the images, especially for isolating reflectant objects, vegetation, buildings, roads, or water. For example, the 3rd place winners obtained good results for predicting the occurrence of water by directly setting a threshold on those indexes from a quick threshold search. The four extra channels (from indexes) are shown below, along with the original image's RGB human-visible channels for comparison:

VESSEL NET- A DENSE BLOCK- U NET FOR RETINAL BLOOD VESSEL SEGMENTATION

Recently many Deep Learning based Retinal vessel segmentation techniques had been proposed. Retinal vessels are tiny structures and retinal vessel structures are often learnt accurately using patch based learning techniques. Dense U nets had been used in scene segmentation. A replacement retinal vessel segmentation framework entirely based on Dense U-net and patch-based training method was proposed through wang et al., (2019). In this method training patches had been received by means of random extraction approach, Dense U-net became adopted as a learning network, and random transformation

Figure 3. UNET architecture

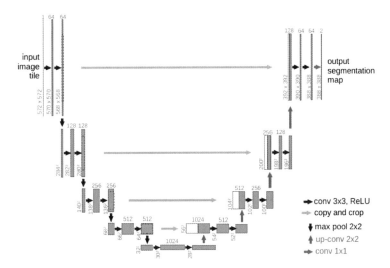

is used in the data augmentation process. In this method of testing, images are divided into patches, patches are predicted with the iearned models, and therefore the segmentation result is often reconstructed through overlapping-patches sequential reconstruction approach. This retinal segmentation method was tested using datasets DRIVE and STARE available in the public domain. The obtained resuls were more accurate than other recent state of art techniques. Sensitivity, Specificity, accuracy and region below each curve had been used as evaluation metrics to verify the effectiveness of this technique. Compared with cutting-edge techniques including the unsupervised, supervised, and convolution neural network (CNN) methods, the obtained result using this Dense UNET confirmed that this system is competitive in those assessment metrics.

ULTRA SOUND IMAGE SEGMENTATION

Ultra sound (US) is one of the imaging modality used in the screening for early prognosis of Breast cancers. Segmentation of ultra sound modality Breast image is very challenging and complex due to huge quantity of shadows, complex and poorly visible boundaries and poor contrast. Additionally US images are inherently corrupted by speckle noise and Rayleigh scattering because of tissue micro systems. Speckle artifacts are tissue dependant and can not be modeled correctly. Many speckle noise elimination and aspect retaining strategies were proposed. Many semi automatic and Interactive techniques were proposed for the segmentation of Breast Ultrasound snap shots. A Dilated fully Convolutional community (DFCN) blended with phase based lively contour approach became proposed by Hu etal., (2018).

The structure of the DFCN is described within the following segment and shown in Fig: 4 (a) & (b). It includes a Convolution layer, Batch Normalization Layer and a Rectified Linear Unit (ReLU). To maintain the dimensions of the function map regular after the convolution operation zero padding turned into executed. ReLU introduces non linearity. M x M x H within the block denotes the kernel size (M) and variety of feature maps (H).Max Pooling layer has a kernel size of 2x2 and a stride of 2. In this DFCN the number characteristic maps successively boom from decrease to better layers. The range of feature maps had been set to 64, 128, 256, 512, 512, and 1024 for layers 1- 6 respectively. A drop out layer become added after Convolution layer 1 at a charge of 0.Five. Subsequently, after the sum layer, a de convolution layer become used to give an up sampling prediction with a component of eight, indicating that the output became eight times the scale of the input. Segmentation end result become improved using dilated convolution since it reduces the effect of shadows within the photo. Max pooling layer make bigger the receptive field and down pattern the characteristic maps; however, in the application of give up-to-stop segmentation, max pooling reduces the decision of the function maps. Dilated convolution, also known as atrous convolution, inserts holes among nonzero clear out taps. Shadows are every now and then close to the lesions, and the distances among them are shortened in characteristic maps with low decision, in Breast extremely Sound pictures. Subsequently, it is difficult to differentiate them the usage of low-resolution feature maps. Dilated convolution can maintain the decision in deep filters and help the feature maps preserve more exact information to differentiate the lesions from shadows.

Figure 4. (a) DFCN architecture; (b) Architecture of a single block used in DFCN

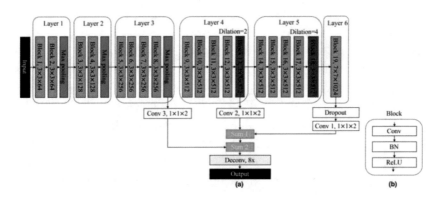

Kidney Segmentation in US Images

Guided segmentation of kidney US images are very hard and quiet time ingesting and quite vulnerable to inter and intra operator variability. Complete automatic segmentation of kidney images are quiet difficult because of various picture intensity distributions and shapes. But kidney limitations have surprisingly homogenous texture styles across pictures. Yin et al., (2019) used pre educated deep neural networks skilled on herbal snap shots to extract excessive degree capabilities from extremely sound images. This approach has been carried out to segmentation of scientific kidney US photos with massive variability in each look and form. The kidney photo segmentation version consists of a switch gaining knowledge of community, a boundary distance regression network, and a kidney pixel wise category community. The switch getting to know community is constructed upon a popular photograph classification network to reuse an photograph type version as a start line for learning high level photograph capabilities from US pictures, the boundary distance regression network learns kidney limitations modeled as distance maps of the kidney obstacles, and the kidney boundary distance maps are subsequently used as enter to the kidney pixel sensible class community to generate kidney segmentation masks. The kidney distance regression network and the kidney pixel sensible class network are skilled based on augmented training data which can be generated using a kidney form-based totally photograph registration.

The imaging facts used on this examine become amassed the use of trendy scientific US scanners at the children's health center of Philadelphia (CHOP). Particularly, 185 first post-natal kidney US images in sagittal view (Zheng et al., 2019; Zheng et al., 2018b), consisting of 85 from 50 kids with congenital abnormalities of the kidney and urinary tract (one from each strange kidney), and 100 from 50 kids with unilateral slight hydronephrosis (one from each kidney). 105 snap shots have been randomly decided on as training facts and 20 pictures as validation records in the present take a look at to educate deep gaining knowledge of-primarily based kidney segmentation models. The kidney segmentation models were then evaluated the usage of the last 60 pics and another set of 104 first submit-natal US images in sagittal view that had been also obtained from the CHOP. All of the pictures were obtained from special kidneys. The images were manually segmented by specialists from the CHOP. Representative kidney US pix are shown in Fig.1. These pix had been resized to have the equal size of 321×321, and their picture intensities have been linearly scaled to [0,255].

Figure 5. CNN based fully automatic kidney segmentation in US images

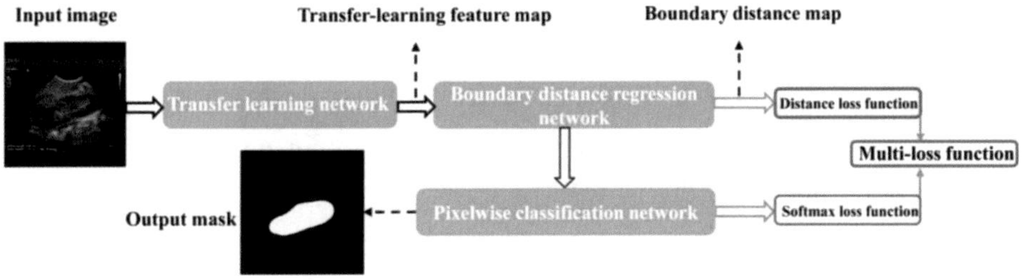

Co-Segmentation in PET/ CT

In clinical envioronment PET/CT scanner which combines PET which is a functional imaging modality and CT which images structural details are widely used. CT images have higher resolution compared to PET scans, but due to the similar intensity range of CT, their ability to distinguish tumor cells from surrounding soft tissues is limited. In the images shown below in figure 6: (a) & (b), the CT image has clear boundary between the tumors (region enclosed by the pink colored contour) and air (region in dark color) but the boundary between the tumor and normal soft tissues are not that much clear. Hence tumor segmentation using CT image alone is difficult. PET images have low spatial resolution, hence the tissue boundaries in them are fuzzy and indistinct, but they, exhibit high contrast, which helps in distinguishing normal soft tissues from tumor cells. Tumor cells in PET image have high standardized uptake value and appear as hot areas. Previous studies indicated that Zhao et al., 2018, proposed a multimodality 3d Deep learning based tumor segmentation co algorithm for ct / pet image. Features extracted from pet and ct were fused in this method use of structural and functional information can improve segmentation performance.

Figure 6. a) Lung CT Image; b) its paired PET Image for one patient

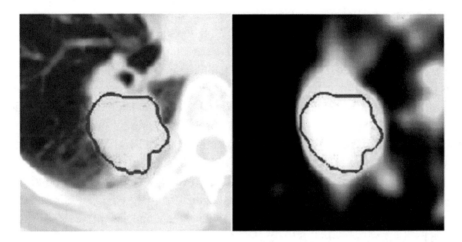

A multimodality 3d convolution neural network tumor co segmentation algorithm for ct/pet image was proposed by zhao et al., 2018. in this method features extracted from pet and ct were fused. two parallel sub segmentation branches consisting of v-net with a weighted loss function followed by a cascaded convolutional block which is used for feature fusion. paired pet/ct images of arbitrary sizes were fed to the network. two v net branches were used to perform feature extraction from pet and ct images respectively. the extracted features were fed to the fusion network having many cascaded convolutional layers a total loss function to achieve feature re extraction and output at the end of the fusion network is the prediction of tumor in the image. this network requires minimal preprocessing and no post processing.

Figure 7. Multi modality Co segmentation network (two parallel feature extraction branch followed by a Feature fusion module)

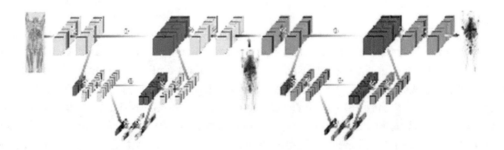

In spite of the availability **of** advanced imaging techniques segmenting and detecting bone lesions is still a challenge. Xu et al., (2018) proposed two architectures namely, V-Net and W-Net for whole body bone lesion segmentation and detection. These DL architectures do not require hand crafted features, which are difficult to identify with regard to multi modal characteristics. In W-Net architectures the structural information is used as regularization in lesion detection. Their preliminary results are based on phantom studies and limited amount of data and observed that by increasing the amount of data performance of the W-Net architecture may be improved.

Combining the sensitivity of PET images in identifying abnormal regions and with precise anatomical localization provided by Tomography images will improve the segmentation and detection of lesions in PET/CT Lung images. Ashnil kumar et al., (2018) proposed a Deep learning architecture which learns to fuse Complimentary information in PET/CT images of lungs. Heart has high FDG (Fluoro Deoxy Glucose) uptake under normal physiological conditions and it needs to be removed based on the anatomical information obtained from Tomographic image. In their study they used Tomographic images with resolution 512 x 512 at 0.98mm x 0.98mm with a slice thickness and inter slice distance of 3mm. and PET images with resolution 200 x 200 at 4.07mm x 4.07mm with a slice thickness and inter slice distance of 3mm. The Convolutional Neural Network proposed by them contains separate encoders for PET and Tomographic images, which extracted visual features relevant to each modality. The stacked convolutional layers of the network have structure similar to Alex Net and VGG Net. Each encoder contains two Convolutional Layer for feature map generation and a Max pooling layer to Down sample the feature Maps. In stacked structure, the change of weights in each layer will influence the subsequent convolutional layers in the network. During training changes happening in a layer will affect the deeper layers and the network has to continuously adapt to these changes. This CNN has inputs from two different imaging modalities hence the co learning and reconstruction will be affected by the cascading weight adjustments from both the encoders. This will reduce the speed of convergence and disturb the learning process. The dead neuron problem can be avoided using leaky RELU. Spatially varying fusion map is derived by the co-learning unit that more precisely integrates functional and anatomical visual features across different regions. Different fusion maps for different input PET- CT images, prioritizing different characteristics at different locations are used.

Figure 8. Co learning on a Single Feature Map

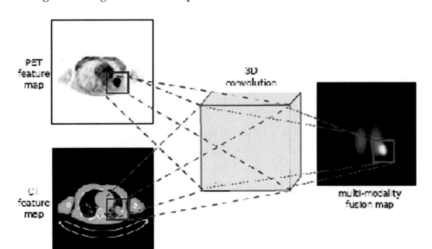

FOETAL ULTRASOUND IMAGE SEGMENTATION USING LINKNET

Imaging modalities like SPECT, PET, Resonance imaging and Tomographic imaging involves radiation hazards, they are invasive and nonportable, whereas Ultrasound imaging is non invasive and free from radiationn hazards. It is a simple portable non invasive imaging capable of diagnosing the causes of infection pain and swlling of internal body parts, used in medication and therapy of internal parts of the body (Rueda et al 2014). Ultra sound imaging is invariably used widely in Obstetrics and gynaecology examination and treatment procedures. Also it's a typical examination procedure during pregnancy which will be used for measuring specific biometric parameters towards diagnostic procedure and estimating fetal age, like the baby's abdominal circumference, head circumference, bi parietal diameter, femur and humerus length, and crown-rump length. Furthermore, the fetal head circumference (HC) is measured for estimating the fetal age, size and weight, growth monitoring and detecting foetal abnormalities.

Figure 9. Link Net architecture for segmentation of Foetal Ultra sound

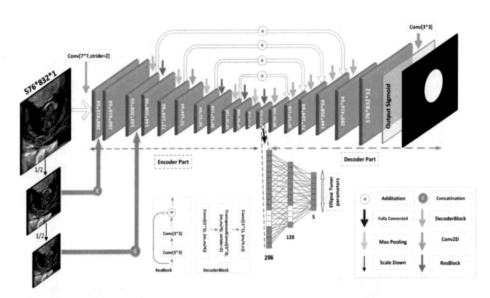

In spite of these advantages, Ultra sound imaging suffers from numerous artifacts like, motion blurring, missing boundaries, acoustic shadows, speckle noise, and low signal-to-noise ratio.

The fetal skull is not visible enough to detect in the first trimester. This makes the US images very challenging to interpret, which requires experts. Zahra Sobhaninia et al., (2019) proposed a multi task CNN based Link Net model originally developed for semantic segmentation. The block diagram of the Multi task Link Net model is shown in Figure 9. This model has two main modules a segmentation network and an Ellipse Tuner. Two different loss functions are defined for the two modules, which improve the training process for the whole network and lead to overall improvement in performance. During the training phase, the network parameters are trained by back-propagation of the combinatorial loss gradients through different entry points. The total loss function is a weighted sum of segmentation loss and Ellipse tuner loss. This network model is a modified version of Link Net with Multi scale inputs and is applied for fetal head segmentation. 2D ultrasound image in three scales, are fed into different layers of this model. The first half of the network contains encoding blocks called ResBlock (Zhou et al., 2018), which consists of convolutional and pooling layers with a residual link. A multi-scale structure is obtained by concatenating first and second feature maps of the network with the down-sampled versions of the input image. The second half of the network consists of decoder blocks that are responsible for up-sampling the feature maps and building up the final segmentation output. Encoders and decoders of similar dimensions are connected by some skip-connections. Feature-map details which might be lost throughout the encoding and decoding process are preserved by these skip connections. Presence of these skips connections leads to more accurate segmentation of the boundaries. The ground truth shapes, provided by the radiologist, have elliptical shapes hence ellipse parameters are investigated for assessment of the fetal head. The extracted features are fed into three Fully Connected (FC) layers for tuning ellipse parameters. The five outputs of the FC network estimate ellipse parameters which represent fetus head location, shape, etc. Furthermore, the proposed FC layers implicitly contribute to segmentation performance and improve the accuracy of the segmentation results, by refining the feature layers to symbolize an ellipse shape. About thousand two-dimensional ultrasound images were used for this study and fetus samples without any growth abnormalities were used. The spatial resolution of the of ultrasound images is 800 x 540 and pixel sizes are in the range 0.052 mm to 0.326 mm. Data augmentation, was done by horizontal and vertical flipping, and rotations from -60 to 60 degrees in steps of 20 degrees. In some cases, a rotation transform destroys the fetal head area by moving the head outside the rotated image. Hence, the corrupted images from the training set are removed and images with a complete fetal head were kept. About Nine thousand augmented US images were generated. The dataset was randomly split into 75%, for training and 25% for testing. 10% of the training data are set as validation data. Dice Similarity Coefficient, Difference, Absolute Difference and Hausdorff Distance were used for the evaluation of this system performance (Vanden et al., 2018). The proposed network was trained on about one thousand images and was evaluated on an independent test set which contains data from all trimesters.

FETAL MIDDLE CEREBRAL ARTERY SEGMENTATION USING MCANET

Recently women facing high-risk pregnancy are increasing at a faster rate because of delayed marriage and child bearing. Foetal growth is affected very much by pregnancy related complications leading to abnormal child births. Risks associated with complicated pregnancy outcomes are assesed using Doppler ultrasonic blood flow signals which reflect the changes of blood circulation in a foetal blood circulation.

The MCA Net architecture for Middle Cerebral Artery segmentation is shown in Fig.10., which consists of a convolution block with the kernel size 7×7, stride 1 and padding 3. .Similar blocks follow the above convolution block. In the following convolutional blocks 2, 3 and 4, the first convolution operation provides down sampling with the kernel size 1×1 and stride 2 and the remaining convolution operations use stride 1. The convolutional blocks 5 and 6 use dilated convolution with dilation 2 and 4. In the susequent layers of the network residual connections were removed and dilated convolution with dilations 2 and 1 is used. Removing residual connections in susequent blocks eliminate gridding artifacts due to dilated convolutions, which results in the improvement of prediction accuracy . Here dense up sampling is done instead of normal up sampling to recover the decoded feature map. Dense upsampling is done using a 3×3 filter kernel and stride 1 which results in the reduction of dimension and computation. Batch normalization operation leads to training with large learning rate and the pixel mixing operation is done to improve the resolution of decoded feature maps. Skip connections combine feature maps in shallow layers of Dilated Residual Network (DRN) and output feature map of Dilated Residual Network. Here an additional 1 × 1 convolution is added in the skip connection. Here the encoder is a Dilated Residual Network pre-trained on Image Net, since the area of the MCA in a raw ultrasound image is relatively small. The resolution of the output feature map of ResNet-34 on ultrasound MCA dataset is 20 × 14, which is only 1/32 of the raw ultrasound image. The output is too coarse for decoder in AlbuNet-34 network to restore fine detailed context information. By removing the max pooling layer Dilated Residual Network overcomes the above said limitations. DRN uses convolution filters for down sampling to remove gridding artifacts caused by the max pooling operation.In Dilated Residual Network dilated convolution helps to preserve spatial resolution, and it produces a comparatively high resolution output feature map. High resolution feature map contains fine details which is helpful in restoring a small object. MCA NET uses the skip connections to effectively combine the local information in shallow layers and semantic information in high layers. The effective small object restoration ability of dense upsampling is used to restore final details in MCA Net. Skip connections are not added to all layers in to maintain the computational efficiency. Feature map of the last layer has both large resolution and sufficient semantic information, which makes it possible to obtain good performance without adding complex skip connections. Furthermore, dense upsampling can directly perform convolution operations on the feature map, which is more computationally efficient than up sampling operations at the same scale. About 4000 qualified raw ultrasound images were selected out of the 5000 images to construct the ultrasound image dataset and experienced radiologists labelled these 4000 ultrasound images. Radiologists manually labelled these ultrasound images with the help of the corresponding color Doppler flow images. Recognizing and locating Middle Cerebral Arteries in raw ultrasound images are complicated tasks, even for experienced Physicians. The labelled dataset is further reviewed by senior radiologists to make sure the correctness of these labels. Jaccard Index, Dice Coefficient and Hausdorff Distance are used as the evaluation metrics to show the performance of each model in different aspects.

Figure 10. MCA Net for Middle Cerebral artery segmentation in an ultrasound image

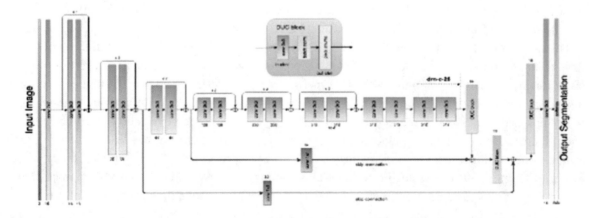

SKULL SEGMENTATION IN FETAL ULTRASOUND IMAGES

Ultra Sound images of the fetal structure have been largely used to assess the growth and well-being of the foetus, and estimating fetal weight. The examination of the fetal skull is a vital part of clinical examination, including head circumference (HC), Bi Parietal Diameter (BPD), and Occipito Frontal Diameter (OFD). But, this measurement is subject to more probability of measurement errors, since this is a manual 2D measurement done from extracted features from specified anatomical planes (i.e., the trans thalamic or trans ventricular planes). This human dependence of the procedure greatly affect the accuracy of sonography based fetal screening, limiting reproducibility, and may affect the early detection of formation of defective foetal structures. Moreover, the early detection of defective formation of cranial structures, such as dolichocephaly, or brachycephaly, requires a thorough understanding of the skull structure and its curved bones, and boundaries, which may be difficult to analyze and make measurement in a single 2D plane. Inspite of this limitations, manual 2D sonography based examination remains the current gold standard in foetal Ultra Sound analysis.

Figure 11. a): Two stage Convolutional Network, b) U-Net architecture

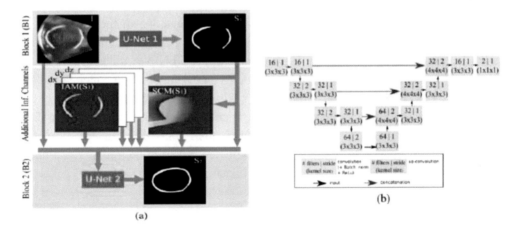

Currently, estimation of foetal head circumference and abdominal circumference (AC are done manually from ultrasound images by well trained radiologists. These parameters are useful benchmarks to gauge gestational age, but the process can be time consuming and laborious. To get around those problems, researchers have developed a machine learning method that takes into account skilled radiologists decisions in order to automate the estimation process. In comparison, earlier efforts to automate fetal structural estimation methods are based on image intensity values. This often led to accurate segmentation of well distinguished anatomical structures but can fail when measuring low-contrast features.

The researchers used a three-stage approach. In the first step, they obtained an initial estimate of the Abdomonal Circumference. Convolutional neural network (CNN) was used to determine the stomach bubble, amniotic fluid and umbilical vein in the ultrasound image, and from these three features, derived an estimate of the Abdomonal Circumference. The novelty of this approach is contained in the second step, where the images together with the Abdomonal Circumference estimates were fed into a second

CNN. This CNN then used these data to estimate the position of bony structures such as the mother's ribs. This information was then used to refine the initial Abdomonal Circumference estimate. In the final step, the researchers passed the final Abdomonal Circumference measurement along with the ultrasound images to a specific class of CNN known as a U-net. The U-net decided whether or not the ultrasound images, together with the Abdomonal Circumference estimate, are accepted or rejected, in a manner that mimics the decision of the clinician. In this way, the machine learns what to look for. Kim and colleagues used 112 images to train each CNN and the U-net, and 62 images to evaluate abdominal circumference. They obtained accurate segmentation of the ultrasound images, including images rejected by the machine because they showed the wrong anatomical plane, in 87.10% of the verification cases. Commercial systems are available to estimate the abdominal cavity volume from ultrasound images. However, these methods often fall short of clinical requirement due their inability to utilize structural information within the ultrasound image, such as shadowing artifacts caused by the ribs. By using a combination of several CNNs and a U-net, researchers have shown that machines have the capacity to learn how to provide this structural information, which can in turn assist clinicians by automatically segmenting images with a good degree of accuracy.

A new two-stage cascade approach as alternative architecture having additional information channels was proposed by Cerrolaza et al., (2018). These new channels are mainly constructed to deal with the difficulties arising in fetal 3D sonographic image segmentation process like shadowing and fading effects, to supply relevant information for the complete reconstruction of the skull. The overall flow diagram of the proposed framework is depicted in Figure 11(a).This segmentation algorithm divides the segmentation process into two parts, using a 3D convolutional U-Net which is a fully convolutional neural network which includes shortcut connections between a contracting encoder and a successive expanding decoder; as shown in Fig. 11(b), which is the basic structural element of the architecture. Taking the original 3D Sonographic image volume as single input channel, and the first block (B1) generates a partial reconstruction of the skull by segmenting the cranial bone visible in the sonographic image. Inspite of the large artifacts and very small signal-to-noise ratio of Sonographic imaging, bone tissue can be partially identified by its hyper echoic characteristics and curved structural pattern. Due to the sequential combination of many convolutional and pooling layers, the proposed CNN-based architecture learns particular filters at different resolutions and able to identify these discriminating features, and provide some incomplete, initial estimation of the skull. Based on this initial estimation, the the second block (B2) generates a full reconstruction of the skull, connecting those gaps and missing portions generated by the combination of fading and shadowing effects. Inspired by the auto-context architecture, the output probability map from B1 is used as an extra input channel for B2, which supplies valuable contextual and anatomical image information. Also, the input to B2 is completed with two extra channels, the Incidence Angle Map (IAM) and the Shadow Casting Map (SCM), both derived from the initial partial segmentation obtained in B1. The IAM and SCM supply necessary complementary information regarding the underlying physics of sonographic imaging, the angle of incidence of the Sonographic wave front, and the shadowing effect caused by the bone structures closer to the probe, respectively.

SEGMENTATION TO ESTIMATE PLACENTAL VOLUME USING OX N NET

Birth weight can be estimated from first trimester placental volume (Pl Vol.) and Pl Vol. measured with B-mode sonogram, which could be used to monitor the growth of foetus. Many studies have proved that a low PlVol between 11 and 13 weeks of gestation can result in adverse pregnancy outcomes, including small for gestational age (SGA) and preeclampsia. Pl Vol. has also shown to be independent of other biomarkers for SGA, such as pregnancy-associated plasma protein A (PAPP-A) and nuchal translucency, a thorough recent study concluded that it could be successfully integrated into a future multivariable screening method for SGA analogous to the "combined test" currently used to screen for fetal aneuploidy. As Pl Vol. is measured at the same gestation as this routinely offered combined test, no extra ultrasound scans would be required, making it more economically appealing to healthcare providers worldwide.

The only available method to determine Pl Vol. for a radiologist is to examine the 3D-sonogram, identify the placenta and manually label it. Software tools, such as Virtual Organ Computer-aided AnaLysis (VOCAL; General Electric Healthcare) and a semi automated random walker–derived (RW-derived) method have been developed to estimate Placental volume. But this method is too time consuming and it's accuracy depend on the radiologist. For Pl Vol. to become a useful measure to monitor the foetal growth, a reliable, real-time, radiologist-independent technique for the estimation of placental volume is needed. A real-time, fully automated technique for estimating Pl Vol. from 3D-Sonograms was proposed by Looney et al., (2018). They also studied the relationship between segmentation accuracy and the size of the training data set, and found the appropriate amount of training data required to maximize segmentation accuracy. Finally, the performance of the Pl Vol. estimates generated by the fully automated fCNN method to predict SGA at term was assessed by them. Deep learning was used with an exceptionally large ground-truth data set to generate an automatic image analysis tool for segmenting the placenta using 3D-Sonogram. This study uses the largest 3D medical image data set to date for fCNN training. In a number of data science competitions, the best performing models have used similar model architectures with poorer performing models but employed data augmentation to artificially increase the training set, suggesting a link between performance and the data set size. The learning curves presented here demonstrate a key finding of the need for large training sets and/or data augmentation when undertaking end-to-end training. The mean squared error learning curves of this model architecture the training and validation curves converged toward 0.275 as the training set size is increased. This was reflected in the monotonic increase across training samples, where the DSC for 1,200 training cases was 0.81 and the DSC was 0.73 for 100. Models are trained end to end on training sets with 100, 150, 300, 600, 900, and 1,200 cases. The mean squared error on the validation set decreased monotonically from 0.039 to 0.030 and increased monotonically from 0.01 to 0.025 on the training set. The median (inter quartile range) DSC obtained on the validation set throughout training increased monotonically from 0.73 (0.17) to 0.81 (0.15). Statistical analysis demonstrated a significant improvement in the DSC values with increasing training set size.

SUMMARY

Deep learning has been used to analyze a wide range of anatomies, including a large number of studies focusing on medical image segmentation. Fully convolutional networks (FCNs) are often used. These networks are closely related to Convolutional Neural Networks, but predict a value for each pixel or voxel, instead of a single prediction for the full image. In this chapter, the roll of various Convolution Neural Network models like Hyper Dense Net, Organ Attention Net, UNet, Link Net, VNet, Dilated Fully Convolutional Network, MCA Net Transfer Learning etc., in the segmentation of different Multimodal

Figure 12. O X N NET Fully Convolutional Network

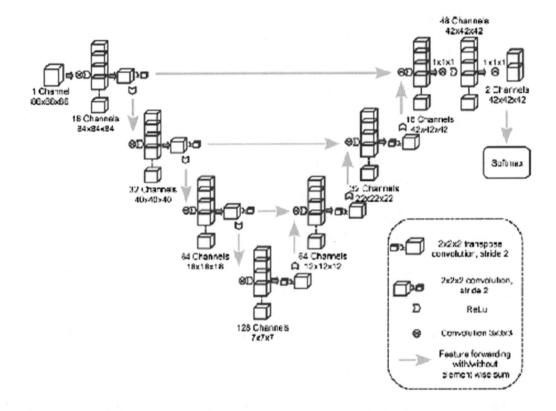

Images are explored in depth. A well-defined dataset, especially large-scale dataset, is essential to deep learning based techniques.

Deep learning, which is a subset of machine learning, is a representation learning approach which will directly process and automatically learn mid-level and high-level abstract features acquired from data (e.g., Tomography images). It holds the potential to perform automatic medical image segmentation tasks in different medical imaging modalities, like lesion/nodule classification, organ segmentation, and object detection. Since Alex Net a deep convolutional neural network (CNN) and a representative of the deep learning method, won the 2012 ImageNet Large Scale Visual Recognition Challenge (ILSVRC), deep learning began to draw attention within the field of machine learning. One year later, deep learning was selected together of the highest ten breakthrough technologies, which further consolidated its position because the leading machine learning tool in various research domains, and particularly generally imaging analysis (including natural and medical image analysis) and computer vision. To date, deep learning has gained rapid development in terms of network architectures or models, like deeper network architectures and deep generative models. Meanwhile, deep learning has been successfully applied to several research domains like Computer Vision, speech recognition, Remote Sensing and medical image analysis, thus demonstrating that deep learning may be a state-of-the-art tool for the performance of automatic analysis tasks, which its use can cause marked improvement in performance. Recent applications of deep learning in medical image analysis have involved various tasks, like traditional diagnosis tasks including classification, segmentation, detection, registration, biometric measurements, and quality assessment, also as emerging tasks including image-guided interventions and therapy. Of these, classification, detection, and segmentation are the three most elementary tasks. they're widely applied to different anatomical structures in medical image analysis involving organs like breast, prostate, liver, heart/cardiac, brain, carotid, thyroid, intravascular, fetus, lymph gland, kidney, spine, bone, muscle, nerve structure, tongue etc., . Multiple sorts of deep networks are involved in these tasks.It's a standard approach to use a CNN model to find out from the obtained data (e.g.,Tomographic images) so as to get hierarchical abstract representations, followed by a softmax layer or other linear classifier (e.g., a support vector machine) which will be used to produce one or more probabilities or class labels. Image annotations or labels are necessary for doing different image analysis tasks, and is called ''supervised learning.'' Unsupervised learning is additionally capable of learning representations from data without proper labels. Auto-encoders (AEs) and restricted Boltzmann's machines (RBMs) are two of the foremost commonly applied unsupervised neural networks in medical image analysis promising improvements in performance. Unsupervised learning has one significant advantage over supervised learning, which is that it doesn't require the use of time-consuming, labor-intensive, and expensive human annotations. Though closely related to semantic segmentation, medical image segmentation includes specific challenges that need to be addressed, such as the scarcity of labelled data, the high class imbalance found in the ground truth and the high memory demand of three-dimensional images. While replacing fully connected layers with convolutions allows for multiple pixels to be predicted efficiently and simultaneously, the resulting outputs typically have a very low resolution. This is caused by the fact that CNNs reduce the size of their input in multiple ways. Convolution layers with filters larger than 1 x 1 reduce the size of their input. This loss of pixels can be prevented by extending the input with a padding of zero-valued pixels.

REFERENCES

Chen, L., Papandreou, G., Kokkinos, I., Murphy, K., & Yuille, A. L. (2018b). DeepLab: Semantic Image 15 Segmentation with Deep Convolutional Nets, Atrous Convolution, and Fully Connected CRFs. IEEE 16. *IEEE Transactions on Pattern Analysis and Machine Intelligence*, *40*(4), 834–848. doi:10.1109/TPAMI.2017.2699184 PubMed

Dolz, J., Gopinath, K., Yuan, J., Lombaert, H., Desrosiers, C., & Ben Ayed, I. (2019). Hyper Dense-Net: A Hyper-Densely Connected CNN for Multi-Modal Image Segmentation. *IEEE Transactions on Medical Imaging*, *38*(5), 1116–1126. doi:10.1109/TMI.2018.2878669 PubMed

Kumar, A., Fulham, M., Feng, D., & Kim, J. (2018). Co-Learning Feature Fusion Maps from PET-CT Images of Lung Cancer. IEEE Transactions on Medical Imaging. Advance online publication. PubMed doi:10.1109/TMI.2019.2923601

Looney, Stevenson, Nicolaides, Plasencia, & Molloholli, Natsis, & Collins. (2018). Fully automated, real-time 3D ultrasound segmentation to estimate first trimester placental volume using deep learning. *JCI Insight*, *3*(10). Advance online publication. doi:10.1172/jci.insight.120178

Pereira, S., Pinto, A., Alves, V., & Silva, C. A. (2015). Deep convolutional neural networks for the segmentation of gliomas in multi-sequence MRI. *International Workshop on Brain lesion: Glioma, Multiple Sclerosis, Stroke and Traumatic Brain Injuries*, 131–143.

Ponomarev, G. V., Gelfand, M. S., & Kazanov, M. D. (2012). A multilevel thresholding combined with edge detection and shape-based recognition for segmentation of fetal ultrasound images. In *Proc. Chall. US biometric Meas. from fetal ultrasound images* (pp. 17–19). ISBI.

Rueda, S., Fathima, S., Knight, C. L., Yaqub, M., Papageorghiou, A. T., Rahmatullah, B., Foi, A., Maggioni, M., Pepe, A., Tohka, J., Stebbing, R. V., McManigle, J. E., Ciurte, A., Bresson, X., Cuadra, M. B., Sun, C., Ponomarev, G. V., Gelfand, M. S., Kazanov, M. D., ... Noble, J. A. (2014). Evaluation and comparison of current fetal ultrasound image segmentation methods for biometric measurements: A grand challenge. *IEEE Transactions on Medical Imaging*, *33*(4), 797–813. doi:10.1109/TMI.2013.2276943 PubMed

van den Heuvel, T. L. A., de Bruijn, D., de Korte, C. L., & van Ginneken, B. (2018). Automated measurement of fetal head circumference using 2D ultrasound images. *PLoS One*, *13*(8). doi:10.1371/journal.pone.0200412

Xu, L., Tetteh, G., Lipkova, J., Zhao, Y., Li, H., Christ, P., Piraud, M., Buck, A., Shi, K., & Menze, B. H. (2018). Automated Whole-Body Bone Lesion Detection for Multiple Myeloma on 68Ga-Pentixa for PET/CT Imaging Using Deep Learning Methods. *Contrast Media & Molecular Imaging*, *23*, 919–925.

Yin, Peng, Li, Zhang, You, Fischer, Furth, Tasian, & Fan. (2018). Automatic kidney segmentation in ultrasound images using subsequent boundary distance regression and pixel wise classification networks. . doi:10.1016/j.media.2019.101602

Zheng, Q., Furth, S. L., Tasian, G. E., & Fan, Y. (2019). Computer aided diagnosis of congenital abnormalities 28 of the kidney and urinary tract in children based on ultrasound imaging data by integrating texture image 29 features and deep transfer learning image features. *Journal of Pediatric Urology*, *15*(1), 75.e71–75.e77. doi:10.1016/j.jpurol.2018.10.020 PubMed

Zheng, Q., Warner, S., Tasian, G., & Fan, Y. (2018b). A Dynamic Graph Cuts Method with Integrated Multiple 2 Feature Maps for Segmenting Kidneys in 2D Ultrasound Images. *Academic Radiology*, *25*(9), 1136–1145. doi:10.1016/j.acra.2018.01.004 PubMed

Zhou, L., Zhang, C., & Wu, M. (2018). D-linknet: Linknet with pretrained encoder and dilated convolution for high resolution satellite imagery road extraction. Proceedings of the IEEE Conference on Computer Vision and Pattern Recognition Workshops, 182–186. doi:10.1109/CVPRW.2018.00034

Chapter 3
Deep Learning Techniques for Biomedical Image Analysis in Healthcare

Sivakami A.

iD https://orcid.org/0000-0001-5075-6365

Bharat Institute of Engineering and Technology, India

Balamurugan K. S.

Bharat Institute of Engineering and Technology, India

Bagyalakshmi Shanmugam

Sri Ramakrishna Institute of Technology, India

Sudhagar Pitchaimuthu

Swansea University, UK

ABSTRACT

Biomedical image analysis is very relevant to public health and welfare. Deep learning is quickly growing and has shown enhanced performance in medical applications. It has also been widely extended in academia and industry. The utilization of various deep learning methods on medical imaging endeavours to create systems that can help in the identification of disease and the automation of interpreting biomedical images to help treatment planning. New advancements in machine learning are primarily about deep learning employed for identifying, classifying, and quantifying patterns in images in the medical field. Deep learning, a more precise convolutional neural network has given excellent performance over machine learning in solving visual problems. This chapter summarizes a review of different deep learning techniques used and how they are applied in medical image interpretation and future directions.

DOI: 10.4018/978-1-7998-3591-2.ch003

INTRODUCTION

In modern years, Deep Learning (DL) (LeCun et al, 2015) has become a large influence on many fields in science and technology. It leads with advances and breakthroughs in audio recognition (Dahl etal,2012) and image recognition (Krishzvesky et al, 2012), it can prepare artificial agents that defeat human players in Go (Silver metal, 2016) and ATARI games (Mnih etal, 2015), and it produces artistic new images (Mordvintsev et al., 2015; Tan etal.,2017) and music (Briot etal.,2017). The final goal is to produce systems that can help in the diagnosis of disease, to automate the difficult and time-consuming tasks of reading and examining medical images, and to promote treatment planning. We require to be capable to employ machine learning methods to automatically distribute medical images (for example, breast x-ray or biopsy images) as healthy (non-cancerous) or not healthy (cancerous). (M H Hesamian et al, 2019). The diagnosis and treatment decisions can be made based on what is learned about the unhealthy images. The goal is that automatic classification and segmentation can be accomplished using distinct and innovative deep learning techniques, and extraordinary levels of accuracy can be accomplished (Shen et al., 2017). The diagnosis of particular image depends on both image acquisition and image interpretation. Image acquisition devices have been developed upto certain extent over the recent few years for getting the high resolution radiological images (X-Ray, CT and MRI scans, etc.) . However, we started to attain the more benefits for automated image interpretation (L B Curial et al, 2019). Computer vision activates the machine learning techniques to detect the image pattern as input and gives the effect in the form of size, colour size etc.Due to the extensive variety of different patient data, conventional learning methods are not guaranteed in the future. Now deep learning has much attention in all the fields especially in medicine. It is supposed to hold a $300 million medical imaging market in future. In 2021, deep learning will show rapid deveolpment in medicine than the other industry. It is the most powerful complex method of supervised learning. DL is particularly used for investigating the psychiatric and neurological disorders noncompulsory of manual feature selection.

DL technology implemented in medical imaging may enhance and increase innovative technology that has observed because of the digital imaging arrived. Over 15 years, most researchers understand that the applications of DL will bring over humans, and not only the diagnosis will be done by intelligent machines but will also help to prognosticate disease, command medicine and control of the disease treatment. The deep learning is transformed in several sectors such as ophthalmology, pathology, cancer detection, radiology etc. Ophthalmology is the first area to be transformed in health care, however, other sectors such as pathology and a cancer diagnosis has gained recognition and sufficient application with proper efficiency at present.

Medical image segmentation, recognizing and resolution of organs from different medical images produced by modern image techniques such as X-ray, ultrasound, CT or MRI. The images produced by imaging techniques has to give critical information about the shapes and volumes of the human organs. Several researchers have been reported (Suzuki etal., 2017; Lakhani et al., 2018; Kim etal., 2018) that the huge automated segmentation systems been developed by utilizing of existing DL methods. More advanced systems were built on common methods like mathematical methods and edge detection filters . Machine learning approaches towards image features have become more influential technique over others for a longer period of time. Due to hardware development, DL methods came into the picture and begun for giving the practical exhibition of image processing tasks in the field of academia and industry. The ability of deep learning procedures has started to good opportunity for image segmentation, and inappropriate for medical image segmentation. DL techniques have earned much attention in an image segmentation (Breininger et al.,2018) for past few years . There are several modern review and research articles on medical image segmentation (Alexander Lundervold et al., 2019; Maier etal., 2019). Figure1 representing the BD and DL in biomedical field. Deep learning is multi-disciplinary research field which includes physics, physiology, signals and systems (ss), mathematics, computer science, biology and medicine.

Figure 1. Diagram of how biomedical information such as biomedical signals and images, genomic sequences and EHRs can be managed by BD and DL in particular domains of healthcare.Reprint produced from L B Curiel etal., Appl. Sci. 2019.

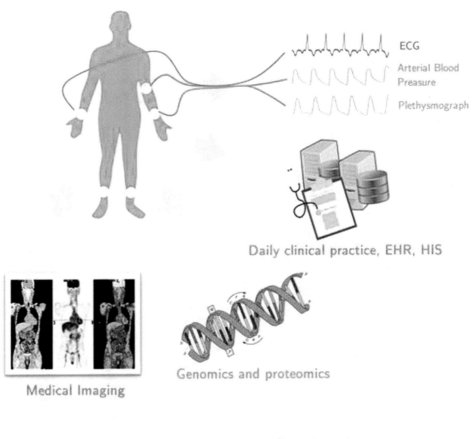

TYPES OF MEDICAL IMAGING

There are various methods of computer-aided detection system was produced and included in the clinical workflow in early 2010. The entirety of the principal responsibilities of medical imaging in radiology applications where the structural abnormalities are identified is classifying them into disease classifications. The different advanced imaging techniques are Microscopic, X-ray, Computer Tomography (CT), and Magnetic Resonance Imaging (MRI) (Ker etal., 2017; Lee etal, 2017; Litjens etal., 2017, Shen etal., 2017)

Radiography Imaging Technique

It is an imaging technique to view the inner parts of the object using ionizing or non-ionizing radiation such as X-rays, gamma rays etc. It employs a wide beam of X-rays to observe non-uniformly composed material present in the organs. The images produced from this technique is used to find the assessment of the presence or absence of disease, damage or foreign object.

MRI-Magnetic Resonance Imaging Technique

MRI is utilizing the high powerful magnets thereby emitting radio frequency pulse at the resonant frequency of hydrogen atoms to polarise and stimulate hydrogen nuclei of water molecules in human tissue. MRI doesn't suggest X-rays nor ionizing radiation. MRI is extensively used in hospitals and regarded as a more suitable choice than a CT scan because MRI attends medical diagnosis without revealing of the body to radiation. MRI scan takes a lot of time and are bigger than other scans.

Ultrasound Imaging Technique

Ultrasound utilizes high range of sound waves for producing the sort of 3D images by tissue. It is more generally associated with fetus image in pregnant women. Ultrasound is also applied for viewing the various inner parts of the human body such as abdominal organs, breast, heart,tendons, arteries, muscles and veins. It presents less anatomical details relative to CT or MRI scans. A significant improvement is ultrasound imaging helps to analyse the use of moving structures in actual time without emitting of radiation. Extremely reliable to practice, can be quickly performed without any unfavourable effects and relatively reasonable.

Endoscopy Imaging Technique

Endoscopy employs an endoscope that is infused directly into the organ to study the hollow organ of the body. The variety of endoscope changes depending upon the site to be observed in the body and can be administered by a doctor.

Thermography Imaging Technique

Thermographic imaging technique produces the graphic visualization with respect to surface variation of temperature of the body by measuring the infrared emissions out from it. The amount of radiation increases with an improvement in temperature. It is intelligent for capturing moving objects in real-time.

Thermographic cameras are entirely valuable. Images of the objects having the information with respect to temperatures might not result in perfect thermal imaging of it.

HARDWARE AND SOFTWARE FOR DEEP LEARNING

GPUs are parallel computing engines, which process the graphical datas per second than central processing units (CPUs). Among modern hardware, GPU deep learning is usually 10–30 times faster than on CPUs. By considering hardware, the motivation behind the most deep learning methods are existence of open-source software packages (OSS). The institutions which provide effective GPU implementation in deep neural network operations. The present most popular packages were detailed below.

1. Theano (Bastien et al., 2012)- provides Python interface
2. Tensorflow (Abadi etal., 2016)- provides C++ and Python and interfaces.
3. Torch (Collobert et al., 2011)- provides a Lua interface and is used by Facebook AI research
4. Caffe (Jia etal., 2014)- provides C++ and Python interfaces

DEEP LEARNING METHODS

DL is one part of machine learning in AI use of multi-layered networks capable of translating the feature from datasets automatically in various tasks such as images, audi, video and text with high accuracy (Shen etal. 2017; Schmidhuber, 2015). DL automatically will learn the complex data representation in a self-taught manner and identify more complex relationships between the input and output (LeCun etal. 2015). Several researchers and data scientists have become interested in using different DL methods to achieve record-breaking performances in various applications (Krizhevsky etal. 2012, Wu et al. 2016).

DL methods are divided into 5 principal groups (shown in Figure 2)

1. CNNs
2. Sequential models (Recurrent Neural Networks (RNNs)
3. Deep neural networks (Auto-Encoders)
4. Pre-trained models
5. Deep generative models (Restricted Boltzmann Machines (RBMs))

CONVOLUTIONAL NEURAL NETWORK'S ARCHITECTURE (CNNS)

It is one type of important DL algorithm to do image classification, recognition faces, object detections and image recognition etc..

CNN architecture consists of 3 different layers and its diagramatic representation as shown in figure 3.

Figure 2. Different existing DL models

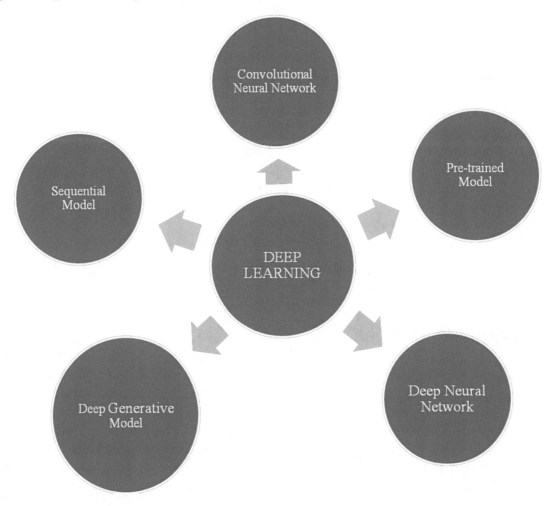

Convolutional Layer (CL)

CLs are major building blocks in the convolutional neural networks.It provides the relationship between pixels by image features using square of input data. The element involved in performing the operation in the first part of the convoultion layer is called Kernel or filter. If it is applied to small regions of an image, the pixel of in this region is converted into single pixel.

Pooling Layer (PL)

A pooling layer is a layer added after convolutional layer and summarizes a region of the feature map generated by CL .Here layer is responsible gradually for reducing the spatial size of the representation and the number of computations and parameters in the network. On comparable with different pooling layers, Max-Pooling is the most important method and widely used in CNN's architecture for analyzing the output of all the preceding layers in detail.

Figure 3. Schematic representation of typical CNN architecture (Khvostikov etal, 2018)

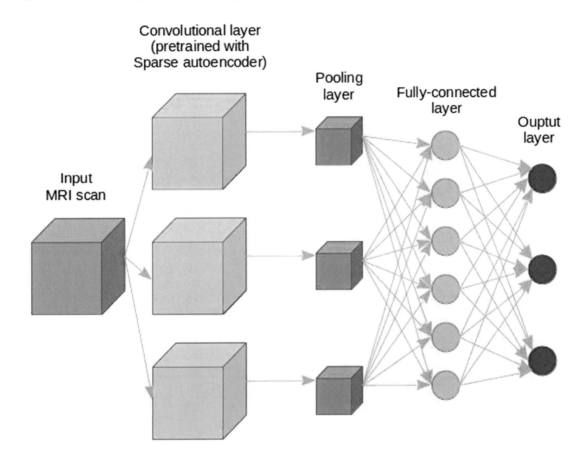

Fully Connected Layer (FCL)

FCLs are much essential component of CNNs which is proved in classifying and recognizing the images in computer vision sucessfully.(van Gerven et al. 2017). It represents the feature vector for the input and also simply called as feed forward neural network. The most powerful modern CNNs employed in the machine learning applications are AlexNet (Krizhevsk et al. 2012), Clarifai (Mamoshina et al, 2016), VGG (Simonyan et al. 2014), and GoogleNet (Szegedy et al 2014). Figure. 4 represents various recommended networks in biomedical applications.

RECURRENT NEURAL NETWORKS (RNNs)

Recurrent neural networks (RNNs) are another type of AI network which is used for sorting the data received from CT,cine MRI and ultrasound images. RNNs should be able to remember the past and process knowledge on image received from background history to progress its existing decision. For example, given a series of images, RNN practices the first image as input, catches the information to make a forecast, and then remember this information which is then appropriated to estimate the next

Figure 4. Most recommended Neural Network architecture used in biomedical fields: (A–C) feed forward neural networks for different depths; (D, E) an Auto-Encoder and a Deep Auto-Encoder,(F, G) descriptions of a restricted Boltzmann machine and a deep belief networks and (H) an AlexNet convolutional neural network (Zemouri et al. 2019)

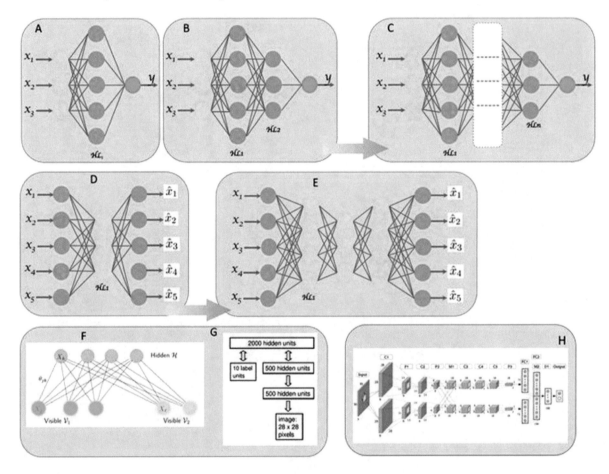

image in the network. The two most widely used architecture in the family of RNNs are gated recurrent unit (GRU) and long-short term memory (LSTM) (Zhou etal., 2015), is capable of modeling long-term memory as displayed in figure 5.

The different deep learning networks with images are presented in Table 1.

DEEP LEARNING USES IN MEDICAL IMAGING

It involves the acquiring, processing and visualizing the functional images of moving objects for both therapaeutic and diagnostic purpose. (Haque etal, 2020). The interpretation of medical image analysis are performed by manually like physicians, radiologists. However, the large variations exist in fatigue of human experts, the computer-assisted interventions are started and benefited. The focus of deep learning applications in bio medical field involves the classifcation and detection of computer-aided images.

Figure 5. Schematic Representation of RNN

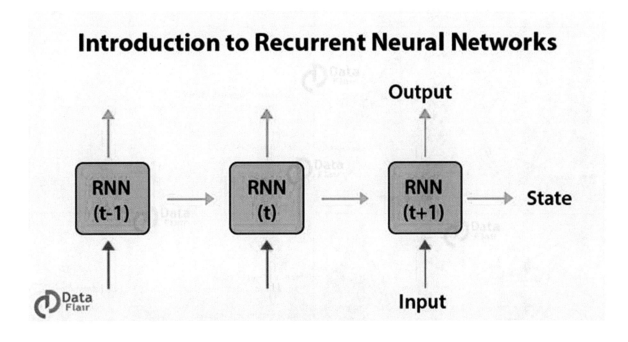

Figure 6. Percentage wise representation of different deep learning methods used in health informatics collected by Google scholar (Ravi etal., 2017).

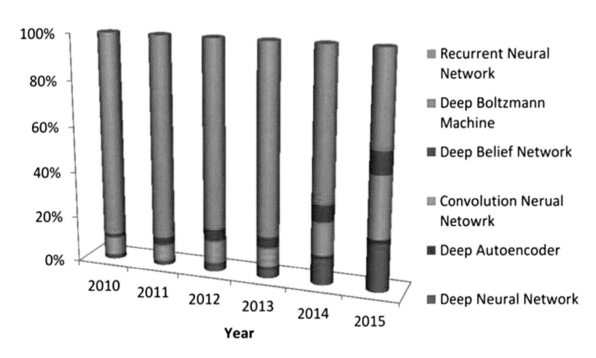

Table 1. Different deep learning architectures represented by embeeded objects (Ravi etal., 2016)

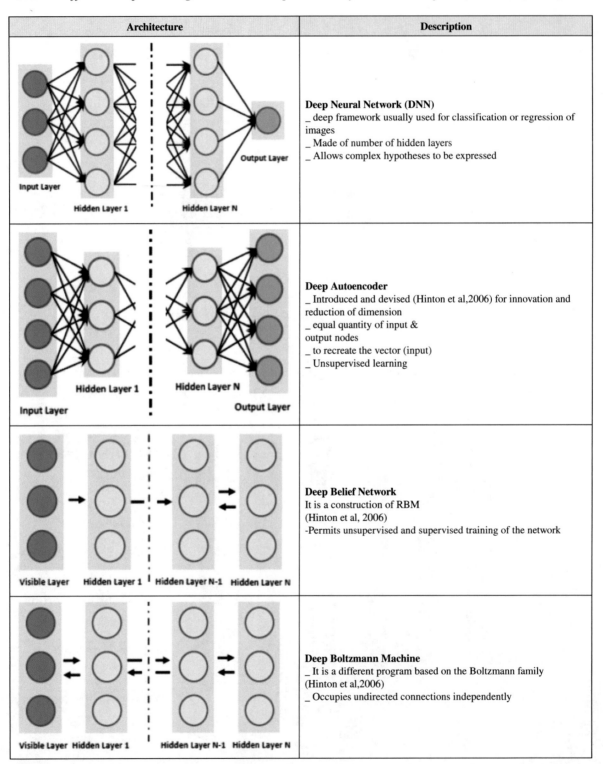

Architecture	Description
	Deep Neural Network (DNN) _ deep framework usually used for classification or regression of images _ Made of number of hidden layers _ Allows complex hypotheses to be expressed
	Deep Autoencoder _ Introduced and devised (Hinton et al,2006) for innovation and reduction of dimension _ equal quantity of input & output nodes _ to recreate the vector (input) _ Unsupervised learning
	Deep Belief Network It is a construction of RBM (Hinton et al, 2006) -Permits unsupervised and supervised training of the network
	Deep Boltzmann Machine _ It is a different program based on the Boltzmann family (Hinton et al,2006) _ Occupies undirected connections independently

continued on following page

Table 1. Continued

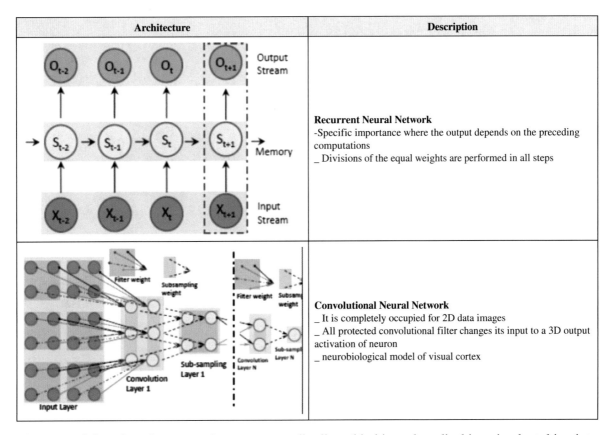

Architecture	Description
	Recurrent Neural Network -Specific importance where the output depends on the preceding computations _ Divisions of the equal weights are performed in all steps
	Convolutional Neural Network _ It is completely occupied for 2D data images _ All protected convolutional filter changes its input to a 3D output activation of neuron _ neurobiological model of visual cortex

The most of deep learning research papers were distributed in bio and medical imaging by taking into account of nuclei image segmentation, localization & classification and detect mitosis in breast histology images by using deep CNN. The biomedical sub- field associated with different biomedical applications are represented in figure 7.

Classification

DL computer vision utilizes a huge number of samples and in most of the cases, transfer learning is used. People in the DL community use the transfer learning technique for several reasons. Two main reasons are: first, use pre-trained weights for feature extraction, and second, use pre-trained weights to an existing network and fine-tune with a new dataset. For example, several studies have been carried out where the trained weights are used to form image net dataset. Out of many studies, some of the researches have reported the best results for medical image classification with transfer learning. Another outstanding work for skin cancer classification where they have achieved dermatologist-level performance for skin cancer disease recognition. In this work, the Google Inception-v3 model used with transfer learning (Suzuki et al, 2016; Shen et al, 2017) .

DCNN approaches are massively applied for neuroimaging such as Brain Tumour, Alzheimer disease segmentation and classification from MRI (Ravi etal., 2016). 2D and 3D convolutional kernels were applied in different models for this purpose. Besides, like Neuroimaging, there are different deep CNN

Figure 7. Schematic representation of different levels of biomedical applications (Zemouri et al,2019)

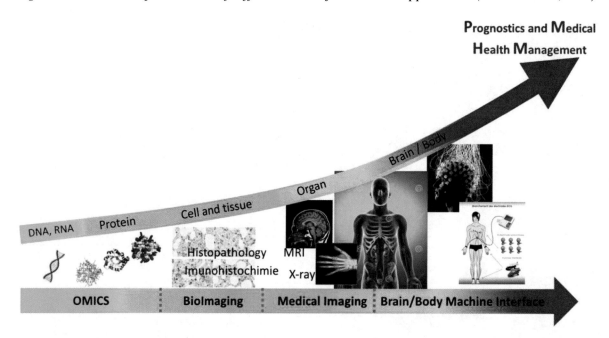

techniques are applied for analysis of detection and classification of lung disease images can be achieved extraordinary performance compared to machine learning approaches.

Segmentation

Numerous DL models have included specifically for the medical image segmentation tasks regarding the disadvantage of data insufficiency and class imbalance problems (Suzuki et al., 2017; Ker etl., 2017). The fast and most precise medical image segmentation called "U-Net". Different types of U-Net models have been introduced for CNN-based data Segmentation of medical image.The effect of skip connections exist in image registration and segmentation have been assessed with U-Net and other residual networks.

Computer-Aided Detection

Medical imaging and data acquisition system have been improved a lot in the last few years. The current CAD system is introduced to achieve two goals: detection and false-positive reduction. The primary objective can be performed based on the detection algorithm, DL methods are dispensing enormous performance on detection tasks. (Haque et al.2020; Liang et al, 2014). On the additional hand, inherited ML procedures such as SVM, PCA are employed for this purpose. Accidentally, traditional ML-based CAD does not function well in clinical practice. However, newly DL based techniques are dispensing excellent representation for false-positive modification responsibilities in different modalities of medical imaging. Newly, there are many CAD methods have recommended for breast cancer detection, lung cancer detection, and Alzheimer 's disease (AD) (Putin et al., 2016).

FUTURE RESEARCH DIRECTIONS

Deep learning is a rapid growing field in the medical application as well as academia. Also, there are many difficulties compared with introduction of several deep learning methods in clinical settings, which provides reliable returns and that are too valuable to abandon. DL performance depends on two main factors such as intensive computation power availability and second is a huge amount of data (Lundervold et al., 2019). In earlier, there are enough high-impact software-systems based on mathematics, physics, chemistry and engineering the entering the daily workflow in the clinic, the reception for other such systems will likely grow. The introduction of bio-sensors in edge computing as wearable devices for continuously observing different disease, an ecosystem of deep learning and other computational medicine-based technologies. AI and DL gradually enabled automated systems, solutions and tools are taking overall business sectors. Finally, there is a bi-directional movement of artificial intelligence between companies and research labs for enhancing the customer experience as well as high speed data analysis.

CONCLUSION

Deep neural network (DNNs) is currently playing the important primary tools in machine learning in all areas especially biomedical imaging, omics, brain and body machine interface and public health welfare. The different DL techniques and its usage in different stages of image analysis were discussed. Among all existing DL network, there is a developing concern in an end-to-end CNN, substituting the traditional handcrafted machine learning methods. This chapter summarized that CNNs are the most well-known and popular deep neural network and how the images are classified and detected. In most circumstances, transfer learning revealing the effect of biomedical imaging applications. Deep learning techniques is a promising interpreter of these data, serving in disease prediction, diagnosis, prevention and therapy. We anticipate that more deep learning applications will be accessible in biomedical areas such as epidemic prediction, disease prevention and clinical decision-making.

REFERENCES

Breininger, K., & Würfl, T. (2018). *Tutorial: how to build a deep learning framework.* https://github.com/kbreininger/tutorial-dlframework

Briot, J., Hadjeres, G., & Pachet, F. (2017). *Deep learning techniques for music generation – a survey.* CoRR abs/1709.01620

Collobert, Kavukcuoglu, & Farabet. (2011). *Torch7: A Matlab-like Environment for Machine Learning.* Academic Press.

Cuirel, L B., Romero, S.M., Cuireses, A., & Alvarez, R. L. (2019). Deep learning and Big data in health care: A double review for critical learners. *Appl.Sci, 9*(11), 2331.

Dahl, G. E., Yu, D., Deng, L., & Acero, A. (2012). Context-dependent pre-trained deep neural networks for large-vocabulary speech recognition. *IEEE Trans Actions Audio Speech Lang Process, 20*(1), 30–42. doi:10.1109/TASL.2011.2134090

Hesamian, M. H., Jia, W., He, X., & Kennedy, P. (2019). Deep Learning Techniques for Medical Image Segmentation: Achievements and Challenges. *Journal of Digital Imaging*, *32*(4), 582–596. doi:10.100710278-019-00227-x PMID:31144149

Hinton, G. E., Osindero, S., & Teh, Y.-W. (2006). A fast learning algorithm for deep belief nets. *Neural Computation*, *18*(7), 1527–1554. doi:10.1162/neco.2006.18.7.1527 PMID:16764513

Hinton, G. E., & Salakhutdinov, R. R. (2006). Reducing the dimensionality of data with neural networks. *Science*, *313*(5786), 504–507. doi:10.1126cience.1127647 PMID:16873662

Ihaque, I. R., & Neubert, J. (2020). Deep leaning approaches to biomedical image segmentation. *Informatics Medicine Unlocked, 18*.

Jia, Y., Shelhamer, E., Donahue, J., Karayev, S., Long, J., Girshick, R., Guadarrama, S., & Darrell, T. (2014). Caffe: Convolutional architecture for fast feature embedding. In *Proceedings of the 22nd ACM international conference on Multimedia*. ACM. 10.1145/2647868.2654889

Ker, J., Wang, L., Rao, J., & Lim, T. (2017). Deep Learning Applications in Medical Image Analysis. *IEEE Access: Practical Innovations, Open Solutions*, *6*, 9375–9389. doi:10.1109/ACCESS.2017.2788044

Kim, J., Hong, J., Park, H., Kim, J., Hong, J., & Park, H. (2018). Prospects of deep learning for medical imaging. *Precis Future Med.*, *2*(2), 37–52. doi:10.23838/pfm.2018.00030

Krizhevsky, A., Sutskever, I., & Hinton, G. E. (2012). ImageNet classification with deep convolutional neural networks. *Proceedings of the 25th International Conference on Neural Information Processing Systems*, 1097-1105.

Krizhevsky, A., Sutskever, I., & Hinton, G. E. (2012). ImageNET classification with deep convolutional neural networks. Advances in Neural Information Processing Systems, 1097–105.

Lakhani, P., Gray, D. L., Pett, C. R., Nagy, P., & Shih, G. (2018). Hello world deep learning in medical imaging. *Journal of Digital Imaging*, *31*(3), 283–289. doi:10.100710278-018-0079-6 PMID:29725961

LeCun, Y., Bengio, Y., & Hinton, G. (2015). Deep learning. *Nature*, *521*(7553), 436–444. doi:10.1038/nature14539 PMID:26017442

Lee, J. G., Jun, S., Cho, Y.-W., Lee, H., Kim, G. B., Seo, J. B., & Kim, N. (2017). Deep learning in Medical imaging: General overview. *Korean Journal of Radiology*, *18*(4), 570–584. doi:10.3348/kjr.2017.18.4.570 PMID:28670152

Liang, Z., Zhang, G., Huang, J. X., & Hu, Q. V. 2014. Deep learning for healthcare decision making with emrs. *Proc. Int. Conf. Bioinformat. Biomed.*, 556–559. 10.1109/BIBM.2014.6999219

Litjens, G., Kooi, T., Bejnordi, B. E., Adiyoso Setio, A. A., Ciompi, F., Ghafoorian, M., van der Laak, J. A., van Ginneken, B., & Sánchez, C. I. (2017). A Survey on Deep Learning in Medical Image Analysis. *Medical Image Analysis*, *42*, 60–88. doi:10.1016/j.media.2017.07.005 PMID:28778026

Lundefvold, A. S., & Lundervold, A. (2019). An overview of deep learning in medical imaging focusing on MRI. *Zeitschrift fur Medizinische Physik*, *29*(2), 102–127. doi:10.1016/j.zemedi.2018.11.002 PMID:30553609

Maier, A., Syben, C., Lasser, T., & Rless, C. (2019). A gentle introduction to deep learning in medical image processing. *Zeitschrift fur Medizinische Physik, 29*(2), 86–101. doi:10.1016/j.zemedi.2018.12.003 PMID:30686613

Mamoshina, P., Vieira, A., Putin, E., & Zhavoronkov, A. (2016). Applications of Deep Learning in Biomedicine. *Molecular Pharmaceutics, 13*(5), 1445–1454. doi:10.1021/acs.molpharmaceut.5b00982 PMID:27007977

Mnih, V., Kavukcuoglu, K., Silver, D., Rusu, A. A., Veness, J., Bellemare, M. G., Graves, A., Riedmiller, M., Fidjeland, A. K., Ostrovski, G., Petersen, S., Beattie, C., Sadik, A., Antonoglou, I., King, H., Kumaran, D., Wierstra, D., Legg, S., & Hassabis, D. (2015). Human-level control through deep reinforcement learning. *Nature, 518*(7540), 529–533. doi:10.1038/nature14236 PMID:25719670

Mordvintsev A, Olah C, Tyka M. (2015). Inceptionism: going deeper into neural networks. *Google Research Blog*.

Putin, E., Mamoshina, P., Aliper, A., Korzinkin, M., Moskalev, A., Kolosov, A., Ostrovskiy, A., Cantor, C., Vijg, J., & Zhavoronkov, A. (2016). Deep biomarkers of human aging: Application of deep neural networks to biomarker development. *Aging (Albany NY), 8*(5), 1–21. PMID:27191382

Ravi, D., Wong, C., Deligianni, F., Berthelot, M., Andreu-Perez, J., Lo, B., & Yang, G.-Z. (2017). Deep Learning for Health Informatics. *IEEE Journal of Biomedical and Health Informatics, 21*(1), 4–21. doi:10.1109/JBHI.2016.2636665 PMID:28055930

Schmidhuber, J. (2015). Deep learning in neural networks: An overview. *Neural Networks, 61*, 85–117. doi:10.1016/j.neunet.2014.09.003 PMID:25462637

Shen, D., Wu, G., & Il Shuk, H. (2017). Deep Learnoing in medical image analysis. *Annual Review of Biomedical Engineering, 19*(1), 221–248. doi:10.1146/annurev-bioeng-071516-044442 PMID:28301734

Silver, D., Huang, A., Maddison, C. J., Guez, A., Sifre, L., Van Den Driessche, G., Schrittwieser, J., Antonoglou, I., Panneershelvam, V., Lanctot, M., Dieleman, S., Grewe, D., Nham, J., Kalchbrenner, N., Sutskever, I., Lillicrap, T., Leach, M., Kavukcuoglu, K., Graepel, T., & Hassabis, D. (2016). Mastering the game of go with deep neural networks and tree search. *Nature, 529*(7587), 484–489. doi:10.1038/nature16961 PMID:26819042

Simonyan, K., & Zisserman, A. (2014). *Very Deep Convolutional Networks for Large-Scale Image Recognition*. arXiv 2014, arXiv:1409.1556

Suzuki, K. (2017). Survey of deep learning applications to medical image analysis. *Med Imaging Technol, 35*, 212–226.

Szegedy, C., Liu, W., Jia, Y., Sermanet, P., Reed, S. E., Anguelov, D., Erhan, D., Vanhoucke, V., & Rabinovich, A. (2014). *Going Deeper with Convolutions*. arXiv 2014, arXiv:1409.4842

Tan, W. R., Chan, C. S., Aguirre, H. E., & Tanaka, K. (2017). ArtGAN: artwork synthesis with conditional categorical GANs. *2017 IEEE International Conference on Image Processing (ICIP)*, 3760–4. 10.1109/ICIP.2017.8296985 van Gerven, M., & Bohte, S. (2017). Editorial: Artificial Neural Networks as Models of Neural Information Processing. *Frontiers in Computational Neuroscience*, *11*, 114–114. doi:10.3389/fncom.2017.00114 PMID:29311884

Wu, G., Kim, M., Wang, Q., Munsell, B. C., & Shen, D. (2016). Scalable High-Performance Image Registration Framework by Unsupervised Deep Feature Representations Learning. *IEEE Transactions on Biomedical Engineering*, *63*(7), 1505–1516. doi:10.1109/TBME.2015.2496253 PMID:26552069

Zemouri, R., Zerhouni, N., & Racoceanu, D. (2019). Deep learning in the biomedical Applications: Recent and Future status. *Applied Sciences (Basel, Switzerland)*, *9*(8), 1526. doi:10.3390/app9081526

Zhou, S., Greenspan, H., & Shen, D. (2017). *Deep Learning for Medical Image Analysis*. Academic Press.

Chapter 4

Heterogeneous Large-Scale Distributed Systems on Machine Learning

Karthika Paramasivam
Kalasalingam Academy of Research and Education, India

Prathap M.
Wollo University, Ethiopia

Hussain Sharif
Wollo University, Ethiopia

ABSTRACT

Tensor flow is an interface for communicating AI calculations and a use for performing calculations like this. A calculation communicated using tensor flow can be done with virtually zero changes in a wide range of heterogeneous frameworks, ranging from cell phones, for example, telephones and tablets to massive scale-appropriate structures of many computers and a large number of computational gadgets, for example, GPU cards. The framework is adaptable and can be used to communicate a wide range of calculations, including the preparation and derivation of calculations for deep neural network models, and has been used to guide the analysis and send AI frameworks to more than twelve software engineering zones and different fields, including discourse recognition, sight of PCs, electronic technology, data recovery, everyday language handling, retrieval of spatial data, and discovery of device medication. This chapter demonstrates the tensor flow interface and the interface we worked with at Google.

INTRODUCTION

In the Google Brain activity started to investigate the utilization of gigantic extent of profound neural frameworks for both disclosure and use in Google items the work of (Karthika P and P. Vidhya Saraswathi, 2017). As a piece of the early work right now, Dist Belief, our unique adaptable arranging and acceptance

DOI: 10.4018/978-1-7998-3591-2.ch004

structure, has been collected and this framework has served us well. We and others at Google have done a wide scope of research utilizing Dist Belief reviewing work for solo language portrayal learning, picture request and article acknowledgement models, video gathering, progression talk acknowledgement, moving spot, stronghold learning, and different regions. Also, in close participation with the Google Brain arrange, in excess of 50 Google groups and other letters in order associations have habitually passed on profound neural frameworks utilizing Dist Belief in a wide scope of things, including Google Search our commercial items, our talk acknowledgement structures, Google Photos Google Maps and Street View, Google Translate, YouTube and numerous others.

Given our association in Dist Belief and a slow comprehension of the appealing structure properties and essentials for the arrangement and utilization of neural frameworks, we have created Tensor Flow, our second-age system for the execution and sending of huge scope AI models. Tensor Flow takes counts spoke to utilizing a dataflow-like model and maps them to a wide scope of gear levels. The range from working wireless determination to, for instance, Android and iOS to humble size frameworks for arranging and induction. This utilizes single machines with one or numerous GPU cards to design monstrous degree frameworks running on a wide range of machines with an enormous number of GPUs. Utilizing a solitary framework that can experience such a wide exhibit of stages fundamentally revamps this current utilization of truth. Computer based intelligence framework, as we have discovered that having separate frameworks for the arranging of gigantic degree and restricted extension structure triggers basic help stacks and split consultations.

PROGRAMMING REPRESENTATION AND BASIC PERCEPTION

A Tensor Flow measurement is represented through a coordinated diagram, which consists of a lot of hubs. The diagram speaks to a data flow estimate, with increments to allow a few types of hubs to maintain and update industrial conditions and to extend and circle organize structures inside the diagram in a method like Naiad. Customers create a logical diagram on a regular basis using single of the maintain front end dialects. A model component for constructing and executing a Tensor Flow chart using the frontage end of Python is revealed in Figure 1 and the calculation diagram below in Figure 1. -hub has at least zero data sources and at least zero yields in a Tensor Flow chart, and speaks to the start of an action. Qualities that stream along the typical edges of the diagram (from yields to inputs) are tensors, self-assertive dimensionality clusters that evaluate or induce the basic component form at the time of graph creation. Extraordinary edges, known as control conditions, can also be found in the chart: no information streams beside such edges, but it specify that the control reliance foundation hub must be executed before the control reliance target hub begins to execute. Since our model integrates variable state, customers may legally use the control conditions to allow the event before connections occur. The execution often integrates control criteria for ordering autonomous operations in any case as a mechanism for monitoring, for example, the use of pinnacle memory.

Activities and Kernels

In recent years we have observed large change sine convoying enema land marketing in particular as a result to fin tenet expansion, globalization, and ubiquitous sin for motion availability. One of the scientific fields which gained result of this was data analysis under various names: statistics, data mining, machine learning, intelligent data analysis, knowledge discovery. Many new data analysis technique re-emerged which exploit avail ability of more and different data from several sources, and increased computational power of nowadays computers. Some examples of these techniques are support vector machines, text analytics, association rules, ensemble techniques, subgroup discovery, etc. These techniques have been accepted into analytics' standard tool box in many disciplines: genetics, engineering, medicine, vision, statistics, marketing, etc.

Meetings

Customer services communicate by having a Session with the Tensor Flow System. The Session interface supports an Extend technique for extending the present diagram supervised by meeting with extra hubs and edges to create a calculation map (the underlying chart is empty when a meeting is made). The other important operation maintained by meeting in which receive a lot of yield forename to processed, just like a discretionary tensor arrangement to be taken care of in the map instead of unique hubs yields. Using the reasons to Fly The use of the Tensor Flow can represent the transitive conclusion of all the hubs that have to be executed in order to record the yields that have been stated, and can then organize the execution of the correct hubs in a request that respects their conditions. Most by far of our Tensor Flow businesses locate a session with the chart once and afterward executes the full graph or a couple of obvious sub-outlines thousands or an enormous number of times through Run Calls.

Figure 1. Corresponding statistical diagram

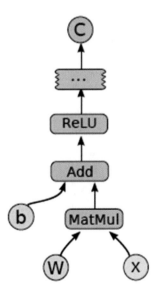

Factors

In many equations on various occasions a diagram is performed. Most tensors don't do a single diagram execution due history. In any case, a variable is a special operation that takes advantage of a handle to an impermanent tensor that gets through a chart's execution. Handles to these relentless alterable tensors can be transferred to a bunch of specific activities, such as Assigning and Assigning Add (proportionate to + =) which changes the referenced tensor. For Tensor Flow's AI uses, the model's parameters are usually displayed in tensors kept in factors and refreshed as a major aspect of the model's preparation chart run.

IMPLEMENTATION

The fundamental segments in a Tensor Flow organization are the customer who utilize the session boundary to speak to the ace with at least one worker type, with every specialist procedure accountable for mediating entrance to at least one device gadget and for implement chart hubs on individuals' gadgets as the ace practiced. It both has neighborhood and Tensor Flow interface use hijacked. Nearby use is used when the client, the ace, and the worker all of a sudden spike in demand for a solitary computer with respect to a solitary working system phase (potentially with specific gadgets if, for example, the machine has added multiple GPU cards). The disseminated use imparts the bulk of the code to the execution of the area, but extends it with the aid of a situation in which the client, the ace, and the staff could all be able to execute different machines in different procedures.

Such separate assignments are compartments in occupations supervised by a group booking system in our conveyed state. These two singular modes are shown in Figure 2. A large portion of the remainder of this field addresses problems that are central to the two executions, while Section 3.3 explores a few specific issues relevant to the proper use.

Gadgets

Gadgets are the heart of Tensor Flow computing. For at least one gadget, each specialist is liable, and each gadget has a form of gadget and a name. Gadget names are made of parts that recognize the type of gadget, the record of the gadget inside the worker, and a distinctive proof of the specialist's operation and undertaking in our conveyed setting. The name of the model gadget is "/job: localhost / computer: cpu:0" or "/job: worker / device:17/device: gpu:3.". Each gadget object is responsible for monitoring portion and gadget memory distribution and for organizing the execution of any bits specified by higher stage in the use of Tensor Flow.

Tensors

A tensor is a composed, multidimensional presentation in our execution. We bolster an assortment of kinds of tensor modules, including checked and unsigned entire numbers running in size of 8 bits to 64 bits, IEEE float and twofold sorts, an incredible number sort, and a string type (an abstract byte appear). Help store of the suitable size is regulated by allocator legitimately to the device the tensor dwells on. Tensor sponsorship shop supports are checked on for reference with are managed if no references remain.

Execution of the Single-Device

Next, helps one to think about the simplest execution situation: a single specialist operation with a solitary gadget. The diagram hubs are implemented in an application that relates to the conditions between hubs. Actually, we track a count for each hub of the quantity of conditions that were not yet implemented in that hub. The hub is ready for execution and is attached to a prepared line when this test drops to zero. In some unknown order, the prepared line is treated which designates execution of the piece for a hub to the object of the gadget. The checks of all hubs that depend on the finished hub are decreased at the point when a hub is wrapped up.

Multi-Device Execution

There are two fundamental intricacies when a system has multiple gadgets: choosing the gadget to position the measurement for each hub in the diagram, and then dealing with the necessary correspondence of knowledge across the limits of the gadget implied by these choices of situation. This chapter looks at those two things.

Node Placement

One of the fundamental responsibilities of Tensor Flow execution, given a calculation map, is to delineate calculation on the arrangement of accessible gadgets. Regarding extensions supported by this estimate, see Section 4.3. One contribution to position calculation is a charge representation that involves assessments the sizes of information and yield tensors for each diagram hub, as well as assessments of the calculation time required for each hub given its information tensors. This charge representation is either evaluated dynamically support on heuristics relevant to specific types of activity, or is calculated dependent on a genuine arrangement of position choices for prior graph execution.

Figure 2. Particular machine and Disseminated Organization Structure

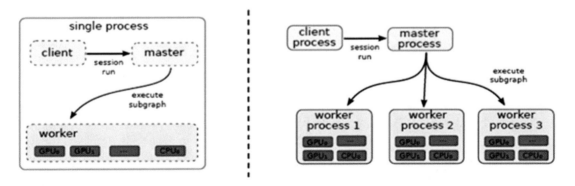

The reproduction is shown below, and winds up selecting a gadget for each hub in the map using avaricious heuristics. The hub to gadget structure created by this reenactment is also used as the real execution situation. The location calculation begins with the measurement chart's wellsprings and recreates the operation on each gadget as it progresses. For each hub that has just arrived, a selection of attainable gadgets is considered (a gadget may not be feasible if the gadget does not give the specific activity a bit). Among hubs with specific attainable gadgets, the situation calculation uses an eager heuristic that looks at the implications of placing the hub on any imaginable gadget for the period of fruition. This heuristic receive into explanation the measured or projected implementation period of procedure on this form of gadget from the charge representation and also involves the expenditure of several communication it will all be familiar with transmitting inputs from various gadgets to the gadget thought.

The gadget where the operation of the hub is to be done as soon as possible is selected as the gadget for that activity and the situation procedure at that stage begins to decide the arrangement choices for the various hubs in the diagram, including downstream hubs currently being prepared for their own re-enactment. A few increases which allow clients to offer insights and fractional imperatives to manage the calculation of the situation. The measurement of the agreement is a region of improvement within the organization.

Communication of the Cross-Device

After enrollment of center point area, the guide is partitioned into many sub diagrams, one for every device. Several cross-contraption edges from x to y is emptied and supplanted by an edge from x to another Send center point in the subsection of x and an edge from a comparable Receive center point to y in the sub diagram. See Figure 4 for an update to this outline.

Figure 3. Previous & after the Send / Receive nodes are inserted

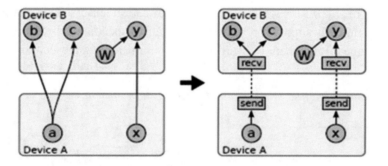

The Send and Receive hubs executions plan for moving information through gadgets at run time to work of (Ganesh Babu, R., & Amudha, V 2016). This helps us to remove all communications within Send and receive executions, thus enhancing the respite of the runtime. Once add Send and the Receive hubs it canonize all customers of a exacting tensor a specific gadget to utilize a particular Receive hub instead of one Receive hub on a specific gadget for each downstream customer. This ensures that the information is only transmitted once between a source gadget for the appropriate tensor! Goal gadget pair, and that tensor memory is only allocated once on the goal gadget, instead of different occasions. By coping with correspondence right now, it is also possible to decentralize the reservation of individual hubs of the map on specific gadgets into the workers: the hubs Send and Receive grant vital communication among different specialists with gadgets only needs to give a solitary one. Scurry demand for each diagram execution to every specialist who has any hubs for the diagram, as opposed to participating in the preparation of each hub or communication with each cross-gadget. It makes the frame considerably more adaptable and allows for much better operations of the granularity center than if the preparation had to be done.

Distributed Execution

Appropriate execution of a diagram essentially is the same as execution of multi devices. A sub graph is rendered per gadget after location of the gadget the work of (P Karthika and P Vidhyasaraswathi, 2017). Send / receive hub matches that use remote communication instruments across specialist types, such as TCP or RDMA, to transfer information across system limits.

Adaptation to Internal Failure

Disillusionments can be found in a variety of areas in a coursed execution. The fundamental ones on which we depend are (an) a bumble in a correspondence between a Send and Receive center point pair, with (b) incidental well-being tests from the as strategy to every master system. At the point when a failure is perceived, the whole execution of the guide is pointlessly ended and restarted with no preparation. Check anyway that Variable hubs allude to tensors that live through chart execution the work of (Antony Selvadoss, Thanamani, Prathap M, 2018). We emphasize automatic check pointing and this state recovery on restart. Each Variable hub is associated to a Save hub in particular form. These Save Hubs are executed intermittently; state each N cycle or every N second the work of (P Karthika and P Vidhyasaraswathi, 2020). The content of the variables are kept in contact with constant stockpiling at the point when they are performed, e.g., a disseminated record structure. Essentially each Variable is connected with a Restore center that is only operated after a restart in the primary cycle. See Section 4.2 for subtleties on how to motivate a few hubs on certain diagrammatic executions.

EXTENSIONS

Image a few more established highlights of the core programming model outlined in section 2.

Gradient Computation

Numerous enhancement calculations, including standard AI calculations such as stochastic angle plummet, show the slope of a cost work related to a lot of sources of data. Since this is a typical requirement, Tensor Flow has helped to calculate the programmed angle.

In the event that a tensor C in a Tensor Flow map relies on some arrangement of tensors fXkg, perhaps through a complex sub graph of activities, there is a worked in function at that stage that will restore the tensors fdC= dXkg. As with different tensors, slope tensors are handled by expanding the Tensor Flow map using the accompanying approach. If Tensor Flow has to process the slope of a tensor C as for various tensors I on in which C depend, it locates the way from I to C first in the calculation diagram. It backtracks from C to I at that point and adds a hub to the Tensor Flow diagram for each retrogressive way process, forming the fractional angles along the regressive way using the chain law.

Figure 4. Gradients in Figure 2 computed for graph

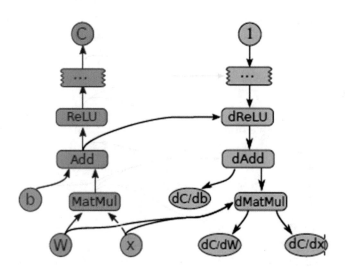

The recently added hub records the forward mode "angle function" for the related activity. Any operation could recruit an angle power.

This capacity takes as information not just the fragmented slants figured starting now and into the foreseeable future along the retrogressive course, yet in addition, on the other hand, the data sources and yields of the forward activity. Figure 5 illustrate slants for a fabricated expense from the Figure 2 occasion. Dull jolts demonstrate potential commitments to tendency works that aren't utilized for explicit exercises. The normal extension to Figure 1 to process these points is: [db,dW,dx] = tf.gradients(C, [b,W,x]) when everything is said to be done an action may have various yields, and C may depend on only a couple. Remote possibility, for instance, that activity O has two yields, y1 and y2, and that C relies upon y2,The essential commitment to O's edge work is set to 0 by then, since dC = dy1=0.

The programmed measurement of the slope entangles changes, in particular the use of memory. When executing "forward" calculation subgraphs, i.e. those unmistakably created by the client, a fair heuristic break connects when selecting which hub to execute next by looking at the request in which the chart was constructed. This for the most part means that not long after being created, impermanent yields are devoured, so that the memory can be reprocess easily. The heuristic is insufficient; the customer can revolutionize the diagram creation request or include control conditions as defined in Section 5.

The client has less control at the point where inclination hubs are naturally added to the diagram, and the heuristics that separate. In particular, in view of the fact that inclinations invert the request for forward calculation, tensors used directly from the bat in the execution of a diagram are needed as often as possible at the end of a slope calculation. These tensors will clutch up a bundle of uncommon GPU memory and boundary the calculation size pointlessly. It effectively was firing executives with changes in memory to handle these cases. Choices include the use of increasingly advanced heuristics to decide the diagram execution request, recalculate tensors instead of keeping them in memory, and swap seemingly perpetual tensors from the GPU memory with progressively ample congregation CPU memory.

Fractional Execution

The consumer usually only has to perform a subgraph of the entire execution diagram. For this reason, when the customer has locate up a calculation diagram in a assembly the run strategy enables to implement a subjective subgraph of the entire diagram and to provide discretionary details along any edge of the diagram. To recover information that streams along every edge of the map. -hub in the diagram has a name, and each hub yield is differentiated the source hub and the hub yield port numbered from 0 (e.g., "bar:0" refers to the first "bar" hub yield, while "bar:1" refers the second yield).

Two disputes to the Run help to identify the particular sub diagram of calculation map to be performed. Second, the Run call recognizes inputs, a discretionary name mapping: port names to "take care of" qualities of tensors. Second, the Run call recognizes yield names, an overview of yield name[:port] information showing the hubs should be executed and, if the port segment is available in a name, the specific yield tensor incentive for the hub should be returned to the customer if the Run call is effectively terminated.

Figure 5. Previous and after the graph transformation for partial execution

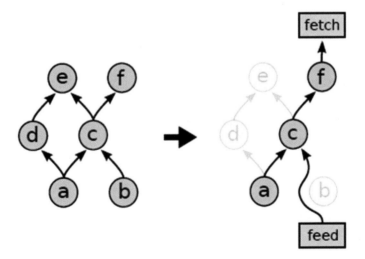

The map is changed depending on estimates of the data and yield sources. That node: the port shown in the inputs is replaced by a feed hub that will get the input tensor from extremely mounted passages in a Rendezvous entity utilizes for the Run identify. In essence, .yield name with a port is connected with a specific hub that masterminds the yield tensor returns with the customer when the identify is completed. To end with the diagram has to change by adding these extraordinary feeds and having hubs, the arrangement of the hubs to be executed can be verified by starting with each of the hubs identified by any yield and working in reverse. The map using the conditions of the diagram to determine the complete structure of the hubs to be applied in the revised diagram to process the yields. Figure 6 displays a separate diagram on the left and a modified diagram resulting from inputs==fbg and production==ff:0 g. Since need to process the hub f yield will not implement d and e hubs, as to have no commitment to f yield.

Device Constraints

Customers of the Tensor Flow can monitor the hubs situation on gadgets by giving a hub fractional requirements about which gadgets it can operate on. For eg, "just spot this hub on a GPU-type gadget," or "this hub can be placed on any gadget in / job: worker / task:17" or "Place this hub with the hub called variable13". The boundaries of imperatives, the arrangement calculation is responsible for selecting a function of hubs for gadgets, which enables the calculation to be performed quickly and also fulfills numerous requirements imposed by the gadgets themselves, for example, by restricting the aggregate quantity of memory wanted on a gadget to perform its subset of chart hubs.

The behind these restrictions necessitate improvements of the estimation of the situation explain in section 3.2.1. To initially process the possible arrangement of gadgets for every hub and then utilize association finds on the limitations diagram to record the diagram segments that need to be compiled. We record the convergence of the achievable gadget sets for each such segment. The possible set of gadgets processed per hub fits effectively into the test system for position calculation.

Control Flow

Despite the fact that data flow charts with no unambiguous control stream are very descriptive, we have observed different situations in which supporting conditionals and circles will prompt ever brief and effective portrayals of AI calculations. Much as in the dataflow-machine approach portrayed, we present a little arrangement of crude control stream administrators into Tensor Flow and sum up Tensor Flow to deal with cyclic dataflow diagrams. For example, elevated level programming constructs if-conditionals and keeps in mind that circles can be effectively organized with these control stream administrators into data flow diagrams. The Tensor Flow runtime, like the MIT Tagged-Token machine the work of (Karthika,P., Ganesh Babu,R., &Nedumaran,A, 2019), updates an understanding of labels and casings adroitly. Any cycle of a circle is interestingly acknowledged by a tag, and a casing speaks to its execution state. The knowledge can join a loop at any point it opens; thus, multiple emphases can be performed simultaneously.

Tensor Flow uses a diffused coordinating instrument to perform control stream maps. All in all, a circle will contain hubs allocated to a wide array of gadgets. Subsequently, addressing a circle's condition becomes a matter of stolen end recognition. The solution to Tensor Flow depends on a reworking of the diagram. Consequently we add control hubs to each section during the diagram distribution. This hubs update a little condition machine that schedules the found and end of each cycle and selects the end of the process. The gadget that claims the circle end predicate sends each taking an interest gadget a little control message for every accent.

As explained above, we train AI models frequently by angle plummet and speak about inclination calculations as a major aspect of data flow diagrams. When a model integrates control-stream activities we will represent them in the measurement of the comparing slope. For eg, the inclination calculation for a model with an if conditional would realize. Part of the restrictive was applying the slope rationale to this division at that point. Similarly slope calculation for a representation with a certain time circle should realize the number of cycles taken and will also depend on the middle of the road estimates processed during this emphasis. The basic strategy is to rework the diagram so as to remember the qualities needed for measuring the inclination. We discard the unpredictable subtleties of this encoding to some degree.

Input Operations

Despite the fact that input information can be given to a calculation by means of feed hubs, another basic instrument utilized for preparing enormous scope AI models is to have uncommon information activity hubs in the chart, which are commonly designed with a lot of yield a tensor enclose at least one models beginning the information put away in that arrangement of documents every time they are accomplish. This permits information to be perused straightforwardly the fundamental stockpiling framework the memory of machine that resolve perform ensuing handling on the information. In setups the customer procedure is independent from specialist procedure, if the information were taken care it regularly would necessitate an additional system bounce.

Queues

Lines are an important element that added to Tensor Flow. It allow special bits of the chart to be executed non-concurrently, perhaps at different cadences, and to distribute information through Enqueue and Dequeue activities. Enqueue operations can hamper before room is obtainable in the line and tasks can hamper waiting an optimal minimum numeral of components is accessible line. One utilize of lines is to permit input information to be prefixed from circle records while the computational section of an AI model still handles a previous group of information. These can also be used for different types of sets, adding multiple inclinations in order to register any increasingly unpredictable mixture of angles over a larger clump or to gaher various information phrases for repetitive language models into canisters of phrases which are roughly similar in length and which can then be prepared even more proficiently. Notwithstanding ordinary FIFO lines, we have also performed a rearranging line that rearranges its components arbitrarily within a massive in-memory cushion. This rearrangement of utility is beneficial for AI calculations that need to randomize the framework in which models are processed.

Containers

A Container is the instrument for monitoring longer impermanent state within TensorFlow the work of (M Prathap, Durga Prasad Sharma, D Arthi, 2010). For a Number, the service store resides in a room. The default holder remains until the procedure is over, but we also require other designated compartments. You can reset a holder by completely freeing it from its content. Using holders, it is possible that even entirely disjoint calculation diagrams linked to different sessions should be shared.

OPTIMIZATIONS

Imagine a portion of the changes in the execution of the Tensor Flow that improve the execution or use of the System properties.

Common Sub Expression Elimination

Since calculation diagrams are frequently created by a wide range of reflections layers in the customer code, calculation diagrams will undoubtedly end up with excess duplicates of a similar calculation. To fix this, we have modified a standard sub expression transfer, such as the calculation which runs larger than the calculation diagram and canonizes multiple duplicates of tasks with indistinguishable sources of information and category of operation to only one of these hubs, and the sidetrack diagram edges suitably represent this canonization.

Data Communication and Controlling the Memory Usage

Cautious preparation of Tensor Flow activities will lead to better execution of the system, specifically with regard to moving information and using memory. In particular, booking will the time during intermediate outcomes should be retained a memory in the middle of activities and consequently the use of pinnacle memory. This decrease is particularly significant for gadgets with GPU where the memory is limited. In addition, organizing knowledge communication through devices will lower disputes over the arrangement of properties. While there are several open doors for planning development, we are concentrating on one that we consider to be particularly important and convincing. This involves Receive hubs planning to peruse remote assets. If no insurance is taken, these hubs will start much sooner than would usually be acceptable, likely at the same time as implementation begins. By playing a count as fast as time permits / as-late-as possible (ASAP / ALAP) as usual in activities are analyzed, we break down the basic charts ways to determine when to create the receive hubs.

Asynchronous Kernels

Excluding traditional synchronous pieces that complete the implementation towards the completion strategy, our structure also wires non-blocking bits. It portions use slightly extraordinary boundary to pass a continuation of the compute strategy that should be conjured when the execution of the bit is finished. This is an improvement in circumstances where having multiple dynamic strings is typically expensive as far as memory use or different assets is concerned, and helps us to refrain from tying up an execution string for infinite timeframes when sitting tight for I / O or various occasions. Instances of non-concurrent pieces combine the component Receive and the sections Enqueue and Dequeue.

Optimized Libraries for Kernel Implementations

For certain tasks we routinely use prior dramatically improved numerical libraries to execute bits. For example, there are numerous enhanced libraries to perform frame duplicates on different hardware, including GPU libraries designed for convolutionary portions of deep neural networks the work of (Antony Selvadoss, Thanamani, Prathap M, 2016). A significant number of our bit executions around such advanced libraries are relatively minor wrappers. For a considerable amount of bit use in the application, we use the open-source Eigen straight polynomial math library the work of (Ekram, H., & Bhargava, V.K. 2007). As one piece of TensorFlow's development, our group extended the open source library Eigen with the help of subjective dimensionality tensoring activities.

Lossy Compression

Many AI measurements, including those usually used to train neural systems, are tolerant of commotion and decreased juggling of the number of accuracies. Similar to the Dist Belief framework, we regularly use loss pressure of higher accuracy inside portrayals when sending information between gadgets (now and then within a similar system, but particularly across machine boundaries). For eg, we also embed extraordinary transformation hubs that turn 32-piece coasting point portrayals into a 16-piece skimming point portrayal.

CONDITION AND EXPERIENCE

The Tensor Flow implementation and a sample use were released publicly under an Apache 2.0 permit, and the software can be downloaded at www.tensorflow.org. The platform provides gritty documentation, various training activities and different models representative how to utilize the organization for a wide array of AI errands. The models integrate templates for grouping digits since the MNIST dataset (the "welcome universe" of AI calculations) the work of (Ganesh Babu,R., Karthika,P., &Aravinda Rajan,V., 2020), arranging images since the CIFAR-10 dataset, demonstrating language periodic LSTM organization, preparing word implanting vectors and that's just the starting point.

The system remembers front-closes for showing Tensor-Flow calculations for expect that additional front-closes should be introduced after a while in view of both Google's inward customers and the larger open-source network. In our past Dist Belief framework the work of (P Karthika and P Vidhyasaraswathi, 2019) we have many AI models which we have switched to Tensor Flow the rest of this area talks about some of the exercises that we have discovered are generalizable any movement of AI representation from one organization to the next, and may be useful to others. In exacting, we concentrate on our exercises to port a best-in-class convolutionary neural network for image recognition called inception. This picture recognition framework characterizes pictures of 224 pixels into one of 1000 marks. When communicated as a Tensor-Flow diagram, a representation contains 13.6 million learnable parameters and 36,000 activities. Running deduction on a solitary photo requires an increase of 2 billion including activities.

Following the construction of all important numerical activities in Tensor Flow, the selection and troubleshooting of each of the 36,000 tasks into the right chart structure proved challenging. Validating accuracy is a worrying undertaking in view of the fact that the system is obviously stochastic and just proposed to carry on with a particular goal in mind in interest — probably evening calculation time. Given these conditions, we consider the accompanying techniques fundamental for transferring the Initiation model to Tensor Flow:

1. Build tools for gathering information in a given model in the specific number of parameters. In an uncertain determination of system design, these apparatuses showed inconspicuous blemishes. Specifically, we had the option to identify activities and factors that began inaccurately because of scheduled broadcasting over a calculation in a numerical operation.

2. Start small, and scale up. A little machine used on the CIFAR-10 knowledge index the work of (M Prathap, Gashaw Bekele, Melkamu Tsegaye, P Karthika, 2020) was the first convolutionary neural network we ported from our past context. Investigating such a method explained unobtrusive edge cases within the AI paradigm in singular behaviours (e.g., max-pooling), which in increasingly complex models would have been unintelligible in all intents and purposes.

3. Continuously ensure that when learning is destroyed, the target (misfortune work) communicates between AI frameworks. The learning to zero enabled us to recognize extraordinary actions by the

way we had haphazardly identified factors in a model. In a complex, preparatory framework such a blunder would have been difficult to recognize.

4. Create a coordinate of solitary computer use before troubleshooting a scattered execution. This method has allowed us to represent and examine the differences between the AI system in planning execution. Specifically, because of race conditions and non-nuclear activities, we identified bugs wrongly considered nuclear.

5. Protect against errors in numbers. Numerical libraries are contradictory in their manner of dealing with nonfinite drifting point esteems. Convolutionary neural systems are especially defenseless to numerical flimsiness and are typically regularly isolated during testing and troubleshooting stages. Guarding against this action by searching for non-finite drifting point esteems helps one to continuously identify blunders rather than discern post-hoc dissimilar behaviors.

6. Running subsections of a neural system on a par with two AI systems offers an exact technique to ensure that a numerical function is indistinguishable between two frames. Given that such calculations are performed with gliding point precision, it is important to predict and understand the extent to which the numerical blunder should all together decide whether a given segment is effectively modified (e.g. realizing "inside 1e-2, amazing!" and "inside 1e-2: why is it so wrong?!").

Figure 6. Parallel processing of the synchronous and asynchronous data

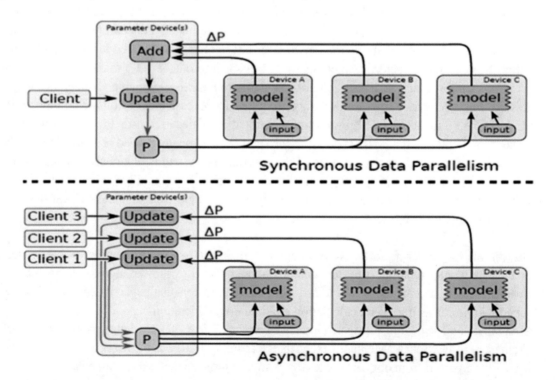

It is very rigorous to consider complex numerical activities in the light of an innately stochastic system. The methodologies outlined greater than proved important in building assurance in the organization and finally in launching the Tensor Flow initiation representation. The final product of these resulted in a 6-higher speed improvement in the preparation time against our current DistBelief model execution and such speed improvements proved irreplaceable in the preparation of another class of larger scope picture recognition models.

7.   The basic dataflow chart model of the common programming idioms tensor flow can be utilize in a multiplicity of behavior for AI applications. This area shows a few systems that we and others have developed to accomplish this and how to understand Tensor Flow these dissimilar methodologies.

This is demonstrate in the peak part of Figure 6. This technique also made non-concurrent the Tensor Flow diagram has several imitations of the diagram bit making up the main part of the model calculation, and each of these reproductions also applies the parameter changes to the parameters of the model non-concurrently. Right now, there is one customer string for each imitation diagram. This portion of Figure 7 non-competitive advance was described.

Figure 7. Model parallel training

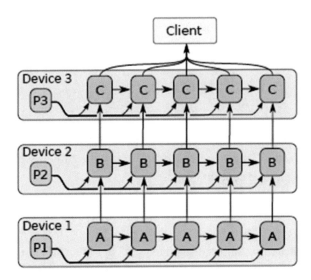

MODEL

Model Equivalent Training

Model equal preparing various parts of the model calculation are done on various computational gadgets at the same time for a similar clump of models, is likewise simple to communicate in TensorFlow. Figure 8 shows a case of an intermittent, profound LSTM model utilized for arrangement to grouping learning, parallelized across three unique gadgets. Simultaneous Steps for Model Computation Pipelining Another regular method to show signs of improvement usage for preparing profound neural systems are to pipeline the calculation of the model inside similar gadgets, by running few simultaneous strides inside a similar arrangement of gadgets.

Figure 8. Simultaneous steps

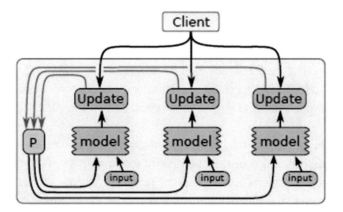

This is appeared in Figure 9. It is fairly like no concurrent information parallelism, then again, actually the parallelism happens inside the equivalent device(s), instead of recreating the calculation diagram on various gadgets. This permits "filling in the holes" where calculation of a solitary group of models probably won't have the option to completely use the full parallelism on all gadgets consistently during a solitary advance.

PRESENTATION

Future adaptation for this paper to complete presentation assessment area of both the single machine and disseminated usage. 9 Tools This segment depicts a few instruments we have built up that sit close by the center Tensor Flow chart execution motor.

Tensor Board

Visualization of diagram structures and rundown insights In request to assist clients with understanding the structure of their calculation charts and furthermore to comprehend the general conduct of AI models, we have fabricated TensorBoard, a partner representation device for TensorFlow that is remembered for the open source discharge. Representation of Computation Graphs Many of the calculation charts for profound neural systems can be very intricate (Rondeau, T.W., &Bostain, C.W. 2009). For instance, the calculation chart for preparing a model like Google's Inception model, a profound convolution for best order execution in ImageNet 2014 challenge, has more than 36,000 hubs Tensor Flow calculation diagram, and some profound intermittent LSTM models for displaying have in excess of 15,000 hubs.

Because of the topology charts, gullible representation procedures frequently produce jumbled and overpowering outlines. To assist clients with seeing the fundamental association of the charts, the calculations in Tensor-Board breakdown hubs into elevated level squares, featuring bunches with indistinguishable structures. The framework additionally isolates out high-degree hubs, which frequently serve accounting capacities, into a different territory of the screen. Doing so lessens visual mess and concentrates consideration on the center segments of the calculation chart. The whole perception is

Figure 9. Tensor Board graph visualization of a convolution neural network model

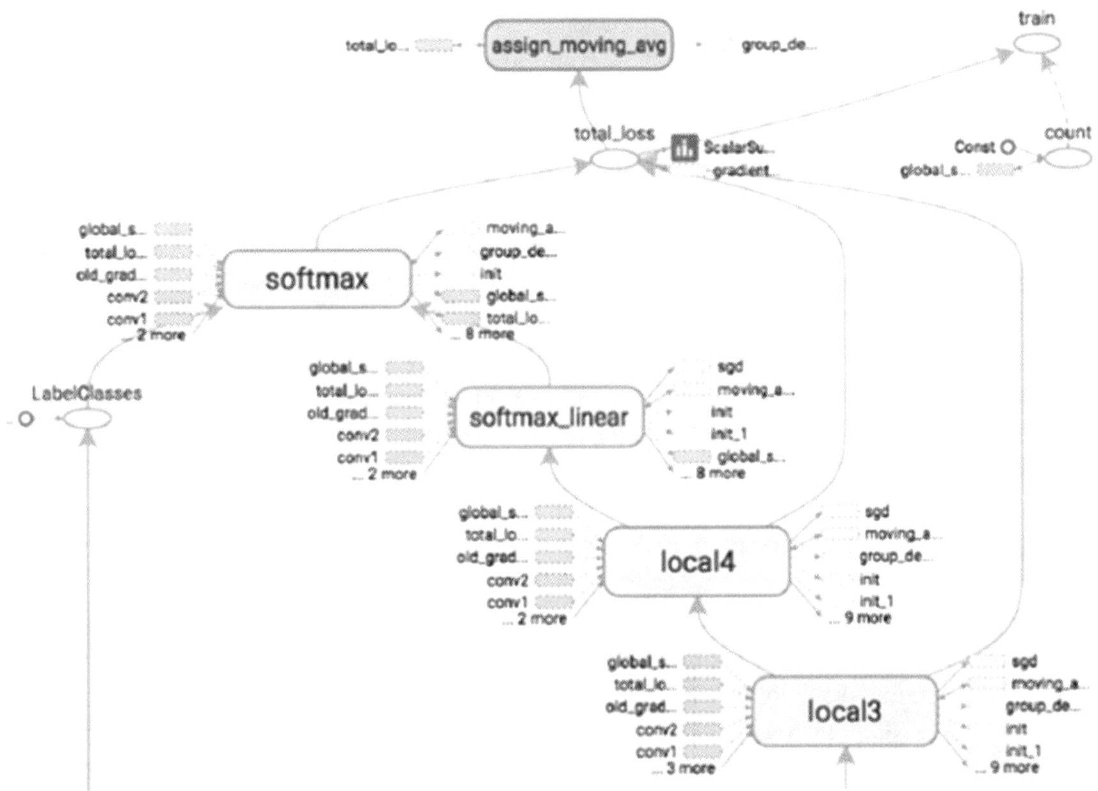

intuitive: clients can skillet, zoom, and extend gathered hubs to bore down for subtleties the work of (P Karthika and P Vidhyasaraswathi 2018). A case of the perception for the chart of a profound convolution picture representation is appeared in Figure 10. Perception of the summary of the data preparing AI models, clients regularly need to have the option to analyze the condition of different parts of the representation to changes after some time. To end of the Tensor Flow backings an assortment of various Summary tasks on embedded diagram,

Figure 10. Tensor Board graphical display of model summary statistics time series data

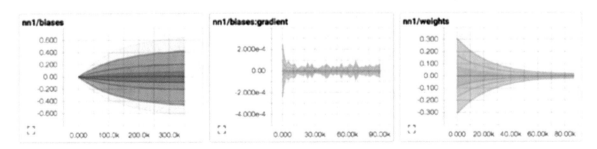

The counting of scalar synopses (e.g., for the general analysis of the model's properties, the measurement of misfortune function finds the mid-value of over a model assortment or the time taken to conduct the calculation diagram), histogram bed rundowns (e.g. the distribution of weight figures in the neural network layer), or picture-based descriptions. Commonly calculation diagrams are set up so Summary hubs are incorporated to screen different fascinating qualities, and from time to time during execution of the preparation chart, the arrangement of synopsis hubs are additionally executed, notwithstanding the typical arrangement of hubs that are executed, and the customer driver program composes the outline information to a log document related with the model preparing with the representation preparing. The TensorBoard system is then built to show this log document for new rundown information, and can display the outline data and how it changes after some. A screen shot of TensorBoard's representation of rundown esteems appears in Figure 11.

Performance Analysis

Any important deferrals due to contact or slows down associated to DMA are differentiated and displayed using bolts in the perception. At first the UI gives a schematic of the whole following, showing only the most notable remnants of the presentation. As the client logically zooms in, subtleties are slowly made to fine goals. A computer EEG experience of a model being prepared on a multi-center CPU stage is shown in Figure 12. The top third of the screen capture shows TensorFlow tasks being dispatched in equal numbers, as shown by the limitations of dataflow the work of (Rondeau, T.W., & Bostain, C.W. 2009). The base section of the following illustrates how the majority of activities are transformed into multiple workplaces simultaneously performed in a string pool. The corner to corner bolts demonstrate where queuing delay occurs in the string pool at the correct hand level. Figure 13 demonstrate another EEG representation, with the mainly occurring calculation for the GPU. The host strings seen interacting with TensorFlow GPU tasks as they become executable, and housekeeping strings can be seen in various shades around processor centers. In reality, bolts display where strings are slowed down to CPU movements on GPU, or where operations encounter enormous delay in queues. Finally, Figure 14 provides a detailed observation that helps to test how Tensorflow GPU administrators are allocate to various GPU streams. At any point in the data flow diagram we seek to uncover the challenging requirements of the GPU gadget utilize streams and stream dependence natives.

FUTURE WORK

It has a few distinct headings for future work. We will keep on utilizing Tensor Flow to grow new and fascinating AI models for man-made reasoning, and over the span of doing this we may find manners by which we should broaden the essential Tensor Flow framework. The open source network can also create new and interesting bearings for the execution of the Tensor Flow. These capacities can become reusable parts in the execution we have designed, even across different front-end dialects for Tensor Flow. So a client may define a functionality that uses the front end of Python, but then use it as a fundamental structure that hinders the C++ frontend from inside. The optimistic that this cross reusability result in a vibrant network of AI specialists sharing entire instances of their analysis, but little reusable sections of their research that can be reused in dissimilar settings.

Figure 11. EEG monitoring of multi-threaded CPU Procedure (x-axis is µs time)

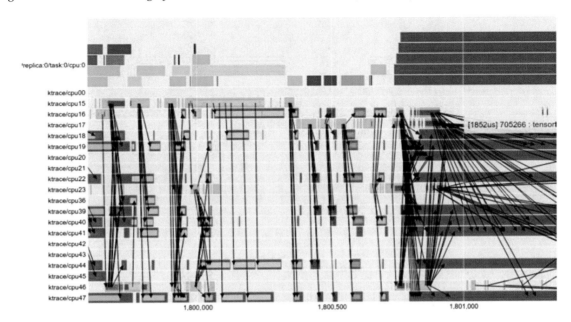

This compiler can understand the semantics of various developments, such as the combination of circles, territorial blocking and tiling, specialization for particular shapes and sizes, and so on. It also envisages an immense area for future work to improve the structure and hub booking calculations used to choose where various hubs will be performed and when they should start to execute. We've modified various heuristics in these subsystems as of now, and we'd rather have the framework find out how to decide on major situational choices.

RELATED WORK

We've modified various heuristics in these subsystems as of now, and we'd rather have the framework find out how to decide on major situational choices. Like Tensor Flow, it supports representative separation, production it easier to characterize and occupation with advancement calculations based on gradients. Tensor Flow, has a core written in C++ that disassembles the sending of models prepared in a extensive range of development environments such as cell phones. The Tensor Flow framework imparts some structure attributes to its preceding system, Dist Belief and later to comparable plans such as Project Adam the work of (M Prathap, Durga Prasad Sharma, D Arthi, 2010) and the Parameter Server venture. Unlike Dist Belief and Project Adam, Tensor Flow allows calculations to be distributed across various computer devices across multiple machines and allows customers to show AI models using relatively high-level representations (Antony Selvadoss, Thanamani, Prathap M, 2016). However, the widely useful dataflow chart model in Tensor Flow is increasingly adaptable and slowly helpful to communicate a wider range of AI models and enhancement calculations, not at all like Dist Belief and Project Adam. It also allows for a crucial change by requiring the declaration of crateful parameter hubs as variables and component updating activities in the diagram that are only extra hubs,

Figure 12. EEG Inception training Hallucination that displays CPU and the GPU operation

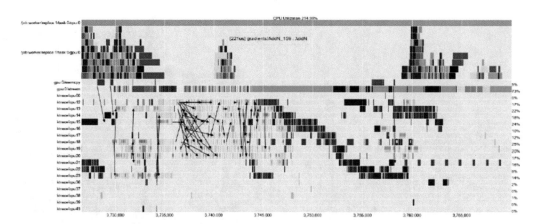

Figure 13. Multi stream GPU execution timeline

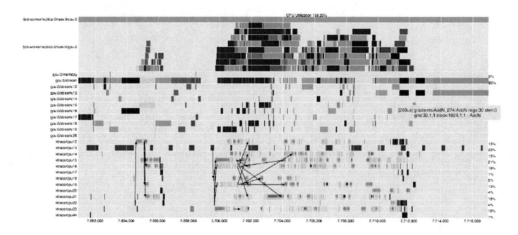

Complete separate server subsystems gave parameter esteems to impart and refresh. The Halide framework for communicating image handling pipelines uses a comparable middle of the road representation to the data flow chart of Tensor Flow. Unlike Tensor Flow, however, the Halide framework has higher-level information about the semantics of its activities and uses this knowledge to produce exceptionally improved bits of code that combine different tasks, taking parallelism and position into account. Halide carries out the calculations on only one computer and not in a distributed environment. We want to extend Tensor Flow with a dynamic assembly framework for comparative cross-activity in future work.

A couple of other articulated frameworks have been developed to execute dataflow charts over a population, including Tensor Flow. This shows how to communicate like a dataflow diagram about a mind-boggling work process. The present non-exclusive support for the subordinate knowledge control stream: CIEL speaks to focus as a DAG that unfurls dramatically, while Naiad uses a process static chart to help emphasize lower-inertness. Sparkle is simplified for calculations that obtain similar information more than once, using "flexible transmitted data sets" (RDDs), which are sensitive pre-calculation state-reserved yields. Dandelion executes data flow charts over a heterogeneous hardware array, such as

GPUs. Tensor Flow uses a model of half breed dataflow that acquires components from each of those frameworks. Its data flow scheduler, the component that selects the following hub for execution, uses a vital calculation similar to Dryad, Flume, CIEL and Spark.

The appropriate architecture is similar to Naiad, since the system uses a single, condensed data flow chart to refer to the whole calculation, and stores data about that diagram on each gadget to minimize overhead coordination. TensorFlow works best, like Spark and Naiad, when there is sufficient RAM in the bunch to carry out work arrangements for calculation. Emphasis in Tensor Flow uses a half-and - a-half methodology: multiple copies of the same dataflow map could be performed without delay for a moment, thus having a similar factor structure. Copies may share information by factors at the same time, or use synchronization systems in the diagram, for example, lines, to work synchronously.

CONCLUSION

The chapter has depicted Tensor Flow, a versatile programming model dependent on stream of data, much the same as a solitary machine and dispersed utilization of this programming representation. The framework is centered on real commitment in coordinating investigation and circulating in excess of 100 AIs over a wide scope of Google things and organizations. We have distributed a Tensor Flow model openly, and we expect a unique network system to be created around the utilization of Tensor Flow. We are eager to perceive how others outside Google utilize Tensor Flow in their own work.

REFERENCES

Ekram, H., & Bhargava, V. K. (2007). *Cognitive Wireless Communications Networks*. Springer Science & Business Media.

Ganesh Babu, R., Karthika, P., & AravindaRajan, V. (2020). Secure IoT Systems Using Raspberry Pi Machine Learning Artificial Intelligence. *Lecture Notes on Data Engineering and Communications Technologies, 44*, 797-805.

Ganesh Babu, R., & Amudha, V. (2016). Cluster Technique Based Channel Sensing in Cognitive Radio Networks. *International Journal of Control Theory and Applications, 9*(5), 207–213.

Ganesh Babu, R., & Amudha, V. (2018c). A Survey on Artificial Intelligence Techniques in Cognitive Radio Networks. *Proceedings of 1st International Conference on Emerging Technologies in Data Mining and Information Security, (IEMIS)in association with Springer Advances in Intelligent Systems and Computing Series*, 99-110.

Karthika, P., Ganesh Babu, R., & Nedumaran, A. (2019). Machine Learning Security Allocation in IoT. *IEEE International Conference on Intelligent Computing and Control Systems*.

Karthika, P., & Vidhyasaraswathi, P. (2017). *Content based video copy detection using frame based fusion technique*. Academic Press.

Karthika, P., & Vidhyasaraswathi, P. (2019). *Image Security Performance Analysis for SVM and ANN Classification Techniques*. Academic Press.

Karthika & Vidhyasaraswathi. (2018). *Digital Video Copy Detection Using Steganography Frame Based Fusion Techniques*. Academic Press.

Karthika & Vidhyasaraswathi. (2020). *Raspberry Pi: A Tool for Strategic Machine Learning Security Allocation in IoT*. Academic Press.

Prathap, Sharma, & Arthi. (2010). *Http server and client performances of multithreaded packets*. Academic Press.

Prathap, M. (2020). Analysis for Time-Synchronized Channel Swapping in Wireless Sensor Network. Academic Press.

Rondeau, T. W., & Bostain, C. W. (2009). *Artificial Intelligence in Wireless Communication*. Artech House.

Selvadoss & Thanamani. (2018). *Investigating the Effects of Routing protocol and Mobility Model and their Interaction on MANET Performance Measures*. Academic Press.

Selvadoss, A., & Thanamani, P. M. (2016). *An Active Surveillance Approach for Performance Monitoring of Computer Networks*. Academic Press.

Vidhya Saraswathi, K. P. P. (2017). A Survey of Content based Video Copy Detection using. *Big Data*, *3*(5), 114–118.

Chapter 5

Impact of Smart Technology and Economic Growth for Biomedical Applications

Dhinesh Kumar R.
Capgemini US, USA

ABSTRACT

Commonly, the sensor installed by these entities collects the citizen and infrastructure data by explicit approval. The performance based on the outline of the citizens could be done through AI depending on the data they offer that could promote advancements in the cities, or it could be utilized for the profit of the corporation. Certain barriers should be crossed in order to succeed in achieving the deployments of the cities. Commonly, the only target is to inform AI through the data gathered by the IoT in the real-time process. At certain times, the generation of data is the main part that is done by several elements, and this could be employed in a deliberate way. Certain challenges are also faced in the economic growth and its impact on biomedical industry. This chapter presents the impact of smart technology and economic growth for biomedical applications.

INTRODUCTION

International Power Institute serves as a mediator in providing guidelines and educational content regarding biomedical industries with the help of International Committee of Power Engineering Professors in International Power Education Academic sector. International power biomedical industries are highly supported with the API, which is considered to be the respectful engineering industry (Beck, Levine, 2012). A survey will be conducted by the API bi-annually based on the transmission, generation and distribution companies (Johansen, 2011). The following requirements are to be satisfied in order to enhance the research innovation in the biomedical Industries:

DOI: 10.4018/978-1-7998-3591-2.ch005

- Most of the power and manufacturing biomedical industries highly depends on the cultural transformation with the constructive engagement among the workforces with the quality leadership and management, which could enhance the rate of workflow along with the innovation
- In order to create an ecosystem with the emerging technologies, greater absorptive capacity is required in the work environment and the thinking ability to collaborate the existing technology with the new one.
- The interconnection of biomedical industries firm promotes greater chances in networking, collaborating and clustering educational institutions, public agencies and research laboratories thereby providing a greater platform for the engineers (Krstevska & Petrovska, 2012).
- In future, the competition among the biomedical industries could be increased with the "boundary-crossing" skills of the teamwork, creative thinking, communication and problem-solving.

Europe is a place known for their second largest industry with highest employment and sixth largest place for the biomedical industries output. Biomedical industries make their highest business expenditure on Research & Development (BERD) with 11% increased annual export earnings (OED, 2012). More than 1 million people in Europe get their job opportunities in more than 47,530 industrial companies.

However, over the past decades, Industrial sectors in Europe had faced many challenges due to the Global Financial Crisis that arises due to the high exchange rates over a long period and also due to the 'fourth industrial revolution', which increases the input costs as well as the energy. Compared to the year 2017 and 2018, the industrial sectors in Europe have survived to these crises in the year 2019. In terms of the supply chains, products, technologies and sectors, International biomedical industries had well balanced in their local and global structural shifts (OCED, 2017).

Figure 1. Rising Healthcare

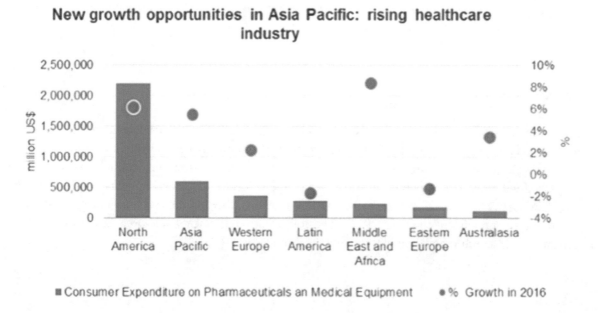

Source. Euromonitor International

In early 2019, RBA forecasted a sharp plunge in the GDP growth of the International economy (+1.7% p.a. till June 2019), grasped by unpretentious growth for the next of upcoming months (+2.6% p.a. till December 2019). Based on the ABS data, there was a sudden break in the global as well as the local economy and the employment growth at the end of the year 2018. Persistent and long-term concerns in the biomedical industries could be bought out in terms of global competition, energy costs, imports and exports by developing a quality workforce with the productivity enhancement, future growth, global connections and local innovations.

The popularity of healthcare and its expenditure is shown in figure 1, 2 and 3. AI technology enhances this process by managing the IoT devices with its ultimate processing and learning abilities that are made possible with the powerful subset known as Machine learning (A.Suresh 2017).

Figure 2. Healthcare Expenditure

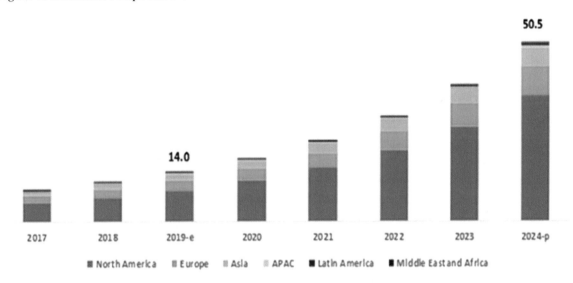

A community-based smart rehabilitation system is employed for sharpening as well as rejuvenating the functional abilities which enhances the physiological care. These systems should interact conveniently apportioning of medical resources based on the requirements of the patient employing IoT based rehabilitation system with an ontology-based automating technique.

IoT technology integrated along with the Artificial Intelligence (AI) seems to unlock huge opportunities in various fields that include adaptive energy systems, autonomous vehicle and smart healthcare(Suresh, A., Udendhran, R., Balamurgan, M. 2019). These examples are considered to be an existing one and still the opportunities with the integration of IoT and AI grows tremendously in all possibilities including the industries and the consumers (Suresh, A. et al 2019). But this combined technology is not found everywhere around the cities although the efficiency of AI along with the IoT nexus is readily available with the deployment of hardware and software. Considering the fact of sustainability, scalability and impact factors of integrating AI along with IoT, significant collaboration and planning is required. In the case of multiple ownerships supporting multiple partners are considered, collaboration and planning is the necessary impact factor.

Figure 3. Rise of Healthcare spendings

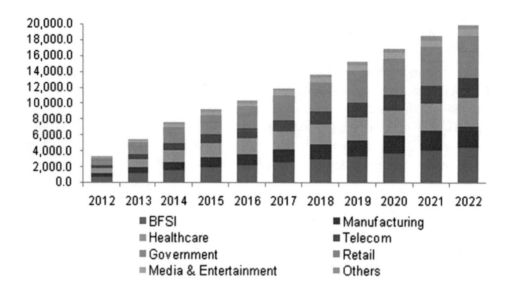

The control could be lost by the cities to the academic institutions, private companies and even citizens in the case where the overarching digital master plan could be confirmed by any city authorities since the solution offered by the IoT are emergent. The question always arises in a way whether the sensors owned by the academia, cities or the business have the computational power in collecting the raw data, filter out the necessary data by applying certain algorithm based on the machine learning process? If this process is done, then what will be the necessity for the process to be carried out?

For example, consider city x that consist of 10 lampposts traffic counting sensors. In this case the physical sensors are being manufactured by company A and the other company B provides the edge computing software and company C promotes the hosting and centralized data management. The planning permission is granted by the Local Authority Y and certain politicians Z based on the analytical decisions.

The strategy should be defined with the use-cases so that the measurements could be customized, planned and considered before the deployment process takes place. Without this strategy, the output will be pool of raw data with the other business content whose use will be a great challenge. The areas that deal with the security, privacy, ethics and legality forms as a core field with the higher priority rather than the technical insight. Moreover, the city authority could get benefit from the machine learning along with the AI. For instance, reduction of accidents due to the traffic, enhancing the transportation and quality of air are the high demands by the agenda. These challenges could be answered when the right information with the intelligence based goes into the hand of the empowered authorities and the necessary actions could be taken by them. IoT generates significant and necessary data and with the AI algorithm the data could be considered at the sufficient volumes and makes use of the AI algorithms. For the authorities to act on the AI/IOT nexus, proper impact should be necessarily considered, and the power should fall on to the right hands. In the context of cities, change of policy will necessarily derive the values or potential decision could be made by the independent decision maker.

SENSOR BASED ALERT MONITORING SYSTEM

An alert system also indicates the operator during the hazardous situation with the audible alarm sound and the visible warning light as shown in figure 4. The principle of alert system operates with the real and reference value comparison, which are obtained with the geometric sensors and pressure measurement. The Central Processing Unit consists of memory unit, where the reference values are stored. Moreover, a low cost controller with the low-cost sensors creates more ideas that could change the definition and the use of the product. Several sensors such as Laser sensor, infrared sensor, ultrasound sensor, camera with feature detection, ultra-wideband (UWB) radar, and induction (Hall effect) sensor could be used. But some sensors among these are said to be highly expensive, produces much noise and not enough to produce high accuracy. Therefore, proper sensors should be utilized to provide high accuracy and then it could be bought under IoT. The system also indicates the operator during the hazardous situation with the audible alarm sound and the visible warning light.

Figure 4. Sensor alert system

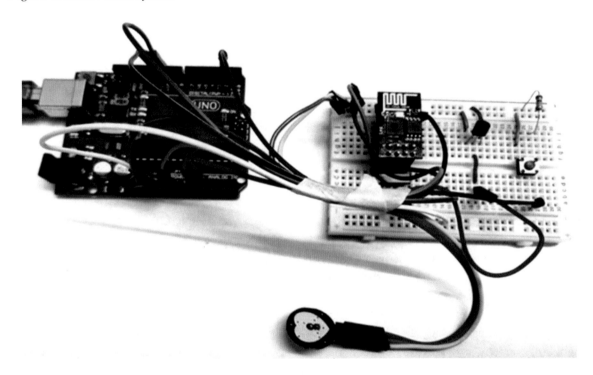

Start: The alert system starts automatically if it was turned off for more than a period of 2 hours. Manually, by pressing key (12) "SEL", the system functions after every patient modifications.

Operation: The system is operation with several functional keys F1,….,F5 in accordance with the key configuration.

Cancellation: All the adjustments made in the patient remain unsorted, when the system is turned off for less than 2 hours. Moreover, an error code E04 will be produced if any sort of unavailable configuration is selected.

- Following steps are required for the proper programming of an alert system setup:
- Set the necessary configuration
- Set the alert configuration
- Set the counterweight configuration
- Set the timing (timings could be increased or decreased)
- Set the outrigger position (outrigger position set to 0%, 50% and 100% and if patient is chosen then static, pick and carry could be selected)
- Confirm the entire procedure for the programming (at the end of the entire process, the selections made are shown in the form of a symbol. The selected symbol will be filled black)

Factors to be Considered in an Alert System

- The operational aid alert system will provide significant warnings to the doctor while the system approaches the overload situation and the hoist condition, which could cause significant malfunction and damage to the entire system.
- The device will not be a substitute for the experience doctor judgment and use of established patient safe functioning procedures.
- The entire functionality of the patient operation and its safety situation depends on the doctor, who is responsible for ensuring all the instructions and the warnings produced should be clearly taken into concern.
- The limitations and the functionality of the patient should be well analyzed with the manual provided from the healthcare end before the doctor works with the patient.
- Proper inspection and observation should be carried out at regular intervals based on the doctor's manual.

SMART BATTERY FOR BIOMEDICAL WEARABLE APPLICATIONS

In Biomedical wearable devices power battery management system is the most important task, which is necessarily carried out to prevent the battery over-charging and over-discharging. The performances of the batteries are mainly decreased due to the aging, elevated temperature and cycling process due to the exchange of ions. Lithium ion batteries is the most used battery component and the battery that promotes high capacity, safety and long life is the lithium iron phosphate (LiFePO4) and the charging method mostly adapted is the current charge- constant voltage method (CC-CV). Therefore in the monitoring process, temperature, total current and voltage are regularly monitored. The cycle count of the batteries is determined by the depth of discharge (DoD) and longer the DoD minimizes the lasting period of the batteries.

The battery monitoring system is estimated to reach US\$8,053.489 million by the year 2024 with the peak of 27.20% when compared to the year 2018, which was around US\$1,901.168 million. The entire process undertaken by the battery monitoring system involves with estimating the state and thereby adjusting the environment. One of the major drivers involves the rising focus on e-vehicles with the BMS market. The government takes initiative in developing the e-vehicles in order to control the air pollution, thereby expanding the growth of BMS.

Figure 5. Smart Healthcare Monitoring

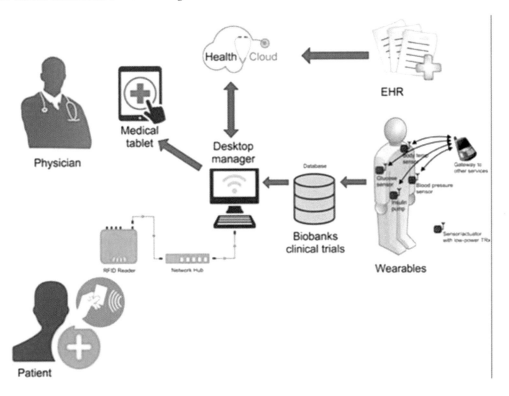

Figure 6 IoT in smart city transportation

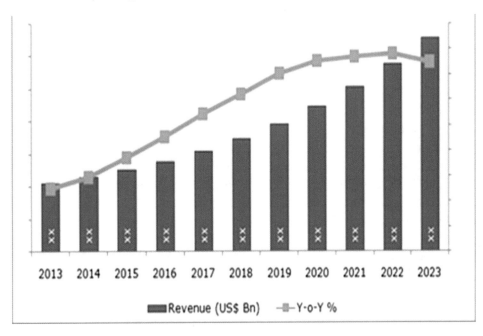

BMS supplier market has been sectioned based on the battery type, components, geography and end users. Moreover, hardware segments in BMS market holds with the fast sampling process, high-resolution recording capabilities and effortless installation. The components also includes with the sensors, data recorders and controllers. Among the other countries, North America holds the global monitoring market with the progressive technologies combined with the data centers expanding investments. This work deals with the study of Battery Monitoring System (BMS) economics and supplier in North America along with the enhancement in the quality. This work deals with the enhancement made with the BMS along the techniques adapted to improve the technology. Detailed analysis had also been carried out with the various temperature sensors implemented for the system. Battery monitoring system:

The most widely used batteries among the e-vehicles includes the Lithium ion batteries due to their exclusive power rating, energy density and charging/discharging coefficient in the pulsed energy flow systems. The problem among the Li batteries is their ability to overcharge that could result in short span and also at the worst cases it could also damages the functioning of the battery systems. Therefore, it is required to have a proper Battery Management system (BMS) thus maintaining a reliable and safe operation for the battery systems [11]. The primary functionality of the BMS system is to provide with a amount of energy that has been utilized by the load, in order to know the regular status of the system. Since the internal resistance and the capacitance of the battery changes with the time, this is determined to be a difficult task. The series-connected battery cells could cause certain unbalancing issues that could be a problem in extending the lifetime of the system, which should also be considered in the BMS system. This could also minimize the battery usable capacity as the low charged cell obtains the end of the discharge, though certain amount of energy are stored in the other battery cells. Lithium ion batteries have a strict voltage limits that still limits the self-recovery of the charge imbalance and it also declines with time. If a single cell from the battery stack has reached the upper limit then it stops the charging process thus making other cells of the battery uncharged. Assuming the cells to have the same capacity (only a few percent of mismatch could be limited), cells that is of varying discharge rate could cause charge imbalance [12]. This could be dealt and obtained by the temperature gradient implemented along with the battery stacks. Therefore, charge equalization technique should be adapted by the BMS system to manage and restore the balanced state.

Critical path analysis (CPA) is a management approach can be used for monitoring, analysis, and prediction purposes as mentioned by Word Health Organization in Critical Path Analysis from Detection to Protection. Critical Path analysis is the latest method which has gained major importance over a short period of time. The main reason for its success is that Critical Path analysis is able to overcome the challenges in traditional method that employs critical path. For many years, critical path was the working the principle for managing, planning applications (healthcare monitoring projects) in the area of management thinking, leading it to deploy in healthcare companies for healthcare monitoring projects management. Critical Path analysis technique reveals the important assets which are required for deploying healthcare monitoring projects activities duties which simplify non-important as well as important tasks such as planning strategic distance from process limited access and period problems. This technique is well suited for applications which consist of tasks that are integrated. The Critical Path analysis assists in control processes as well as planning based on theory of constraints which reduces waste including six sigma reduce variations.

Critical path methodology includes (i). Identifying every activity which employs work breakdown structure to separate every activity that is retains as well as manages high-level activities. (ii) Establish Dependencies (iii) Draw the network Diagram which employs Critical Path Analysis Chart (iv) determine Activity Completion Time (v). Determine the critical path the advantage of critical path includes proper visual representation of the healthcare monitoring projects activities and provides effective time management for completion of overall healthcare monitoring projects with maintaining tracking of important tasks.

In-order to obtain the critical path, these steps should be followed:

- Create Schedule Network Diagram
- Obtain activities
- Obtain all Possible Paths
- Determine the each path duration

Traditional healthcare monitoring projects Management Methods: These methods such as waterfall method provide an effective scheduling for worst-case task durations with its main focus on progress of tasks. These methods includes faster reaction towards to uncertainty by modifying priorities, expediting, as well as planning fresh schedule. The major disadvantages in traditional healthcare monitoring projects tools can be overcome by employing critical path and chain technique.

Steps for effective healthcare project integration:

Healthcare project integration management plays an important role in a project as well as co-ordinate it.

- Develop Project Charter: The project charter is important for the healthcare monitoring applications since it formally authorizes the project and it provides high-level description of the project objectives and the craved product. Moreover, project charter determines the project manager and provides authority to request as well as maintain resources.
- Develop Project Management Plan: The project management plan is also important since it will plan all necessary requirements including the healthcare- technology budget, schedule, database. The project management plan provides consolidation of the other management plans which leads to quick overview of the intended healthcare monitoring applications .
- Direct and Manage project work: This process forms the backbone of the healthcare monitoring applications which controls the technical aspects of the project such as back-end and front-end programming language, thus enabling reliability, status report and work-flow of the application.
- Monitor and Control Project Work: In-order to direct and management, the healthcare should be monitored in aspects such as customer-care after medication, any modification in the project.
- Perform integrated change control: This process overlooks the modifications which may affect the performance of the healthcare-project.
- Project Close: This process should be incorporated since it reviews the several processes employed as well as rate the process whether it is successful or not

CONCLUSION

Industry 4.0 is said to be an evolution that combines manufacturing process and digital technology with the consumer life. This is known as the 'Fourth Industrial revolution' and Europe is known to be the highly potential country for these future biomedical industries (reported by the World Economic Forum). The workplace will also be filled with the complex organizational and complex structure that relates the coordination, decision-making, quality control, sales, and distribution and support services. Additionally, workforce also impacts higher demands in terms of complexity, problem-solving, connectivity and abstraction.

REFERENCES

Beck, T., & Levine, R. (2012). Industry growth and capital allocation: Does having a market- or bank-based system matter? *Journal of Financial Economics*, *64*(2), 147–180. doi:10.1016/S0304-405X(02)00074-0

Financial Services Institute of Europe. (2016). *A discussion paper on Regulating Foreign Direct Investment in Europe*. Author.

Franklin, A., & Douglas, G. (2010). *Comparing ▫nancial systems*. MIT Press.

Jevcak, A., Setzer, R., & Suardi, M. (2010). Determinants of capital flows to the new EU member states before and during the financial crisis. European Commission. *Economic Papers*.

Johansen, S. (2011). Statistical analysis of cointegrating vectors. *Journal of Economic Dynamics & Control*, *12*(2-3), 231–254. doi:10.1016/0165-1889(88)90041-3

Krstevska, A., & Petrovska, M. (2012). The economic impacts of the foreign direct investments: Panel estimation by sectors on the case of Macedonian economy. Journal of Central Banking Theory and Practice, 55-73.

Mitrovic, N., Petronijevic, M., Kostic, V., & Bankovic, B. (2011). Active Front End Converter in Common DC Bus Multidrive Application. XLVI Proc. of Inter. Conf. ICEST 2011, 3(2), 989-992.

OECD. (2012). Glossary of foreign direct investment terms and definitions. Retrieved from http://www.oecd.org/investment/investmentfordevelopment/2487495.pdf

OECD. (2017). Economic survey – Europe. doi:10.1787/eco_surveys-aus-2017-en

Paul, A. K., Banerje, I., Snatra, B. K., & Neogi, N. (2009). Application of AC Motors and Drives in Steel Industries. *XV National Power System Conference*, 6(2), 159-163.

Petronijevic, M., Veselic, B., Mitrovic, N., Kostic, V., & Jeftenic, B. (2011, May). Comparative Study of Unsymmetrical Voltage Sag Effects on Adjustable Speed Induction Motor Drives. Electric Power Applications, IET, 5(5), 432–442. doi:10.1049/iet-epa.2010.0144

Suresh, A. (2017). Heart Disease Prediction System Using ANN, RBF and CBR. *International Journal of Pure and Applied Mathematics*, *117*(21), 199–216.

Suresh, A., Udendhran, R., & Balamurgan, M. (2019). A Novel Internet of Things Framework Integrated with Real Time Monitoring for Intelligent Healthcare Environment. *Springer-Journal of Medical System*, *43*, 165. doi:10.100710916-019-1302-9

Suresh, A., Udendhran, R., & Balamurgan, M. (2019). Hybridized neural network and decision tree based classifier for prognostic decision making in breast cancers. Springer - Journal of Soft Computing. doi:10.100700500-019-04066-4

Udendhran, R. (2017). A Hybrid Approach to Enhance Data Security in Cloud Storage. ICC '17 Proceedings of the Second International Conference on Internet of things and Cloud Computing at Cambridge University, United Kingdom. Doi:10.1145/3018896.3025138

ADDITIONAL READING

Darwish, A., & Hassanien, A. E. (2011). Wearable and Implantable Wireless Sensor Network Solutions for Healthcare Monitoring. *Sensors (Basel)*, *11*(6), 5561–5595. doi:10.3390/s110605561 PubMed

Hulko, M., Hospach, I., Krasteva, N., & Nelles, G. (2011). Cytochrome C Biosensor—A Model for Gas Sensing. *Sensors (Basel)*, *11*(6), 5968–5980. doi:10.3390/s110605968 PubMed

Chapter 6
Importance of Automation and Next-Generation IoT in Smart Healthcare

Ahmed Alenezi

Taibah University, Al-Ula Campus, Madhina, Saudi Arabia

M. S. Irfan Ahamed

Taibah University, Al-Ula Campus, Madhina, Saudi Arabia

ABSTRACT

Generally, the sensors employed in healthcare are used for real-time monitoring of patients, such devices are termed IoT-driven sensors. These type of sensors are deployed for serious patients because of the non-invasive monitoring, for instance physiological status of patients will be monitored by the IoT-driven sensors, which gathers physiological information regarding the patient through gateways and later analysed by the doctors and then stored in cloud, which enhances quality of healthcare and lessens the cost burden of the patient. The working principle of IoT in remote health monitoring systems is that it tracks the vital signs of the patient in real-time, and if the vital signs are abnormal, then it acts based on the problem in patient and notifies the doctor for further analysis. The IoT-driven sensor is attached to the patient and transmits the data regarding the vital signs from the patient's location by employing a telecom network with a transmitter to a hospital that has a remote monitoring system that reads the incoming data about the patient's vital signs.

INTRODUCTION

Healthcare became technically advanced with IoT and ubiquitous computing by leveraging devices like connected sensors as well as wearable devices which patients can get send the information for real-time monitoring by doctors. Therefore, IoT and ubiquitous computing in healthcare can be considered an important life-saving technology in the healthcare domain which is predominately employed for gather-

DOI: 10.4018/978-1-7998-3591-2.ch006

ing data from the bedside devices, viewing patient information, as well as diagnosing in real-time the entire system of the patient care.

Ubiquitous Computing (UbiComp) is characterized by the use of small, networked and portable computer products in the form of smart phones, personal digital assistants and embedded computers built into many devices, resulting in a world in which each person owns and uses many computers. The consequences include enhanced computing by making computers available throughout everyone's daily life while those computers themselves and their interaction are 'invisible' to the users. The term 'invisible' in this context is used to mean interaction between computer and user in a more natural manner such as speech and physical interaction, with the computer itself automatically capturing its external parameters while concurrently communicating with other computers. UbiComp proposes many minute, wireless computers that can monitor their environments, and communicate and react to monitored parameters (Emiliano Miluzzo et al 2008). However, the challenge which prevails in healthcare is that it causes loss of data and even fault in diagnosis and most of confidential healthcare data is stored in cloud. Even a minute in treating patients is life-saving, therefore doctors should save precious minutes which can be done by employing monitoring of medical assets and less manual visiting each patients through remote diagnosis IoT and ubiquitous computing enabled remote diagnosis (Firouzi, F et al 2018).

Most of our daily life has become part of the internet because of their faster communication as well as capabilities. It is estimated by 2020, more than 90 percent of healthcare industry will integrate IoT technology which will in return enhance the efficiency of healthcare and provide quality care in the modern society (Gubbi, J et al 2013). Quality healthcare is based on the speed as well as accuracy and supporting many people with huge range of devices which are connecting with IoT (Hank, P et al 2013). Since providing healthcare to every person plays a major role in developing a country. Another compelling reason for adapting healthcare is that there is more increase of patients which leads to a smaller number of doctors. Hence, most of the diagnostics are delayed because of this reason since it is time consuming and even patients ignore because of the expense and rely on the doctors. Many health problems are not detected because of non-availability of doctors as well as not accessing the healthcare systems. The only solution to this problem is to integrate healthcare with IoT for real-time monitoring of every patients as well as analyze the data and provide real-time healthcare(Kumar, L et 2012).

Generally, the sensors employed in healthcare are used for real-time monitoring of patients, such devices are termed as IoT driven sensors. These type of sensors are deployed for serious patients because of the non-invasive monitoring, for instance physiological status of patients will be monitored by the IoT-driven sensors which gathers physiological information regarding the patient through gateways and later analyzed by the doctors and then stored in cloud which enhances quality of healthcare and lessens the cost burden of the patient. The working principle of IoT in remote health monitoring systems is that it tracks the vital signs of the patient in real-time and if the vital signs are abnormal then it acts based on the problem in patient and notifies the doctor for further analysis. The IoT-driven sensor is attached to the patient which transmit the data regarding the vital signs from the patient's location and employing a telecom network with a transmitter to a hospital which consists of remote monitoring system that reads the incoming data about the patient's vital signs. In some cases, the sensor will be implanted into patient's body which transmits data electronically (Kuo, A and M.-H., 2011). This confidential information will be encrypted and then decrypted for further analysis when the need arises.

The popularity of Internet of things is shown in figure 1. AI technology enhances this process by managing the IoT devices with its ultimate processing and learning abilities that are made possible with the powerful subset known as Machine learning (A.Suresh 2017).

Figure 1. Popularity of Internet of Things

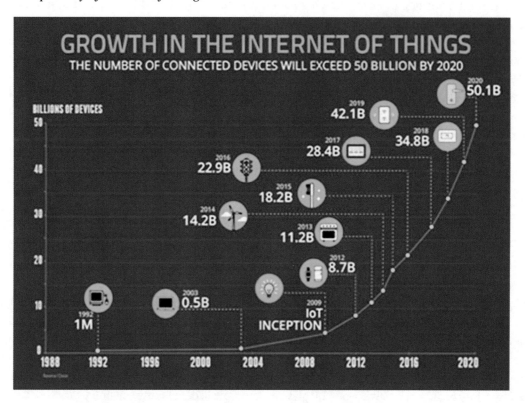

A community-based smart rehabilitation system is employed for sharpening as well as rejuvenating the functional abilities which enhances the physiological care. These systems should interact conveniently apportioning of medical resources based on the requirements of the patient employing IoT based rehabilitation system with an ontology-based automating technique as shown in figure 2.

IoT technology integrated along with the Artificial Intelligence (AI) seems to unlock huge opportunities in various fields that include adaptive energy systems, autonomous vehicle and smart healthcare(Suresh, A., Udendhran, R., Balamurgan, M. 2019). These examples are considered to be an existing one and still the opportunities with the integration of IoT and AI grows tremendously in all possibilities including the industries and the consumers(Suresh, A. et al 2019). But this combined technology is not found everywhere around the cities although the efficiency of AI along with the IoT nexus is readily available with the deployment of hardware and software. Considering the fact of sustainability, scalability and impact factors of integrating AI along with IoT, significant collaboration and planning is required. In the case of multiple ownerships supporting multiple partners are considered, collaboration and planning is the necessary impact factor.

The control could be lost by the cities to the academic institutions, private companies and even citizens in the case where the overarching digital master plan could be confirmed by any city authorities since the solution offered by the IoT are emergent. The question always arises in a way whether the sensors owned by the academia, cities or the business have the computational power in collecting the raw data, filter out the necessary data by applying certain algorithm based on the machine learning process? If this process is done, then what will be the necessity for the process to be carried out?

Figure 2. Applications of Internet of Things

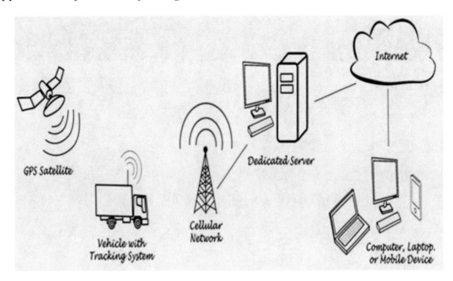

More commonly, the sensor installed by these entities involves in collecting the citizens and the infrastructure data either by the approval explicitly. The performance based on the outline of the citizens could be done through AI depending on the data they offer that could promote advancements in the cities and also the citizens or either it could be utilized for the profit of the corporate. Certain barriers should be crossed in order to succeed in achieving the deployments of the cities. Commonly, the only target is to inform AI through the data gathered by the IoT in the real-time process. At certain times, the generation of data is the most main part that is done by several elements and this could be employed in a deliberate way. Certain challenges are also faced in the urban service sectors (transportation, healthcare sector) by promoting AI with the IoT

For example, consider city x that consist of 10 lampposts traffic counting sensors. In this case the physical sensors are being manufactured by company A and the other company B provides the edge computing software and company C promotes the hosting and centralized data management. The planning permission is granted by the Local Authority Y and certain politicians Z based on the analytical decisions.

The strategy should be defined with the use-cases so that the measurements could be customized, planned and considered before the deployment process takes place. Without this strategy, the output will be pool of raw data with the other business content whose use will be a great challenge. The areas that deal with the security, privacy, ethics and legality forms as a core field with the higher priority rather than the technical insight. Moreover, the city authority could get benefit from the machine learning along with the AI. For instance, reduction of accidents due to the traffic, enhancing the transportation and quality of air are the high demands by the agenda. These challenges could be answered when the right information with the intelligence based goes into the hand of the empowered authorities and the necessary actions could be taken by them. IoT generates significant and necessary data and with the AI algorithm the data could be considered at the sufficient volumes and makes use of the AI algorithms. For the authorities to act on the AI/IOT nexus, proper impact should be necessarily considered, and the power should fall on to the right hands. In the context of cities, change of policy will necessarily derive the values or potential decision could be made by the independent decision maker

Figure 3. Internet of Things and its applications

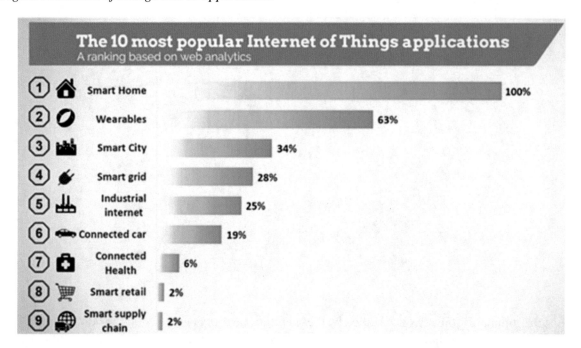

TECHNICAL BACKGROUND

Citizen-focused and test-driven approaches make sure that the end scenario is one that has been tested with the public along the way, and not just imposed upon them. After this significant experimentation, the theories could be tested with the stakeholders and then implementing it in the small scale without any huge investments. This could give a massive lead along the technology as well as the techniques adapted and further deals with the effective discussion of citizens

Several benefits could be attained by integrating cloud computing with the medical sector that drives the technology. Though there are certain difficulties, certain developing countries like Malaysia, Ghana and South Africa witnessed to improve the healthcare with the cloud computing technology. To enhance the outcomes of the patients, Cloud technology embraces with the healthcare that forms a great answer to all healthcare organizations. The interaction among the patients, doctors and the ICT environment through the internet to the cloud. This process starts with the patients and their visit to the doctor who makes a verification of the patient's details in the eHealth (Cloud) system. The doctor analysis the patient's health with the details provided and offers certain medicine for their health with the possible medication. Through this system the interactions could be very soon and the details of the patients could once again be updated in the cloud environment. In the world of healthcare cloud computing is said to be an emerging technology and seems to be very effective when compared to the current health-based information storage systems.

The quick development of the device, wireless and cloud computing technologies alter a wise healthcare that supports the consistent remote watching on the physical conditions of patients, older individuals or babies, and the efficient process of the massive sensing knowledge sets. Such a smart healthcare will enhance the standard of life significantly. However, as investigated by the Cloud Standards Customer

Figure 4. IoT Communications in multi-domains

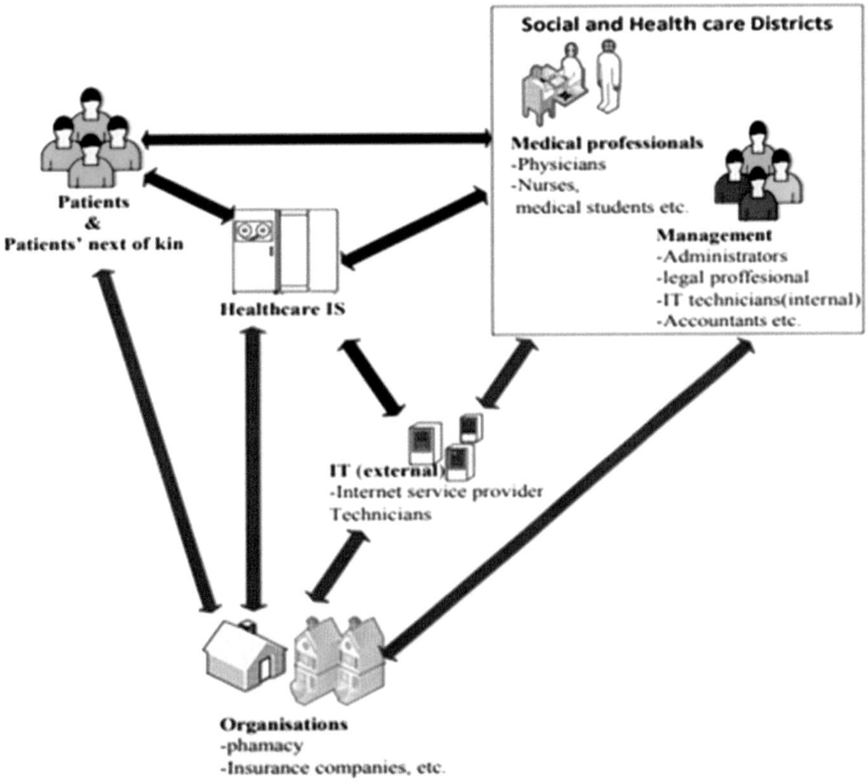

Council, the healthcare institutions don't seem to be keen on building sensible healthcare systems supported the IT technologies, particularly in developing countries. The under-utilization of IT technologies prevents the wide information sharing and processing in healthcare industry. Cloud computing paradigms alter a virtual mechanism for IT resource management and usage. Sensor-cloud infrastructure could be a technology that integrates cloud techniques into the WSNs. It provides users a virtual platform for utilizing the physical sensors in a clear and convenient approach. Users will manage monitor, create, and check the distributed physical sensors while not knowing their physical details, however simply using a few of functions. Cloud computing provides the IT resources for WSNs, that supports the storage and quick process of a large amount of detector information streams. The connection between WSNs and clouds is enforced by two kinds of gateways: sensor gateways and cloud gateways, where sensor gateways collect and compress data, and cloud gateways decompress and process data.

Sensor-clouds are utilized in various applications, like the disaster prediction, atmosphere watching and healthcare analysis. Sensor-clouds collect knowledge from totally different applications and share and process the information based on the cloud computational and storage resources. An interface is provided to the users to manage and monitor the virtual sensors. In Sensor-clouds, the sensor modeling language (SML) is utilized to explain the data of the physical sensors, that is processed as the metadata of the sensors. The standard language format allows the collaboration among sensors in several networks and platforms. The XML encryption additionally provides a mapping mechanism for transforming commands and information between virtual and physical sensors.

Compared with the traditional sensor networks, Sensor-clouds have the following advantages: (1) the capability of dealing with numerous data types; (2) scalable resources; (3) user and network collaboration; (4) data visualization; (5) data access and resource usage; (6) low cost; (7) automated resource delivery and data management; (8) less process and response time.

Figure 5. IoT in smart city transportation

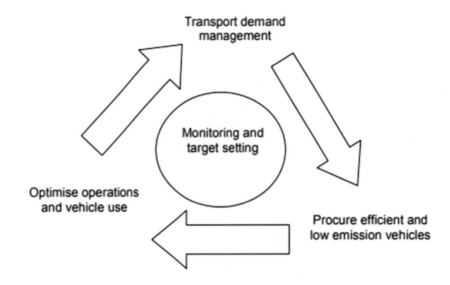

IoT-BASED ELECTRIC VEHICLES FOR URBAN PLANNING AND SMART CITY

The toxic gases that are emitted by these vehicles are as follows:

- Carbon monoxide (CO): Oxygen plays a key role in our body. With the inhalation of carbon monoxide it could block the oxygen carrying capacity that could affect the most important organs of our body. Extreme emission of CO could be dangerous and should be controlled. But a limited emission could save the normal people but could affect those people suffering from the heart disease.
- NOx (oxides of nitrogen including nitrous oxide and nitrogen dioxide): This could extremely affect the lung functionality causing illness in breathing. Extreme exposure to these gases could cause asthma and other allergies to the lungs. Nitrogen dioxide (NO2) could affect the environment that could cause acid rain and ozone depletion.
- Particulate matter: According to Dr Matthew Loxham, a research fellow in air pollution toxicology at the University of Southampton says that these fine particles are extremely dangerous that could cause heavy respiratory disorders and cardiovascular disorders. More than 29,000 deaths have occurred in United Kingdom due to this fine particle emission.
- Hydrocarbons: This could be extremely carcinogenic and leads to death. In fact, this could also result in the formation of greenhouse gases.

The above pollutants are caused by either diesel or petrol vehicles. Plug-in Electric vehicles could save us money as well as reduce in the pollutants causing factors. There are many advantages of using EVs which could react quickly that the internal combustion engines.

Certain advantages are caused due to the operation of fleet rather than fuel vehicle:

- Electric vehicles (EVs) react quickly. It has a very good torque and extremely responsive.
- EVs charging are user-friendly. Just like plugging a smart phone the vehicles are connected. Moreover, we can able to monitor the amount of charge taken by our vehicle.
- The cost of the fuel is highly controlled using IoT based electric vehicles. Since we could charge the vehicle using any form of renewable sources from solar and wind.
- Electric train for smart urban transport: Considering an electrical driven train that seems to be simpler that the conventional one. Since the components of the electric vehicle are reduced say for example, it has only electric motor. They do not require any replacement of fuel. In fact, features like regenerative braking reduce wear and tear on basic vehicle components. Recent survey says that a typical fleet that uses electricity reduces 50 percent of its cost for the maintenance purposes.

A programmable logic controller (PLC) plays a vital role in the field of microprocessor. It's very simple and easy for the users to handle since it could consist of hardware and software components which could handle all the industrial components, and the work of the engineer is to program the PLC using ladder logic which could do certain type of automation and control in the industrial equipment. The greatest advantage of using PLC is that they can be reprogrammed depending on the user who handles it. They have an astounding effect on the automation since these PLC components are highly flexible and reliable at their implementation stage. Since the functionalities of the micro-controllers have been increased as well as their reduced cost had increased their scope and made it to be used in the variety of fields.

CONCLUSION

The advancements in healthcare are not implemented in developing countries because of the poor healthcare infrastructure. The solution for this problem is integrate health sensing devices with portable devices such as smart phones and deploy them in the cloud. By following this technique, poor people can make use of healthcare by employing smart phones which are cheap these days. However, healthcare should enhance the reliability by deploying real-time monitoring for patients and analyzing patient data by providing smart healthcare monitoring devices since IoT has turned into an effective communication paradigm.

REFERENCES

Firouzi, F., Rahmani, A. M., Mankodiya, K., Badaroglu, M., Merrett, G. V., Wong, P., & Farahani, B. (2018). Internet-of-Things and big data for smarter healthcare: From device to architecture, applications and analytics. *Future Generation Computer Systems*, 78(2), 583–586. doi:10.1016/j.future.2017.09.016

Gubbi, J., Buyya, R., Marusic, S., & Palaniswami, M. (2013). Internet of Things (IoT): A vision, architectural elements, and future directions. *Future Generation Computer Systems, 29*(7), 1645–1660. doi:10.1016/j.future.2013.01.010

Hank, P., Muller, S., Vermesan, O., & Van Den Keybus, J. (2013). Automotive ethernet: in-vehicle networking and smart mobility. In *Proceedings of the Conference on Design, Automation and Test in Europe*. EDA Consortium. 10.7873/DATE.2013.349

Kumar, L. D., Grace, S. S., Krishnan, A., Manikandan, V., Chinraj, R., & Sumalatha, M. (2012). Data ŏltering in wireless sensor networks using neural networks for storage in cloud. In *Recent Trends In Information Technology (ICRTIT), 2012 International Conference on*. IEEE.

Kuo, A. M.-H. (2011). Opportunities and challenges of cloud computing to improve health care services. *Journal of Medical Internet Research, 13*(3), e67. doi:10.2196/jmir.1867 PMID:21937354

Miluzzo, Lane, Fodor, Peterson, Lu, Musolesi, Eisenman, Zheng, & Campbell. (2008). Sensing meets mobile social networks: the design, implementation and evaluation of the cenceme application. In *Proceedings of the 6th ACM conference on Embedded network sensor systems*, (pp. 337–350). ACM. 10.1145/1460412.1460445

Suresh, A. (2017). Heart Disease Prediction System Using ANN, RBF and CBR. *International Journal of Pure and Applied Mathematics, 117*(21), 199 – 216.

Suresh, A., Udendhran, R., & Balamurgan, M. (2019). A Novel Internet of Things Framework Integrated with Real Time Monitoring for Intelligent Healthcare Environment. *Springer-Journal of Medical System, 43*, 165. doi:10.100710916-019-1302-9

Suresh, A., Udendhran, R., & & Balamurgan, M. (2019). Hybridized neural network and decision tree based classifier for prognostic decision making in breast cancers. *Springer - Journal of Soft Computing*. doi:10.100700500-019-04066-4

Udendhran, R. (2017). A Hybrid Approach to Enhance Data Security in Cloud Storage. *ICC '17 Proceedings of the Second International Conference on Internet of things and Cloud Computing at Cambridge University, United Kingdom*. Doi:10.1145/3018896.3025138

ADDITIONAL READING

Ben-David, A. (1995). Monotonicity maintenance in information-theoretic machine learning algorithms. *Machine Learning, 19*(1), 29–43. doi:10.1007/BF00994659

Bernardi, E., Carlucci, S., Cornaro, C., & Bohne, R. (2017). An analysis of the most adopted rating systems for assessing the environmental impact of buildings. *Sustainability, 9*(7), 1226. doi:10.3390u9071226

Ghaffarianhoseini, A., Berardi, U., AlWaer, H., Chang, S., Halawa, E., Ghaffarianhoseini, A., & Clements-Croome, D. (2016). What is an intelligent building? Analysis of recent interpretations from an international perspective. *Architectural Science Review, 59*(5), 338–357. doi:10.1080/00038628.2015.1079164

Chapter 7
A Comparative Study of Popular CNN Topologies Used for Imagenet Classification

Hmidi Alaeddine

(iD) https://orcid.org/0000-0002-2417-3972

Laboratory of Electronics and Microelectronics, Faculty of Sciences of Monastir, Monastir University, Monastir, Tunisia

Malek Jihene

Higher Institute of Applied Sciences and Technology of Sousse, Sousse University, Sousse, Tunisia

ABSTRACT

Deep Learning is a relatively modern area that is a very important key in various fields such as computer vision with a trend of rapid exponential growth so that data are increasing. Since the introduction of AlexNet, the evolution of image analysis, recognition, and classification have become increasingly rapid and capable of replacing conventional algorithms used in vision tasks. This study focuses on the evolution (depth, width, multiple paths) presented in deep CNN architectures that are trained on the ImageNET database. In addition, an analysis of different characteristics of existing topologies is detailed in order to extract the various strategies used to obtain better performance.

INTRODUCTION

Machine learning is a field that is interested in the development of technology that allows the interpretation and prediction of data in an automatic way. It was originally a result of pattern and object recognition research that seeks to make algorithms able to learn automatically from a huge amount of data. During the years of the field of machine learning, applications in this domain have gone beyond the scope of pattern and object recognition where they are widely used in various tasks such as filtering content on networks of social media or web searches. In the early days, the technique of carrying out this process relies on methods developed by experts who have knowledge in the field purpose of interpreting the

DOI: 10.4018/978-1-7998-3591-2.ch007

data. Today, the current methods are based on networks of artificial neurons that have the ability to perform the same role at the top level compared to the first techniques and without the need for human intervention. In the course of a long history, deep learning (Y. LeCun, 2015), (Schmidhuber, 2014) has undergone alternating periods of excitement and forgetfulness, under which it was known under various names and many researches. The first period of work on deep learning was for nearly 20 years from 1940 until 1960 and described simple neural networks with a supervised and unsupervised learning process (Rosenblatt, 1958). The second period began 19 years after the end of the first period (1979 to 1990) with fukushima's work Neocognitron (Fukushima, 1979) which can be considered as the first work that deserves to be described "deep" (Schmidhuber, 2014). In addition, this period has seen the application of back propagation to artificial neural networks (Werbos, 1981), it recalls us that the back propagation is an effective method of gradient descent for supervised learning has been developed in the years 1960 and 1970. The most modern period represents the golden period of deep learning started from 2006 as shown in the following works (G. E. Hinton, 2006), (Y. Bengio, 2007), where we can notice that in the previous last years, deep learning (Y. LeCun, 2015), (Schmidhuber, 2014) achieve huge success in different areas especially in computer vision. In addition, deep networks have won numerous competitions in image classification (Alex Krizhevsky, 2012), (Matthew D Zeiler, 2013), (Zisserman, 2015), (al, 2015), (He X. Zhang, 2015), (Komodakis, 2016), (Saining Xie, 2016), (Gao Huang, 2017), (Huang, 2016), (P. Sermanet, 2014), (Bengio, 1994) and object detection (P. Sermanet, 2014) such as the annual ILVRC competition Russakovsky. It is based on ANN consisting of various processing layers to learn representations of large sets of data using the back propagation algorithm to update the local parameters of each layer. Among various deep learning algorithms [1,2], Convolutional Neural Networks (CNN) has become the popular and frequently used neural model in deep learning it where has achieved a revolution in the tasks of artificial vision thanks to these capacities persising in which resist for many years. In addition, it offers good performance in terms of precision (Alex Krizhevsky, 2012), (Matthew D Zeiler, 2013), (Zisserman, 2015), (al, 2015), (He X. Zhang, 2015), (Komodakis, 2016), (Saining Xie, 2016), (Gao Huang, 2017), (Huang, 2016), (P. Sermanet, 2014), (Bengio, 1994).

This paper is organized as follows: the second section describes the background. The third section presents the comparative study of popular deep CNN topologies used for imageNET classification and the conclusions are drawn in the last section.

Background

Recent work on deep neural network approaches has improved the performance of these visual recognition systems. In section, we focus on the convolutional neural network, supervised learning and the ImageNet database.

Motivations

CNNs are described as a special type of artificial neural networks that is based on convolution, which is defined as a specialized type of linearity operation. They benefit from a capacity that can be controlled by varying these two elements: the depth and the width. In addition, they have a lower density of connections and parameters compared to the multilayer perceptron (MLP) which makes them easier to train.

Supervised Learning

To Supervised learning (Bishop, 2006) is based on the exploitation of labeled data. It makes it possible to create a link between the input and the label for correctly associating each input E_i with its output S_i. For each input sample E_i from the incoming data of size n $E_i = \{0,1,2..n-1\}$, the neurons that are located in the last layer of the network will be labeled, so that for one input E_i the activated neuron after the learning phase will be assigned by the label which represents the expected output Si for this input.

$$S_i = f(E_i) + \Delta \tag{1}$$

The principle of supervised learning is to evaluate the error (Δ) for each input and update the local parameters of the neurons in order to minimize the error in the next iterations. This means that the training data must be iteratively provided to the ANN in order to compare the prediction of each data sample with the associated desired label and analyze the error for each input and then make a modification of the local parameters of the system (called: weights, they are real numbers) will be realized to improve the prediction. By repeating this operation on a set of data, the system will converge to a state where its predictions become more and more accurate. Among the most widely used supervised learning algorithms are the gradient retro-propagation algorithms (D. Rumelhart, 1986). The major disadvantage of this learning class is sometimes the difficulty of finding a database in the desired domain and the cost of building the new learning database and finally the complexity of these learning algorithms.

ImageNet

ImageNet is an image database organized according to the WordNet hierarchy and intended for computer vision research work. It is considered the standard reference for evaluating the performance of different image classification and object detection algorithms. It integrates more than fifteen million high resolution and variable size images from approximately 22,000 categories. ImageNet consists of variable resolution images collected from the web and labeled by "Amazon Mechanical Turk". This foundation was first presented at the Conference on Computer Vision and Pattern Recognition (CVPR, 2009) during poster sessions. Since 2010, an annual competition has been organized. ILSVRC (Russakovsky, 2015) uses a subset of ImageNet with approximately 1000 images in each of the 1000 categories. In total, there are approximately 1.2 million training images, 50,000 validation images and 150,000 test images with variable resolutions.

The Bases of Convolutional Neural Networks (CNN)

The different CNN models use many layers that typically include multilayer convolution stacks, combined with grouping and normalization layers, to extract simple and hierarchically useful images. The stacking of these several layers ensures the construction of the CNN. Convolution is the main step in convolutional neural networks and is described as the core of the convolutional network. A convolution is defined as the scalar product of a matrix of pixels with a weight matrix or core that represents the size convolutional filter (L x C) where L and C are generally odd and equal values. CNNs perform a 3D convolution (figure 1) to extract the spatial and temporal characteristics of the data (Dou Q, 2017), (N. C. Camgoz, 2016). The principle of 3D convolution consists of convolving an image cube that represents several images stacked in a successive manner by a set of filters that forms a 3D core. The value of the position (x, y, z) in the three-dimensional convolution for a fifth image in the first layer is determined using the following mathematical equation:

$$Conv(Pixel_{ij}^{XYZ}) = B_{ij} + \sum_{N}\sum_{P=0}^{P_i-1}\sum_{Q=0}^{Q_i-1}\sum_{R=0}^{R_i-1} W_{ijN}^{PQR} \times Pixel_{(i-1)\times N}^{(X+P)(Y+Q)(z+R)} \tag{2}$$

Knowing that R denotes the length of the size of the 3D convolution kernel.

Figure 1. Representation of a 3D convolution

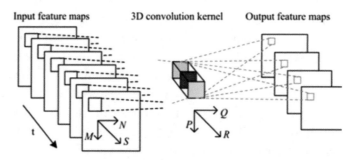

The choice of the activation function is very important since it will be applied to each incoming pixel. In addition, the choice of the appropriate activation function can speed up the learning process. The figure 2 summarizes the popular activation functions and the most used ones:

Figure 2. The most popular nonlinear activation functions

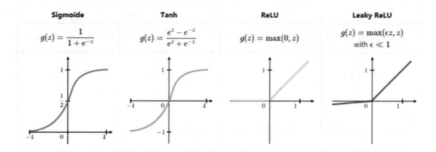

The Relu function is the most used function; it represents the identification of positive arguments and assigns zeros to others. In most early works, the hyperbolic tangent and the sigmoid function are often the most used. However, today these functions are no longer exploited because they require complex arithmetic and they present difficulties in the material implantation. In addition, the hyperbolic tangent does not represent a solution for the problem of vanishing gradient problem on the contrary to the Relu, which allows to reduce and relieve this problem (Clevert & Unterthiner, 2016). The pooling operation is an interesting local operation like convolution. This operation consists of generating the dominant response by summarizing similar information near the receptive field in the local region. These layers make it possible to reduce the spatial dimensions of each input (generally the inputs are images) by replacing a group of pixels by a single pixel representing either the maximum value (Max pooling) or the average of these pixels (average pooling). Although pooling layers reduce spatial resolutions, they keep the same number of incoming images. This reduction in the size of feature maps regulates the complexity of the network and contributes to the fight against over-adjustment. Fully connected layers are sometimes used as the last layers of CNN or the penultimate. These layers are localized after many convolution and pooling layers. The fully connected layers are identical to the multilayer perceptrons (MLPs). They are mainly used at the end of the network for classification. In this layer, each neuron is connected with all the combinations of the outputs of the neurons in the previous layer, which leads to a density of connections that is very high compared to those in the convolution layers (Dou Q, 2017), (N. C. Camgoz, 2016). The fully (fully) connected layer is described as a vector of the neurons of a surfacing of different volumes received from the previous layer (Dou Q, 2017), (N. C. Camgoz, 2016).

Batch Normalization

This layer is introduced in 2015 by Google researchers (Szegedy, 2015). It is usually located after fully connected layers or convolutional layers, and before non-linearity. It alleviates the problems associated with internal covariance lag in feature maps by normalizing the output distribution of the layer to zero mean (unit variance). The use of this layer makes the networks more resistant to bad initialization; moreover, it eliminates the need for the use of the Dropout layer.

Dropout

This is a popular (Salakhutdinov, 2014) and simple method to prevent neural networks from over-adjusting. During training, these layers eventually improve the generalization by randomly jumping a select percentage of their connections. Neurons that are "abandoned" in this way do not contribute to the propagation and do not participate in the back propagation, At the time of the test, all the neurons are used but their outputs are multiplied by the probability (Generally: 0.5 the most used). This layer introduces regularization within the network. An extremely efficient regulation technique complements

Figure 3. Dropout layer

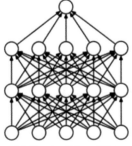 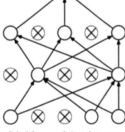

(a) Standard Neural Net   (b) After applying dropout.

the L1, L2 regularization methods that are used to control the ability of neural networks to prevent over-learning. The disadvantage of this layer is that it doubles the number of iterations needed to converge.

Case Study: CNN Topologies for Image Classification

In scientific research, a comparison between the different architectures proposed is required. This is what is required in the following where we will try to represent its different architectures of convolutional neural networks. We will first look at the different architectures that have won first places in the annual competition of ImageNET. Various CNN models have been offered since AlexNet's outstanding performance on the ImageNet database in 2012. These innovations were mainly due to the reorganization of different layers, the design of new blocks or the operation of multiple paths. The table 1 summarizes the number of layers for each topology, including the convolution layer, the pooling layer, and the fully connected layer. We note that the activation function exploited by these different topologies is the ReLU function.

Table 1. Number of layers for each topology

Topology	References	Num of layer		
		Convolution	Pooling	FC
AlexNet	(Alex Krizhevsky, 2012)	5	3	3
GoogleNet	(al, 2015)	57	14	1
VGG16	(Zisserman, 2015)	13	5	3
VGG19		16	5	3
ResNet50	(He X. Zhang, 2015)	49	2	1
ResNet101		100	2	1
ResNet-152		151	2	1

Table 2. The number of Multiply Accumulate operation (MACC) and the number of parameters for the different architecture

Topologies	References	Convolution		FC		Total	
		MACC	Param.	MACC	Param.	MACC	Param.
AlexNet	(Alex Krizhevsky, 2012)	666 M	2.33M	58.6M	58.6M	724M	62M
GoogleNet	(al, 2015)	1.58G	5.97M	1.02M	1.02M	1.58G	6.99M
VGG16	(Zisserman, 2015)	15.3G	14.7M	124M	124M	15.5G	138M
VGG19		19.5G	20M	124M	124M	19.6G	144M
ResNet50	(He X. Zhang, 2015)	3.86G	23.5M	2.05M	2.05M	3.86G	25.5M
ResNet101		7.57G	42.4M			7.57G	44.4M
ResNet-152		11.3G	58M			11.3G	60M

The different architectures of the deep neural networks proposed until these days mainly focused on the improvement of the precision. Although precision figures have been steadily increasing, the use of resources from winning models has not been properly accounted for. The table 2 summarizes the number of Multiply Accumulate (MACC) operations and the number of parameters for the different architecture where most MACC operations are consumed during convolutional layers.

AlexNet

Introduced by Alex et al in 2012. It is described among the first CNN for image classification. He won the ILSVRC in 2012, consisting of 5 convolutional layers followed by 3 fully connected layers. This architecture (figure 4) is of significant importance in the new generation of CNN and has opened a new line of research on CNN. AlexNET (Alex Krizhevsky, 2012) is the first architecture that has used the non-saturated ReLu activation function (linear rectified unit) for the non-linear part, instead of a Tanh or Sigmoid function that is the norm for traditional neural networks. This function is placed after each convolutional and fully connected (FC) layer. The use of this function was to improve the convergence rate by mitigating the problem of the disappearance gradient.

This architecture (Alex Krizhevsky, 2012) uses five convolution layers with a Local Response Normalization Layer (LRN) after the first, the second convolution layer, and a max-pooling layer after the first, the second, the fifth convolution layer. An advantage in this architecture is that it solved the over-adjustment problem by using a Dropout layer after each FC layer. This network requires 62.4 million parameters and offers an ImageNet TOP 5 error rate equivalent to 16.4%. It is noted that this model is formed for about 90 cycles using a stochastic gradient descent (SGD) with:

- Momentum of 0.9.
- Decrease in weight: 0.0005.
- Batch size: 128.

The weights are initialized from the mean Gaussian zero distribution with a standard deviation of 0.01, N (0, 0.012), the conv1 and conv3 biases are initialized to zero, the other biases are initialized to 1. The Learning was initialized at 0.01 and reduced 3 times before the end where it divided by 10 when the validation error rate has stopped improving with the current learning rate.

Figure 4. AlexNet architecture

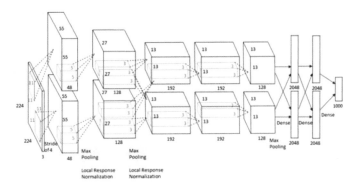

ZF Net

In 2013, Zeiler and Fergus introduced a multilayer deconvolutional neuron network known as (ZF NET) (Matthew D Zeiler, 2013), where they proposed a new visualization technique that provides a quantitative overview of network performance. The idea of visualizing network activity is to discover the contribution of different model layers to performance by monitoring the learning pattern during training and using these results to extract the problem associated with the model. Experimentally validating this idea was done on AlexNet (Matthew D Zeiler, 2013). Experimental results showed that only a few neurons were active, unlike other neurons that were inactive in the first two layers of the model. Based on these results, the authors modify the CNN topology and optimize CNN learning by reducing both the size of the first layer size (11 x 11) filters (Stride: 4) to the size (7 x 7) (Stride: 2) In addition, they increased the numbers of the kernel of the 3rd convolutional layer from 384 to 512 kernel, the 4th convolutional layer from 384 to 1024 kernel, the 5th layer of convolution of 256 to 512 kernel. This readjustment of the CNN topology resulted in an improvement in performance where this architecture achieved an Image Net "TOP 5" error rate equivalent to 11.7% instead of 16.4% for (Y. LeCun, 2015). Feature visualization can be exploited to extract design problems and adjust settings in a timely manner.

VGG Net

In 2014, the VGGNET architecture (figure 5) proposed in (Zisserman, 2015) won part of the ILSVRC 2014 challenge with an error rate equivalent to TOP 5 equivalent to ImageNET 7.4% and nearly 138 million parameters knowing that it uses fully connected layers. There are several topologies with different numbers of layers ranging from 11 to 19 layers (the VGG-16 topology with very regular structure of 16 layers). This architecture replaced the 7x7 convolution filters with a pitch of 2 by three (3) convolution filters of sizes 3x3 (stride 1, pad 1) in order to keep almost the same resultant dimension of convolution 7x7 while decreasing the number of parameters (calculation complexity) and increasing the number of activation functions used. In addition, this architecture uses a maximum sub-sampling of size 2 × 2. VGG16 (Zisserman, 2015) is based on convolution filters of sizes 3x3 and a network width which starts from 64 and increases by a factor of two every two layers of convolution. It is noted that VGG (Zisserman, 2015) is the most expensive architecture in terms of both computational requirements and number of parameters. This model is formed for about 74 epochs (370K) using a stochastic gradient descent (SGD) with:

- Momentum: 0.9.
- Decreased weight: 0.0001.
- Batch size: 256.

The learning rate was initialized to 0.01 and reduced three times before the end where it is divided by 10 when the validation error rate stopped improving with the current learning rate.

GoogLeNet

Introduced by Christian Szegedy et al, (al, 2015). She won the 2014-ILSVRC contest. The GoogLeNet contains 22 layers and achieves an ImageNet TOP 5 error rate equivalent to 6.67% while requiring only

Figure 5. VGGNET model

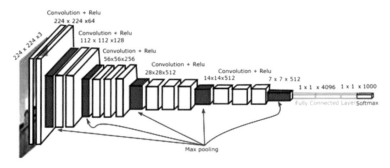

about 7 million parameters. These savings are realized through a complex architecture that uses a sub-architecture in the network called "inception" modules (figure 6).

Figure 6. Architecture du module «inception»

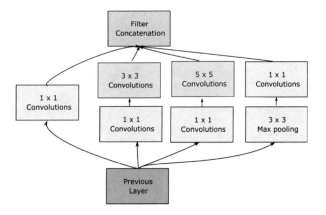

The main target of this architecture was to reduce the cost of computation while offering high accuracy; this can be noticed from the 1 × 1 convolution layer before the 3x3 and 5x5 layers in order to reduce the number of channels. Reducing the channel size automatically causes a decrease in the number of MACC parameters and operations. In addition, a global average-pooling (AVG) layer is used at the last layer, instead of using a fully connected layer (FC) to reduce the density of connections. The authors of the Google NET Architecture [8] have added auxiliary classifiers connected to intermediate layers in order to increase the gradient signal that is propagated backwards (Retro-gradient propagation) and accelerate the convergence rate. At the time of testing, these auxiliary networks are discarded and at the time of training, the losses of auxiliary classifiers are balanced to 0.3. This model is formed for approximately 600K iteration using a stochastic gradient descent (SGD) with Momentum of 0.9. The learning rate decreases by 4% each eight time.

ResNet

After the invention of VGG (Zisserman, 2015) and Google Net (al, 2015), it was thought that the more layers architecture stacks, the accuracy will be better. However, deeper neural networks with more than 20 layers generally are more difficult to form where very deep network formation requires slow training time and present several problems such as the disappearance of gradients in the backward propagation. In (He X. Zhang, 2015) He et al, presented a residual learning bloc (figure 7) to facilitate the formation of considerably deeper networks than (Zisserman, 2015) and other networks by including a link around each two convolutional layer, summing both the original diverted and its function. This solution is called the "residual block". Each "residual block" integrates two layers of convolution (3x3). ResNet (He X. Zhang, 2015) was proposed by He et al and offers a result that ranked ILSVRC 2015 as the leading candidate for residual networks with maximum depth (152 layers) nearly 8 times deeper than (Zisserman, 2015). The error is equivalent to 3.57%. It is noted that even with a deeper architecture, ResNet (He X. Zhang, 2015) has a computing complexity lower than that (Zisserman, 2015). This architecture is similar to GoogLeNet (al, 2015) in terms of the use of a global sub-sampling followed by the classification layer, so this architecture does not use the "Dropout" layer. This model is formed for approximately 600K iteration using a stochastic gradient descent (SGD) with:

- Momentum of 0.9.
- Decreased weight 0.0001.
- Batch size 256.

Figure 7. Residual Block

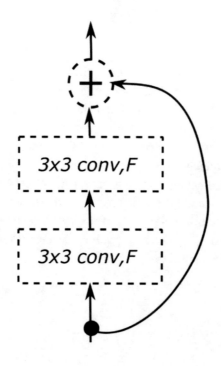

The learning rate was initialized to 0.1 and divided by 10 when the validation error rate stopped improving (with the current learning rate). The increase of a fraction in the precision requires an exploitation of the new tens layers. Although most of these convolutional networks are very deep with hundreds of layers and lead to a reduction in error rates, it also introduces several challenges such as vanishing gradients (Bengio, 1994), decreased reuse of features during the forward propagation and the long learning time. For this, several approaches are proposed to reduce this problem such as Batch Normalization (Szegedy, 2015), cautious initialization [19] but the shorter networks always remain the best solution where they have the advantage that the information can spread in back and forth efficiently and can be easily trained in a very acceptable learning time. Several works have convinced this principle as (Huang, 2016) proposed by Huang et al. Where they introduced the concept of stochastic depth. This concept allows a contradictory configuration to form short networks during learning and to use deep networks during the test time. This concept of stochastic depth offers the possibility of increasing the depth of residual networks even beyond 1000 layers while achieving a reduction in training time and test error.

Wide ResNet

Wide ResNet (Komodakis, 2016), proposed by Zagoruyko and Komodakis. This work (Komodakis, 2016) supports the principle: "Residual blocks are important factors in architecture and not depth which is an additional effect". This principle is validated when you notice that a Large-ResNet 50 layers wide surpasses the original ResNet 152 layers on ImageNet.

Figure 8. Wide Residual Block

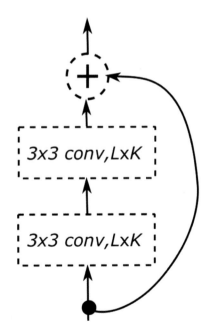

Wide ResNet (Komodakis, 2016) has exploited a large residual block (figure 8) that integrates (L x K filters instead of L filters like (He X. Zhang, 2015), where k is defined as an additional factor, which controls the width of the network. shown that the use of a larger residual block represents an effective solution for improving accuracy than deepening residual networks.

ResNeXt

In ResNeXt (Saining Xie, 2016), is similar in spirit to the "Inception" module of GoogLENET (al, 2015). It is another architecture that is proposed by the designers of the ResNET (He X. Zhang, 2015)in 2017 where they have increased the width of the residual block through multiple parallel channels. This strategy put in sight a new dimension, called "cardinality". The results of the test showed that increasing cardinality is a solution to improve the classification accuracy. Moreover, it is more effective than deepening networks. This architecture used the deep homogeneous topology of VGGNET [7] and the simplified heterogeneous architecture of GoogleNet (al, 2015). ResNeXt (Saining Xie, 2016) obtained an error rate of 3.03% (Top-5) and won the 2nd place in the 2016 ILSVRC classification task. The ResNext building block (Saining Xie, 2016) is shown in Figure 9.

Figure 9. ResNext

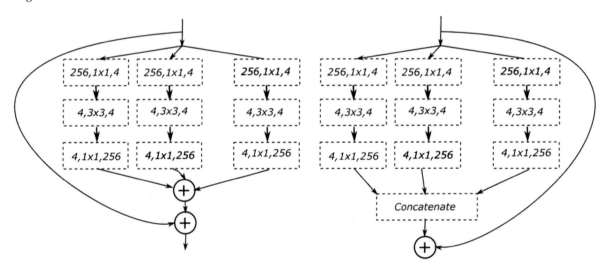

DenseNet

It is widely known that CNN can be much deeper, and more efficient during training if they contain shorter connections between the near-entry layers and those close to the classification layer.

Convinced by this principle, the authors proposed In (Gao Huang, 2017) a DenseNET architecture based on dense bloc (Figure 10) where each layer is connected to all other layers in "feedforward" mode. Thus, feature maps of all previous layers were used as inputs in all of the following layers parameters, which enhances the propagation of features, and encourages their reuse. This established network L (L + 1) / 2 direct connections since it connected each layer to a layer on two retroactively. DenseNET (Gao

Figure 10. DenseNet

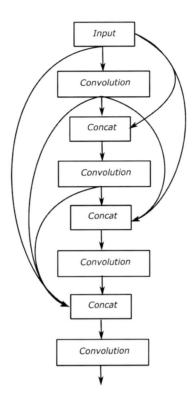

Huang, 2017) offers several advantages such as mitigating the vanishing-gradient problem (Glorot, 2010), encourage the propagation and reuse of feature maps which causes the reduction of parameters exploited.

CONCLUSION

In recent years, research on convolutional neural networks has been rapidly developed and conducted to improve CNN's performance in vision-related tasks. In this paper, we have studied CNN extensively among all the different types of DNN, or provided insights into the improvements and changes made to the topologies of various deep CNN architectures. (Depth, width, multiple path ...) which are trained on the database ImageNet. In the recent literature, it has been observed that the first CNN performance improvements were obtained because of the use of small filters, and the increase in channel numbers. The increase of the precision, the reduction of the size of the convolution filters is more favorable since it makes it possible to reach a level of precision a little high with an additional advantage related to a low complexity of computation and parameters compared to the same one architecture that uses a higher convolution filter size. This is what we can see from the comparison between AlexNet (Y. LeCun, 2015)and ZFnet (Matthew D Zeiler, 2013). It is noted that despite the use of small size filters and its simplicity, VGGNET requires a high cost of calculation and a heavy calculation load (VGG16: 138 M, VGG19: 140 million parameters). After the release of ResNet (Alex Krizhevsky, 2012) in 2015, the new trend in research is to replace convolution layers with modules. These modules play an important role in improving CNN's performance.

REFERENCES

Alex Krizhevsky, I. S. (2012). ImageNet Classification with Deep Convolutional Neural Networks. *Advances in Neural Information Processing Systems 25(2).*

Bengio, P. L. (2007). *Greedy layerwise training of deep networks.* NIPS.

Bengio, Y. (1994). Learning long-term dependencies with gradient descent is difficult. *IEEE Transactions on Neural Networks*, 157-66. doi:10.1109/72.279181

Bishop, C. M. (2006). *Pattern recognition and machine learning.* Springer.

Camgoz, S. H. (2016). Using Convolutional 3D Neural Networks for User-independent continuous gesture recognition. *International Conference on Pattern Recognition (ICPR)*, 49-54. 10.1109/ICPR.2016.7899606

Clevert, D.-A., & Unterthiner, T. H. (2016). *Fast and Accurate Deep Network Learning by Exponential Linear Units (ELUs).* ICLR.

Dou, Q. C. H., Chen, H., Yu, L., Qin, J., & Heng, P.-A. (2017). Multilevel Contextual 3-D CNNs for False Positive Reduction in Pulmonary Nodule Detection. *IEEE Transactions on Biomedical Engineering*, *64*(7), 1558–1567. doi:10.1109/TBME.2016.2613502 PMID:28113302

Fukushima, K. (1979). Neural network model for a mechanism of pattern recognition unaffected by shift in position. *Neocognitron*, 658–665.

Glorot, X. B., & Bengio, Y. (2010). *Understanding the difficulty of training deep feed forward neural.* http://proceedings.mlr.press/v9/glorot10a/glorot10a.pdf

He, X., & Zhang, S. R. (2015). *Deep residual learning for image recognition.* ArXiv preprint arXiv:1512.03385

Hinton, G. E., Osindero, S., & Teh, Y.-W. (2006). A fast learning algorithm for deep belief nets. *Neural Computation*, *18*(7), 1527–1554. doi:10.1162/neco.2006.18.7.1527 PMID:16764513

Huang, G., Liu, Z., van der Maaten, L., & Weinberger, K. Q. (2017). *Densely Connected Convolutional Networks.* https://arxiv.org/abs/1608.06993v5

Huang, G. S. (2016). Deep networks with stochastic depth. *European Conference on Computer Vision*, 646–661.

Ioffe, S., & Szegedy, C. (2015). *Batch normalization: Accelerating deep network training by reducing internal covariate shift.* http://proceedings.mlr.press/v37/ioffe15.pdf

Komodakis, S. Z. (2016). *Wide residual networks.* BMVC.

LeCun, Y., Begio, Y., & Hinton, G. (2015). Deep learning. *Nature*, *521*(7553), 436–444. doi:10.1038/nature14539 PMID:26017442

Rosenblatt, F. (1958). The perceptron: A probabilistic model for information storage and organization in the brain. *Psychological Review*, *65*(6), 386–408. doi:10.1037/h0042519 PMID:13602029

Rumelhart, D., Hinton, G. E., & Williams, R. J. (1986). Learning representations by backpropagating error. *Nature*, *323*(6088), 533–536. doi:10.1038/323533a0

Russakovsky, O. D., Deng, J., Su, H., Krause, J., Satheesh, S., Ma, S., Huang, Z., Karpathy, A., Khosla, A., Bernstein, M., Berg, A. C., & Fei-Fei, L. (2015). ImageNet Large Scale Visual Recognition Challenge. *International Journal of Computer Vision*, *115*(3), 211–252. doi:10.100711263-015-0816-y

Salakhutdinov, N. S. (2014). Dropout: A Simple Way to Prevent Neural Networks from Overfitting. *Journal of Machine Learning Research*, 1929–1958.

Schmidhuber, J. (2014). Deep learning in neural networks: An overview. *Neural Networks*, *61*, 85–117. doi:10.1016/j.neunet.2014.09.003 PMID:25462637

Sermanet, D. E. (2014). *OverFeat: Integrated Recognition, Localization and Detection using Convolutional Networks*. ICLR.

Shuiwang Ji Wei Xu, M. Y. (2012). 3D Convolutional Neural Networks for Human Action Recognition. *2nd International Conference of Signal Processing and Intelligent Systems (ICSPIS)*.

Szegedy, C., Liu, W., Jia, Y., Sermanet, P., Reed, S., Anguelov, D., Erhan, D., Vanhoucke, V., & Rabinovich, A. (2015). Going deeper with convolutions. *IEEE Conference on Computer Vision and Pattern Recognition (CVPR)*, 1 - 9. doi:doi:10.1109/CVPR.2015.7298594

Werbos, P. J. (1981). Applications of advances in nonlinear sensitivity. *Proceedings of the 10th IFIP Conference*, 762-770.

Xie, S., Girshick, R., Dollar, P., Tu, Z., & He, K. (2016). *Aggregated Residual Transformations for Deep Neural Networks*. https://openaccess.thecvf.com/content_cvpr_2017/papers/Xie_Aggregated_Residual_Transformations_CVPR_2017_paper.pdf

Zeiler, M. D., & Fergus, R. (2013). *Visualizing and Understanding Convolutional Networks*. https://cs.nyu.edu/~fergus/papers/zeilerECCV2014.pdf

Zisserman, K. S. (2015). *Very deep convolutional networks for large-scale image recognition*. ICLR.

Chapter 8
A Novel Framework on Biomedical Image Analysis Based on Shape and Texture Classification for Complex Disease Diagnosis

Reyana A.
Nehru Institute of Engineering and Technology, India

Krishnaprasath V. T.
Nehru Institute of Engineering and Technology, India

Preethi J.
Anna University, India

ABSTRACT

The wide acceptance of applying computer technology in medical imaging system for manipulation, display, and analysis contribute better improvement in achieving diagnostic confidence and accuracy on predicting diseases. Therefore, the need for biomedical image analysis to diagnose a particular type of disease or disorder by combining diverse images of human organs is a major challenge in most of the biomedical application systems. This chapter contributes an overview on the nature of biomedical images in electronic form facilitating computer processing and analysis of data. This describes the different types of images in the context of information gathering, screening, diagnosis, monitor, therapy, and control and evaluation. The characterization and digitization of the image content is important in the analysis and design of image transmission.

DOI: 10.4018/978-1-7998-3591-2.ch008

INTRODUCTION ON THE NATURE OF BIOMEDICAL IMAGES

Bio-medical image analysis is the visual view of human body interiors for medical investigations and interventions in order to reveal hidden structures of skin, bone, organs, tissues to diagnose diseases and abnormalities. It uses technologies like imaging, radiology, endoscopy, photograph, tomography, etc. The medical imaging is interpreted for radiology as it involves videos and pictures thus acquiring quality image for diagnostic is still a challenge. Many scientific investigations and research contribute radiology as relevant area of medical science. The various medical investigations include cardiology, psychology, neuroscience etc, Magnetic resonance imaging instrument are powerful to polarize the human tissue water molecules enabling detectable signals resulting the human body image.

Although the instruments are capable there arises more health risks associated to the exposure of radio fields. The other applications are detecting of tumour and assessing vascular disruption effects since computers are considered as integral components of medical imaging performance, image generation, and analysis and data acquisition are the tasks preferred. Though there are increases in modalities of medical imaging the nature increase in complexity and their associated problems lead to the need for advanced solutions. Thus use of medical imaging has improved the diagnosis of specific diseases on discovering a comprehensible visual display. Developing an algorithm for medical image analysis is the variability features like signals, images and systems. Bio medical image analysis techniques incorporates techniques for i) quantification, ii) computer aided diagnosis and iii) evaluation and validation techniques. Quantification is the method radiologist use to measure and extract objects from images by segmentation. Computer aided diagnosis measures the features of a diagnosis procedure for accuracy and efficiency. A common medical image analysis requires prior knowledge on the disease symptoms, ability to classify the features by matching the model to sub images, description of their shapes.

A digital image is the signal sampled in space or time quantized based on amplitude. However, any medical image can be processed i) manually, ii) semi – automatic or iii) automatic analysis. Thus it has become important in health care the use of digital image processing for medical diagnosis. Picture image are composed of pixel values assigned with discrete brightness or colours. With the emergence of advanced imaging techniques the bio-medical images are processed efficiently and quickly. The "biomedical image processing" covers areas like image capturing, image visualization and image analysis. Both of these achieved by applying any of the technique like sampling quantization, matrix representation, blur and spread function or image resolution as described by authors Alvarez, Lions & Morel (1992). Further the analysis of image shape with respect to factors like compactness, moments, chord length structure improves the quality of the image. The analysis of texture based on directional filtering, Gabor filtering and finally pattern oriented classification and diagnostic detection will provide efficient measurement on diagnostic accuracy.

Today most of the diagnosed images suffer loss of quality due to artifacts and practical limitations. The presented methods and mathematical derivations shall provide a better understanding on biomedical image analysis. Each system in human body composed of many physiological processes that are complex including neural, hormonal or control simulations. Investigating such systems requires sophisticated techniques and the chapter provides methods related to analysis of such real time specimens and their diagnosis. Several imaging procedures and techniques have limitations on the quality and information content. Therefore special imaging techniques are required to facilitate visualization. The case study is on the image segmentation for tumour analysis. However some malignant tumours have smoothed shapes some have rough shapes. The main features are summarized and the algorithm measures the segmentation error, intrinsic structure, time efficiency, low memory consumption is proven.

Hence image analysis has many algorithms and specific functions those can be integrated with multiple application domains. The other levels of analysis included edge, texture, region, and object and scene level. Lower level techniques of image processing are pixel, texture, and edges based providing very minimal abstractions. Whereas higher level include object, texture, region and scene levels. Hence processing of high level biomedical images is complex and is difficult to formulate as it involves high level cognitive interpretations of physicians. Thus automatic image analysis for medical prescription should never go negative as it leads to harmful results.

Challenges on Characterization and Digitization of Image Quality

There are certain difficulties the investigator faces in applying these techniques is the limitation on the quality and the information content. The following paragraphs illustrate the difficulty in biomedical image acquisition and analysis. There are various medical imaging modalities that play a vital role in disease diagnostics, planning procedures and treatment and further follow up studies. Figure 1 depicts the various sorts of medical imaging modalities.

Figure 1. Imaging Modalities

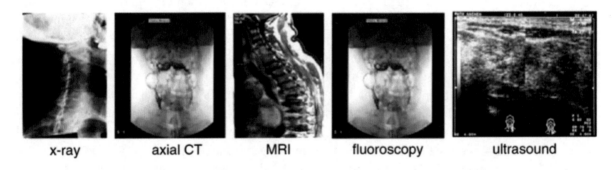

x-ray   axial CT   MRI   fluoroscopy   ultrasound

These modalities are based on spin, sound, radiation, temperature, reflection and transmission. There are plenty of imaging modalities and here X-ray imaging, CT, MRI, and ultrasound are considered under study. There are also modalities like photography, microscopy, and endoscopy belonging to optical modalities. Despite the use of such devices the images obtained are affected by severe artifacts as per Ando (2000).

Considering the radiation from x-rays passing through human body, the photon interaction leads with nucleus leads to absorption and scattering. These causes damage to living cells when photon releases all its energy to absorbing material, but obtaining of contrasted image is mandatory for diagnosis. Similarly secondary photon from scattering lowers SNR ration contributing at wrong location. However scattering radiations are filtered using scatter rasters from lead. The thickness of imaged material decides the radiation level as high radiation leads to harmful effects stated by Angenent, Haker & Tannenbaum

(2003). The reconstruction process in CT captures signals from various lines and gradient fields. Also, the strong currents in devices like MRI to be on and off very quickly.

Ultrasound works on reflection in sound waves and it travels in the tissues, reflected or transmitted partly depending on transducer. The strength in echo based on the material properties and the sonographic view of organs and bony structures through water-based gel that performs air-free coupling in the human body. There is a linear increase in attenuation with sound frequency where as high frequency is mandatory for spatial resolution. Diagnostic scanners operate at frequency range of 2-18 MHz with trading off in imaging depth and special resolution. Digitization based on the image's discrete nature and not on film based radiograph or altering on the image. Digitization includes sampling and quantization based on value range. The maximum number of gray scale for each image defines the quantization. It is either 8bit or 24 bit for both full color and gray scale images. The continuous brightness worsens the quality of the image and assistive noise alteration enhanced by adding gray scales. Quantization worsens quality of the image on continuous brightness. The quantization noise is depicted in Figure 2.

Figure 2. Quantization Noises

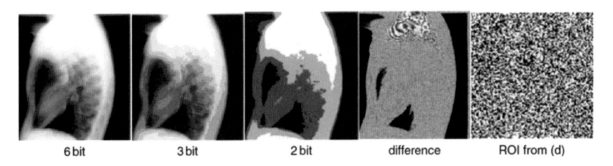

6 bit	3 bit	2 bit	difference	ROI from (d)

Similar to in-sufficient quantization, sub-sampling information suppressed inside the representation of digital image. On falsely adding quantization the sampling theorem not fulfilled. More important is aliasing effects. Thus there exist still a challenge in the usage of biomedical images as measurements of human body on different scales for clinical tasks (e.g. a diagnosis) leading to large impact on decision making of physicians.

Apart from these there are challenges related to the quality of the image like:

i) Image restoration
ii) Image enhancement
iii) Accurate segmentation
iv) Accurate registration
    iv) Fusion of modality images
    v) Classification and characterization and
    vi) Quantitative measurements

The role of imaging in therapeutic use in before, during and after treatment is the key challenge in the image guided therapy and surgery. The major methods involved in medical imaging are i) segmentation, ii) registration, iii) visualization and iv) simulation. Segmentation is the method of creating specific models from images of patient's anatomy. Registration aligns each other with multiple data sets. Visualization helps to simulate and evaluate the strategies and planned treatments. There are many imaging techniques that are complementary and offering different insights.

Image Quality Analysis

In mathematical models the data extracted from images helps in achieving clinical, biomedical, experimental progress in scientific research. Arrays of data samples arranged geometrically quantify the physical phenomenon of time variation in haemoglobin deoxygenating on neuronal metabolism, i.e. diffusion of water molecules inside tissues. The scope of organizing observations in biophysical world lead to increase in the emergence of new processing technology combining complex and sophisticated mathematical models to physiological dysfunction by Bloembergen, Purcell & Pound (1948) and Canet (1996). Shape analysis is detecting similarly shaped objects to automatically analyze and represent it in a digital form. The objects are modelled either by scanning or by extracting 2D or 3D images and simplified for a shape description. The simplified representations are easy to handle and reconstruct the original object. Segmentation in an image gives lots of data. Shape descriptors are required to evaluate to the same (vector or scalar) value for all sizes, positions, and/or orientations of any given shape. The various ways of analyzing a biomedical image by means of shape considered under study is given below:

Spatial Domain Shape Metrics

This is the broadly used metric to the measures aspect ratio, compactness, regularity on describing the shape of a feature. The metric is considered as their boundary points. If K boundary points define distance then for kth point described in equation (1)

$$r_k(\phi) = \sqrt{(x_k - x_c)^2 + (y_k - y_c)^2} \tag{1}$$

Were the associated angle 9 is as follows (equation (2))

$$\tan \phi = \frac{(y_k - y_c)}{(x_k - x_c)} \tag{2}$$

This representation converts the perimeter value r_k the sequence, where $0 < r < k$. The aspect ratio, A is given in equation (3):

$$A = \frac{\max r_k}{\min r_k} \tag{3}$$

The cluster shown in part having a zero-valued background, black is the segmentation process and the lines highlighted in white are the perimeter. The dashed lines that are thick are the diagonal lines with features approximated. In diagram B the centroid is the intersection of the diagonal lines and the

Figure 3. Sample Cluster

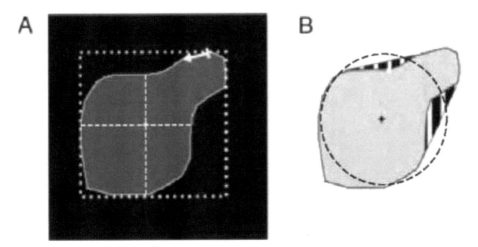

concave regions are filled with striped lines to obtain convex hull. The compactness of the shape is described in equation (4) and (5):

$$C = \frac{1}{N}\left[\sum_{k-1}^{k-1}\sqrt{(x_{k+1}-x_k)^2+(y_{k+1}-y_k)^2}\right] \tag{4}$$

$$C = \frac{\sqrt{4\pi N}}{\sum_{k=0}^{k-1}\sqrt{(x_{k+1}-x_k)^2+(y_{k+1}-y_k)^2}} \tag{5}$$

Here N the pixel numbers belonging to the area, x_k and y_k are the pixel coordinates with boundary points $x_K = x_0$, $y_K = y_0$ a closed contour.

The value for an ideal circular shape is normalized to 1 and for irregular shapes its zero. Since the images are discrete there will be deviation of small circles from theoretical values. Compactness C is the metric to increase both eccentricity and irregularity. The outline for irregularity is quantified well with the coefficient variation V for r_k is mentioned below in equation (6):

$$v = \frac{\sqrt{\sum_{k=0}^{k-1}r_k^2 - (1/k)\left(\sum_{k=0}^{k-1}r_k\right)^2}}{\sum_{k=0}^{k-1}r_k} \tag{6}$$

Figure 4. Separation of features with different shapes

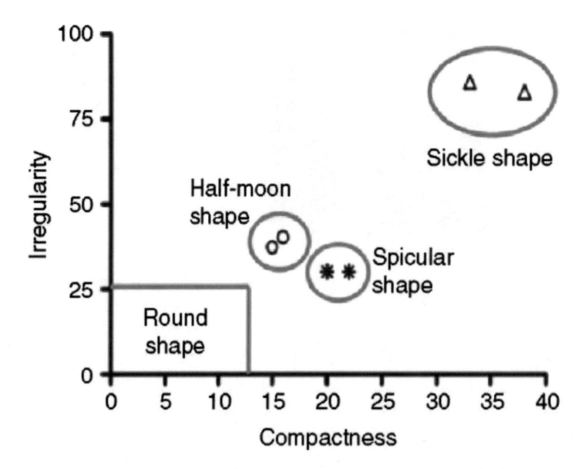

Compactness, irregularity, and aspect ratio are the metrics to shape scaling, translation, and rotation of pixel limits discretization. Vector used group features have a similar shape. The feature vector with irregularity and compactness are the considered elements.

The major and minor axis is aligned with the bounding rectangle. The extent of the shape within the same axis defines the bounding rectangle as described in figure 4. The zero crossings and aspect ratio computed with average radius is defined in (equation (7)) terms of area ratio parameter (ARP):

$$ARP_P = \frac{1}{N r_{mean}} \sum_{i=0}^{N-1} [r_k - r_{mean}] P \tag{7}$$

The operation indicates indicate threshold at zero for a > 0 and a < 0. The value of p is the aspect ratio parameter to emphasize specular outliers. The metric is not a scale-invariant. The descriptors are based on convex hull and tangent to all edges in the boundary pixel. The actual perimeter length is the convexity metric and the examples of these are malignancy of skin lesions with melanocytic is defined under r_k and the nth boundary moment m_n is given in equation (8) to (12):

$$M_n = \frac{1}{N} \sum_{i=0}^{N-1} \left(r_k - m_1 \right)^n \tag{8}$$

$$m_n = \frac{1}{N} \sum_{i=0}^{N-1} r_k^n \tag{9}$$

M_n is defined through

$$\bar{m}_n = \frac{m_n}{M_2^{n/2}} = \frac{(1/N) \sum_{i=0}^{N-1} r_k^n}{[(1/N) \sum_{i=0}^{N-1} (r_k - m_1)2] n / 2} \tag{10}$$

$$\bar{M}_n = \frac{M_n}{M_2^{n/2}} = \frac{(1/N) \sum_{i=0}^{N-1} (r_k - m_1)n}{[(1/N) \sum_{i=0}^{N-1} (r_k - m_1)2] n / 2} \tag{11}$$

The feature including translation, rotation, and scaling of invariant elements are F1, F2, and F3 (equation (13) to (15)) computed from the four lowest-order moments:

$$\bar{M}_n = \frac{M_n}{M_2^{n/2}} = \frac{(1/N) \sum_{i=0}^{N-1} (r_k - m_1)n}{[(1/N) \sum_{i=0}^{N-1} (r_k - m_1)2] n / 2} \tag{12}$$

$$F_1 = \frac{\sqrt{M_2}}{m_1} \tag{13}$$

$$F_2 = \frac{M_3}{(M_2)^{3/2}} \tag{14}$$

$$F_4 = \frac{M_4}{(M_2)^2} \tag{15}$$

Statistical Moment Invariants

The statistical moment describes the texture and is defined to describe the shape. Here the pixel value is assumed to be unity and the shape moments are not influenced by intensity variations. The feature computed from all the pixels is only boundary points.

M_k specified as

$$M_k = \frac{1}{N_F} \sum_{y=F} \sum_{x=F} (x - x_c)^k (y - y_c)^k \tag{16}$$

where x, y e F indicates the equation, the statistical moments are generally not rotation-invariant.

$$\varphi_2 = (\mu_{2,0} - \mu_{0,2})^2 \tag{17}$$

$$\varphi_6 = (\mu_{2,0} - \mu_{0,2})[(\mu_{1,2} - \mu_{3,0})^2 - (\mu_{2,1} - \mu_{0,3})^2] \tag{18}$$

φ_1 through φ_7 form a feature vector that can be used to classify the shape.

$$M_{p,q} = \sum_{Y=F} \sum_{x=F} (x - x_c)^P (y - y_c)^q \tag{19}$$

$$\mu_{p,q} = \frac{M_{p,q}}{M_{0.0}^{\gamma}} \tag{20}$$

$$\gamma \left[\frac{p+q}{2} \right] + 1 \tag{21}$$

where $p + q > 2$. Hu defined the feature metrics (equation (22) to (25)) to like rotation-invariant described as:

$$\varphi_1 = \mu_{2,0} + \mu_{0,2} \tag{22}$$

$$\phi_2 = (\mu_{2,0} + \mu_{0,2})^2 + 4\mu_{1,1}^2 \tag{23}$$

$$\phi_3 = (\mu_{3,0} - 3\mu_{1,2})^2 + (3\mu_{2,1} - \mu_{0,3})^2 \tag{24}$$

$$\phi_4 = (\mu_{3,0} + \mu_{1,2})^2 + (\mu_{2,1} - \mu_{0,3})^2 \tag{25}$$

Normalization by SD contains the co-ordinates x_c, y_c as centroid given by equation (26) and (27):

$$x_c = \frac{1}{N_F} \sum_{x=F} x \tag{26}$$

$$y_c = \frac{1}{N_F} \sum_{y=F} y \tag{27}$$

Normalized moments are less sensitive to noise. The computational invariants stored in tables having compactness, size, irregularity, and aspect ratio described in the work by Nishimura (1996) and Schempp (1998).

Topological Analysis

Topology is the connectivity among shapes that are under transformations. When shape stretched becomes topologically invariant. Consider a shape as a rubber sheet, it remains in the same topology when stretched and distorted. The description of connectivity in terms of topology is for shape analysis. The notion of graph leads to define set of nodes connected by edges. Euler proved in 1735 the Euler number

e=v-n+l

Here v - vertices, n – edges and l indicates the loops. Also v, n, l, and e are invariant to scaling, translation, rotation. The invariance belongs to affine transformations. The vertices graph and connections are extracted utilizing Skeletonization, shapes to medical axis a thinning process. The median axis is the set of centres inside a maximum diameter nearing the boundary shape and tangent without intersecting the boundary. Any two background pixel having same minimum distance belongs to medial axis. Pixel joined with three or more edges called as vertex. And the edge connected by link, edge connected to one vertex called branch and the other end is the endpoint.

The medial axis is the set of centres inside a maximum diameter nearing the boundary shape and tangent without intersecting the boundary. Loop is the sequence of vertices and links. The branches, loops, vertices and links are topological invariants describing the network connectivity based on scaling, rotation and position metrics. Invariants are called offline transformations. Every minor changes of binary image reflected in network is shown in figure 5 described by Liang & Lautenberg (1999). The shape characterization is based on sensitivity and skeleton needs analyzation to indicate the presence

Figure 5. Skeletonization Process

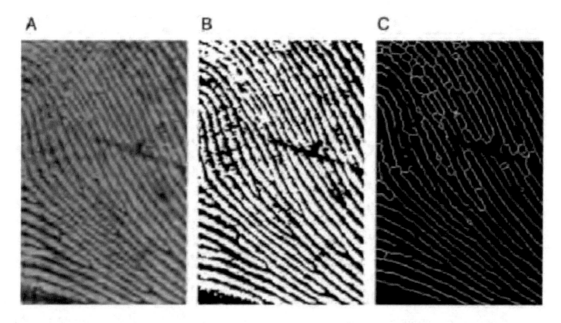

of loop. If loops are connected then each loop contributes one branch. These metric combined together describes the feature vector of image and its shape.

Case Study

Partitioning the image into various regions for the purpose of some specific task described to be image segmentation/labelling problem. For example detection of brain tumour from CT/MR image. Before interpretation and image analysis segmentation is the first step. Further based on features utilizing, dif-

Figure 6. A- 4 Connected Pixels, B- 8 Connected Pixels

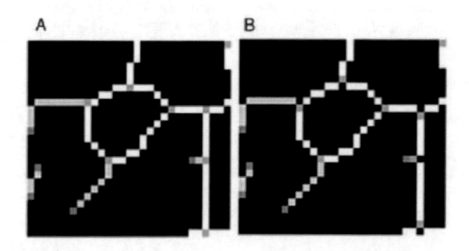

ferent techniques the images underneath for classification. Features like texture, edge information, pixel intensity, color, etc., are considered and broadly classified into statistical and structural methods. Here we consider the structural method on spatial properties like regions and edges for classification. The boundaries from different brain tissues are extracted using edge detection algorithm. These algorithms are sensitive to noise and artifacts. In this the image is divided into many small regions called seeds and boundaries adjacent examined. Weak boundaries rejected whereas strong boundaries are accepted. The process is iteratively carried out until weak boundaries get rejected. The performance depends based on the selection of the seed and well defined region. Consider the segmentation of a 3D computed tomography data.

Requisites of the 3D Segmentation Algorithm: Every pixel in 2D data is characterized by color domain vector values. The third dimension partition for segmentation in 3D space is color. The voxel have 12 bit scalar in HU domain and extension to 3D has certain disadvantages in terms of time and memory. The neighbourhood of voxel is elongated on the scanning direction. The image processing assuming regular shapes on pixels are not considered here for study.

The main features are summarized and the algorithm measures the segmentation error, intrinsic structure, time efficiency, low memory consumption.

2D to 3D CT Imagery

Consider the statistical region merging algorithm build from statistical considerations. The feature of the algorithm to satisfy the segmentation requirement and grows with region emerging statistical test. The statistical region growing is linear and fast as per the number of pixels exploiting low memory

Figure 7. Segmentation results

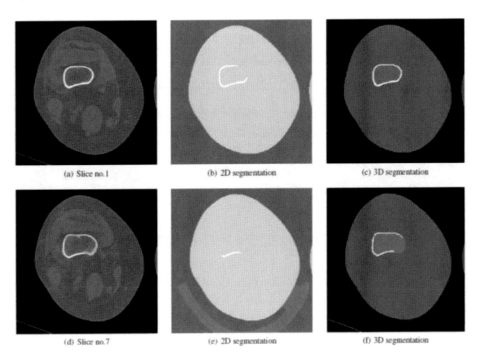

consumption and time complexity. The algorithm could use 3D pixels belonging to neighbouring slices on considering memory occupancy.

The result of 2D segmentation SRM compared to extended 3D on merging the regions of neighbouring slices is shown in figure 7. These extensions are used in most cases and the dataset is robust. Whereas loading the entire dataset in main memory is not feasible for many real cases. For this, one can utilize multiple slices spanning to overcome the memory issues.

CONCLUSION

Developing computational techniques like reconstruction analysis, segmentation, physiological signal processing, multimodal imaging, etc., are the major areas where research contributions required. However the chapter proposed helps on the basic study of image processing and its importance in processing image and helping medical descriptor on diagnose of a specific disease. Some fundamental concepts of medical image processing are discussed.

REFERENCES

Alvarez, L., Guichard, F., Lions, P.L., & Morel, J.M. (1992). Axiomes et ´equations fundamentals du traitement d'images. *Paris 315. C. R. Acad. Sci.*, 135–138.

Alvarez, L., Lions, P. L., & Morel, J. M. (1992). Image selective smoothing and edge detection by non-linear diffusion. *SIAM Journal on Numerical Analysis*, 29(3), 845–866. doi:10.1137/0729052

Ando, S. (2000). Consistent gradient operators. *IEEE Transactions on Pattern Analysis and Machine Intelligence*, 22(3), 252–265. doi:10.1109/34.841757

Angenent, S., Haker, S., & Tannenbaum, A. (2003). Minimizing flows for the Monge-Kantorovich problem. *SIAM Journal on Mathematical Analysis*, 35(1), 61–97. doi:10.1137/S0036141002410927

Bloch, F. (1946). Nuclear induction. *Physical Review*, 70, 460474.

Bloembergen, N., Purcell, E. M., & Pound, R. V. (1948). Relaxation effects in nuclear magnetic resonance absorption. *Physical Review*, 73(7), 679–712. doi:10.1103/PhysRev.73.679

Canet, D. (1996). *Nuclear Magnetic Resonance: Concepts and Methods*. John Wiley & Sons.

Hashemi, R. H., & Bradley, W. G. (1997). *MRI: The Basics*. Baltimore, MD: Williams & Williams. www.cis.rit.edu/htbooks/mri/inside.htm

Liang, Z.P., & Lautenberg, P.C. (1999). *Principles of Magnetic Resonance Imaging: a signal processing perspective*. IEEE Press.

NessAiver, M. (1997). *All You Really Need to Know about MRI Physics*. Baltimore, MD: Simply Physics.

Nishimura, D.G. (1996). *Principles of Magnetic Resonance Imaging.* Academic Press.

Schempp, W. G. (1998). *Magnetic Resonance Imaging: Mathematical Foundations and Applications.* John Wiley and Sons.

Chapter 9
Deep Learning Techniques for Prediction, Detection, and Segmentation of Brain Tumors

Prisilla Jayanthi

https://orcid.org/0000-0002-4961-9010

K. G. Reddy College of Engineering and Technology, India

Muralikrishna Iyyanki

Defence Research and Development Organisation, India

ABSTRACT

In deep learning, the main techniques of neural networks, namely artificial neural network, convolutional neural network, recurrent neural network, and deep neural networks, are found to be very effective for medical data analyses. In this chapter, application of the techniques, viz., ANN, CNN, DNN, for detection of tumors in numerical and image data of brain tumor is presented. First, the case of ANN application is discussed for the prediction of the brain tumor for which the disease symptoms data in numerical form is the input. ANN modelling was implemented for classification of human ethnicity. Next the detection of the tumors from images is discussed for which CNN and DNN techniques are implemented. Other techniques discussed in this study are HSV color space, watershed segmentation and morphological operation, fuzzy entropy level set, which are used for segmenting tumor in brain tumor images. The FCN-8 and FCN-16 models are used to produce a semantic segmentation on the various images. In general terms, the techniques of deep learning detected the tumors by training image dataset.

INTRODUCTION

Deep Learning (DL) techniques include machine learning methods, where learning can be supervised, unsupervised, semi-supervised and reinforcement learning. DL performance is based on neural network (NN) modeling for huge data and requires faster processors for better outcomes. These techniques are very powerful in segmentation and detection of tumors in the healthcare. DL has the potential in transforming

DOI: 10.4018/978-1-7998-3591-2.ch009

healthcare centers. DL in healthcare systems comes with improving accuracy and increasing efficiency. The excellent approaches of DL proved in classification, segmentation, recognition and detection with automated algorithms. DL systems develop and evolve by assisting humans with those tasks at which humans are not good. In this chapter, the approaches of ANN, CNN and DNN in healthcare are highlighted.

DEEP LEARNING TECHNIQUES

This approach is used to build and train neural networks (NN), with high propitious decision-making nodes. An algorithm is said to be deep if the input data is passed through a series of layers before it gives output. The state-of-the-art of DL is focused on training NN models using the backpropagation algorithm. Deep Neural Networks (DNN) is one among the DL technique and all the networks that have more than a two or three layers are DNN. The most popular DL techniques are shown schematically in figure 1 (Brownlee, 2019):

Figure 1. Techniques of Deep Learning

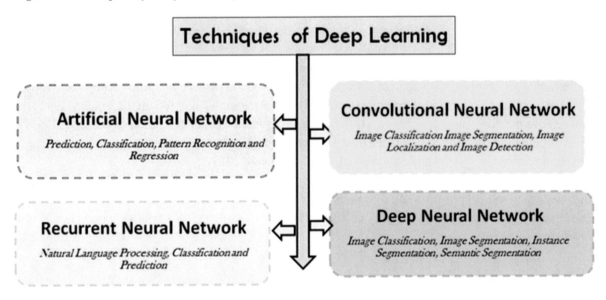

DL has been dominant in the medical imaging, in which different techniques are used for various analysis purposes. One of such technique of DL named ANN works on the numerical datasets of the patient in healthcare whereas CNN deals with the medical imaging that includes magnetic resonance imaging, digital images and so on.

The four tasks of DL are:

Image Classification, that includes two types of classification namely binary and multiclass. It allocates a label to an image.

- Labeling an x-ray comes under binary classification whereas
- Naming a face in the photo is multiclass.

Image Classification with Localization (Object Localization) includes class labelling and displaying the object location by a bounding box.

- Classifying different objects like birds and drawing a rectangular box known as bounding box around the birds.
- Labeling a digital image in MRI as cancer with a box around the region.

Object detection is the task of image classification with localization. Whenever an image has multiple objects and different objects are to be detected; it requires localization and classification.

- Labeling each element in the satellite image with a bounding box.
- A bounding box and labeling each vehicles, and pedestrians on a road.

Image segmentation is the technique of drawing a line around the object detected in the image. The process of dividing an image into segments is called as Image Segmentation. The segmentation identifies the specific pixels in the image that belong to the object image with a fine-grained localization (Brownlee, 2019).

In short, one can classify what are present in an image using classification. Image localization points to the single object position in an image whereas object detection detects the multiple objects in the image. Lastly, image segmentation produces pixel-wise masking on each object in the images. Thus, one can achieve the different shapes of objects in the image using image segmentation (Sharma, 2019). The need of bounding box reduces the range of search for the object features and thereby conserves computing resources: memory allocation and processing time (Zhao, 2017).

In DL, ANN model is a potent tool for ML, propitious to reform the future of AI. DL aims at models that computes and is collection of multiple processing layers to learn data representations with multi-levels abstraction. For solving complex problems, DL provides new sophisticated approaches to train DNN architectures.

ARTIFICIAL NEURAL NETWORKS

It is known as computing system built with many interconnected processing elements, which processes the information by their dynamic state response to external inputs. ANN comprises of a group of algorithms used in ML and DL. ANNs are a type of models in ML, and called as Multilayer Perceptron (MLP) and its great potential lies in the high-speed processing provided for a massive parallel implementation (Abiodun, 2018).

ANN is one such model that is acquainted by innovations like AI and big data into the healthcare. Huge quantity of raw information is transformed into valuable information which is needed to take decisions for treatment and care. According to Ferguson (2019), many prognostics methods are turning to ANN to discover new insights into healthcare future.

Many wide applications are found to be supported by techniques of AI and intelligent systems (Teodorrescu, 1998). NN was found to be very useful in diagnosing diseases in the pathological study and in the areas of biomedical. ANNs are powerful non-linear modelling, that causes and effect relationships inherent in complex processes (Shi, 2004), typically for controlling quality (Moghimi & Wickramasinghe, 2014). In the medical information system with uncertain growth of health databases, the system requires manual data analysis coupled with efficient computer-assisted analysis techniques (Lavrac, 1999).

The Working of ANN Model

ANN is organized in layers with the series of neurons; where each neuron is inter-linked through a weighted link with neurons in the next layer. In simple terms, ANN is designed by input layer, one or more hidden layers (HL) and an output. The model complexity is obviously decided by the number of neurons and the layers present in it. The input layer takes the data and transfers the neurons through the weighted links in the first HL. The data are processed and the outcomes are reassigned to the neurons in the next layer, the network's output is achieved in the last layer shown in figure 2 (Amato, 2013). ANNs named as connection-oriented networks (Sarasvathi & Santhakumaran, 2011) includes processors performing parallel with the set of inputs taken at a time and producing output based on processing algorithm (Gharehchopogh, 2013).

Figure 2. NN model with hidden layer

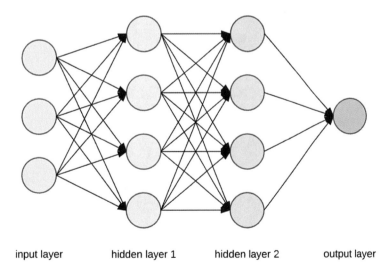

input layer   hidden layer 1   hidden layer 2   output layer

The n inputs values are represented by $x_1, x_2, ..., x_n$ in figure 3. Each of the inputs have the weights $w_1, w_2, .., w_n$. The input values are multiplied by the weight and summed as follows in the Eq. (1) (Bishop, 1995). A single layer NN is called perceptron (Bishop, 1995) whereas a MLP is called Neural Networks

Figure 3. Perceptron learning process

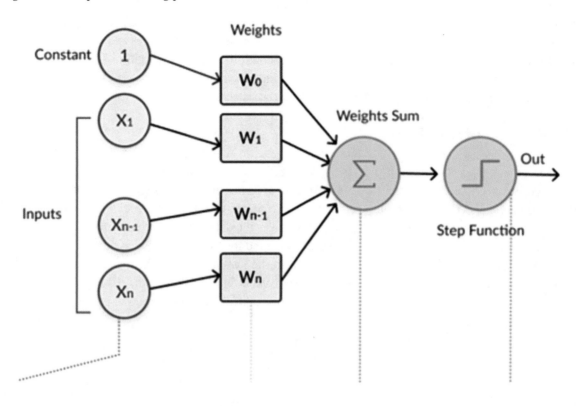

$$S = w_1\,x_1 + w_2\,x_2 + \ldots + w_n\,x_n = \sum_{i=1}^{n} w_i\,x_i \qquad (1)$$

The output function *z* of the weighted sum is known as activation function, $z = f(s) + b$, where *b* is known as bias parameter (see Eq. (2)) (Hinkelmann, 2016)

$$z = b + \sum_{i=1}^{n} w_i\,x_i \qquad (2)$$

In ANN, the activation function of a node describes the outcome, given an input or set of inputs. There are several activation function of which one is named as logistic function in Eq. (3) and is known as the sigmoid function. The function takes the value ranging from -1 to +1 in Eq. (4) (Bishop, 1995).

$$\sigma(a) = \frac{1}{1 + \exp(-a)} \qquad (3)$$

The mathematical equation of an activation function decides the outcome of each element i.e. perceptron or neuron in NN. Each neuron input is taken and transformed into an output, the value is in between 1 and 0 or between -1 and 1.

$$f(a) = \begin{cases} +1, & a >= 0 \\ -1, & a < 0 \end{cases} \qquad (4)$$

In ANN model, the two types of architecture used are feed forward and feed backward. In feed forward architecture, it does not have any connections with the results and input neurons and does not

Figure 4. Feed forward network (Agatonovic - Kustrin & Beresford, 2000)

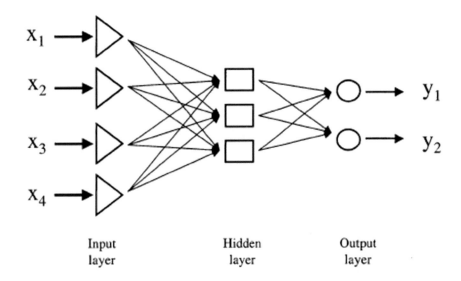

Figure 5. Feed backward network (Agatonovic - Kustrin & Beresford, 2000)

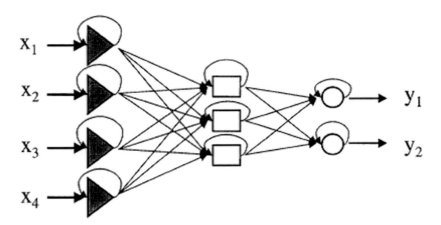

maintain a record of its previous result values as shown in figure 4. Whereas in figure 5, feed backward architecture has the connections from the results obtained with input neurons. For each neuron, an additional weight is given as an input, that helps in minimizing the training error (Agatonovic - Kustrin & Beresford, 2000). In the areas of remote sensing applications, multilayer feed-forward NN algorithm is used for classification. Back Propagation NN (BPNN) is used only for image classification (Saravanan & Sasithra, 2014).

The applications of ANN's are wide in healthcare. For processing ANN, the inputs are taken from the patient's health reports which are maintained in electronic health records (EHR). EHR includes patient health history details such as diagnostic tests, medications and treatment plans, immunization records, radiology images, electroencephalogram EEG, laboratory and test-lab reports e.t.c. shown in figure 6. This information gives a complete opinion of a patient condition or/and disease and then improves the quality of the obtained inference (Ravi, 2016).

Figure 6. ANN's applications in Healthcare

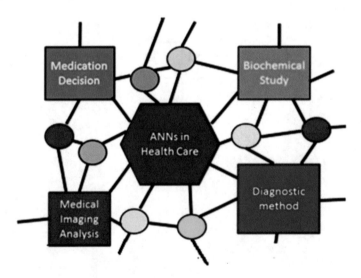

ANN model in brief has an input layers with nodes based on the number of input values fed into it; the weights are multiplied with the output of the first layers and along with the bias, it is passed for the next layer and continues the process based on the number of layers between the first and final layer which are known as HLs. The final layer is called output node/layer.

Prediction of Brain Tumour using ANN

In this study, ANN model is trained with the input datasets to get the predicted output with minimum error rate. The brain tumor (BT) symptoms readings of few patients are presented in table 1. The reports were obtained from Gandhi Hospital, Secunderabad and Omega Hospital, Hyderabad, India. The symptoms are headache, vomiting, seizure, altered behavior, decreased vision, difficulty in speech and loss of consciousness that are represented from S1 to S7. These symptoms are considered as inputs for

Table 1. Patient's brain tumor symptoms readings

Age	Sym1	Sym2	Sym3	Sym4	Sym5	Sym6	Sym7	Tumor
71	1	2	0	0	3	0	0	1
18	3	3	1	0	0	0	0	1
60	0	0	2	0	0	0	0	1
35	0	0	0	0	0	0	0	1
53	3	0	0	0	0	1	0	1
25	3	3	0	0	0	0	0	1
60	1	0	0	0	0	1	0	1
6	2	2	0	0	0	0	0	1
30	1	0	0	0	0	0	2	1
38	0	0	1	0	0	0	0	1
48	0	3	0	1	0	0	0	1
9	1	3	0	0	3	0	0	1
65	3	0	0	3	0	0	0	1
46	0	0	1	0	0	0	3	1
40	1	0	1	0	0	1	0	1
56	0	0	0	1	0	0	0	1
40	3	2	0	1	0	1	0	1
42	1	2	0	0	0	0	0	1
60	3	2	0	0	0	0	0	1
30	1	1	0	0	0	0	0	1
45	2	0	0	0	0	0	0	1
40	0	0	1	0	0	0	0	1
51	0	0	0	0	0	0	0	1
70	0	0	0	0	0	1	0	1
13	1	2	0	0	0	0	0	1
65	3	0	0	0	0	0	0	1
22	1	2	0	0	0	0	0	1
16	2	3	0	0	0	0	0	1
4	1	0	1	0	0	0	0	1
55	1	0	0	0	0	0	0	1
38	1	3	3	0	0	0	0	1
60	3	0	0	0	0	3	0	1
35	1	3	1	0	0	0	0	1
65	1	0	0	0	0	0	0	1
21	3	3	0	0	0	0	0	1
45	0	0	1	0	0	0	0	1
25	1	0	3	0	0	0	0	1
56	1	2	0	0	2	0	0	1
71	0	0	0	0	0	0	0	1
65	0	2	0	0	0	0	0	1
50	1	0	1	0	0	0	0	1

*Very Low = 0, Low = 1, Medium = 2, High = 3

Table 2. The significance F's value

	SS	MS	F	Significance F
Regress	3.6673	0.4584	3.7911	0.00069
Res.	11.1246	0.1209		
Total	14.7920			

the ANN input layer and the outputs of the input layer are in turn passed to the next layers as input, and so on to obtain an output.

Using Leven-berg Marquardt back-propagation algorithm, various symptoms are trained through NN. Thus, NN model is obtained by adjusting the weights and number of HLs. Using MATLAB R2012b, the outputs of the network fitting model and performance plot are shown in figures 7 and 8 respectively. The best validation performance is 0.0159 achieved at epoch 2 when NN was trained with 101 inputs, 44 HLs and 1 output. The next validation performance was obtained 0.0181 at epoch 2 for 62 HLs shown in figures 7a and 8a respectively.

Figure 7a. Performance plot for NN model with 44 HLs

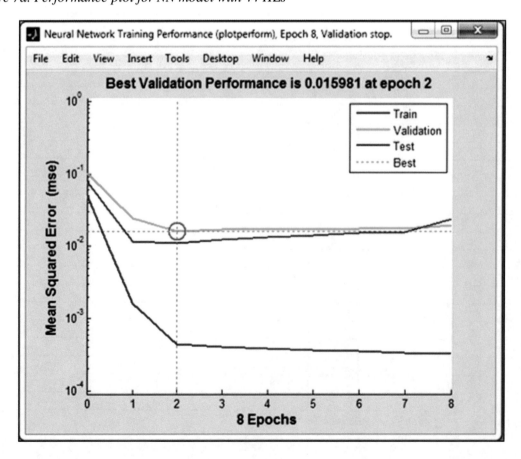

Figure 8a. Performance plot for 62 HLs in NN model

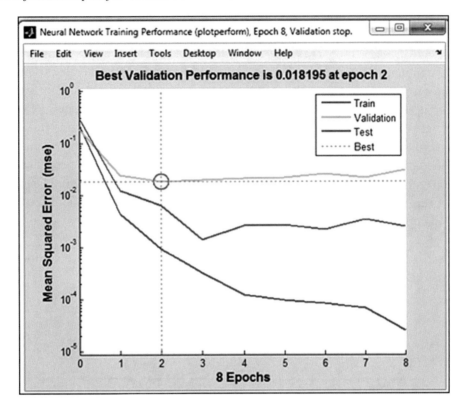

Table 2 shows the significance F's value calculated from the table 1. Significance F's value was found to be 0.00069 which is less than 0.05, if the value is greater than 0.05, it's good to stop using the set of independent variables.

In performance plot, drastically dropping of mean squared error can be seen. In this plot, the blue line demonstrates error declining on the data training. The validation error is indicated by green line and as the error in validation declines, the training stops and the red one displays the test data error

Figure 7b. NN model with 101inputs, 44 HLs and 1 output

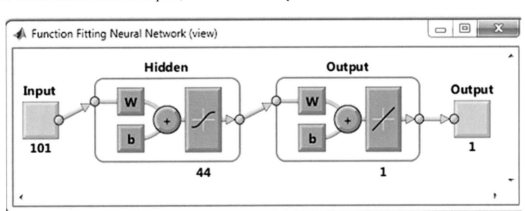

that signifies how thoroughly the net will generalize to the new data. The standard network is used for function fitting is n-layer feed forward network, with a sigmoid transfer function in the HL and a linear transfer function in the output layer.

Let's say the total hidden neurons are set to 44. With the input and target vectors are randomly distributed as follows: For training, 65% will be used; 20% for validating the net and generalizing and training stop before overfitting and the last 15% will be used as a completely independent test of network generalization. Each training set purpose is used to fit the models; the validation set indicates the prediction error for model selected, the test set is used for the generalization error assessment of the final model chosen.

NN has generalization ability i.e. a trained net would classify data from the same class as the learning data. The learning stops with the least validation error set. At this point the net generalizes the best. The performance of the model decreases with overtraining and when the learning is not stopped, and the training data error is reduced. After completing the learning phase, the net should be finally checked with the third dataset, known as the test set. The two reasons for using NN are

1. The dependence between input and output data is non-linear and NN have ability to model non-linear patterns.

Figure 7c. Regression model using NN for 44 HLs

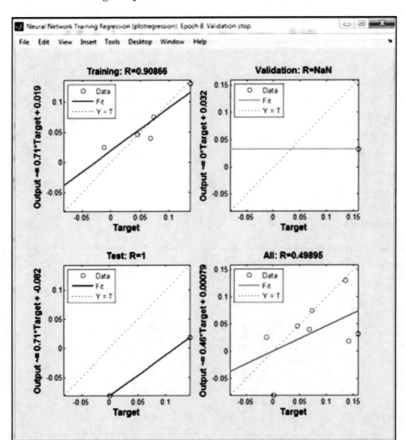

Figure 8b. NN Model with 101 inputs, 62 HLs and 1 output

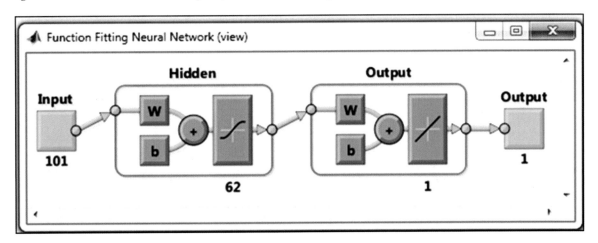

Figure 8c. Regression model using NN with 62 HLs

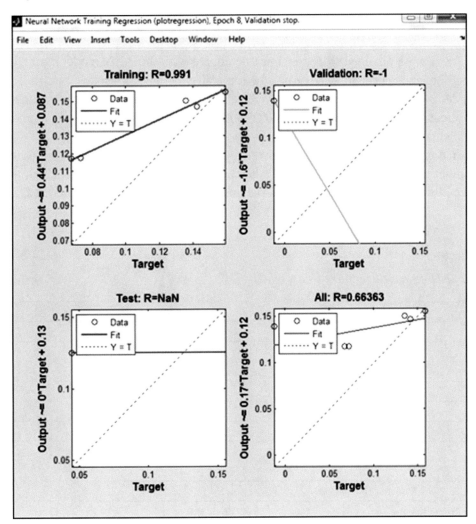

2. ANN is more reliable at predicting. When huge data is available, NN trains this data by adjusting the weights and predicts output with minimal error when working on new data with similar characteristics of the input data.

Table 3. Summary output of patient's indicators P-value

Symptoms of Tumor	Coefficients	P-value
Intercept	0.5226	1.9150E-05
Age	0.0014	0.5038
Headache	0.1355	0.0021
Vomiting	0.0730	0.0478
Seizure	0.1420	0.0079
Altered behavior	-0.0119	0.9052
Decreased vision	0.0452	0.3579
Difficulty in speech	0.0677	0.4965
Loss of consciousness	0.1586	0.0395

Tables 2 and 3 are obtained by regression to compare with the ANN model regression values in figure 7c and 8c; and are found to be nearly the same. The p-value is displayed in the Table 3 for all the independent variables and its coefficient values. The indicators as headache; vomiting, seizure and loss of consciousness have p-value less than 0.05 and are measured to be the most common symptoms of all the patients.

The prediction of tumor (BT) analysis is carried out by the symptoms reading in numerical form; and applied to input layer along with varying HLs and adjusting the weights for the model, the validation performance plot is obtained. Based on the curve line, the performance can be analyzed.

Table 4. Various geometric features (Masood, 2017).

Sno.	Features
i.	Lt ey –Nos / Rt ey-Nos
ii.	Lt ey-Mou / Rt ey-Mou
iii.	Lt ey-Nos / Lt ey-Mou
iv.	Rt ey-Nos / Rt ey-Mou
v.	Lt ey-Rt ey / Nos-Mou
vi.	Lt ey-Rt ey / Lt ey-Nos
vii.	Lt ey-Rt ey / Lt ey-Mou
viii.	Rt ey-Nos / Nos Ht
ix.	Nos Ht / Nos-Mou

*Left = Lt, *Right = Rt, *Eye =ey, *Nose =Nos, *Mouth = Mou, *Height =Ht

Classification of Human Ethnicity using ANN

In a case study of facial features were extracted from the images carried out by training a feedforward ANN using backpropagation algorithm (Masood, 2017). Extraction of skin color and calculation of normalized fore head color. In table 4, all the geometric features are shown and the results for different classes P1, P2 and P3 and the calculated precision, recall, f –measure and accuracy are shown in the table 5. The best result was achieved for 17 HLs with 150 epochs with learning rate 0.17.

Table 5. ANN model for facial features (Masood, 2017)

Attr	S1			S2			S3		
	P1	**P2**	**P3**	**P1**	**P2**	**P3**	**P1**	**P2**	**P3**
Pre	80.0	82.8	78.1	80.6	80.0	83.3	80.6	83.3	83.3
Re	80.0	77.4	83.3	83.3	77.4	83.3	83.3	80.6	83.3
F	80.0	80.0	80.6	81.9	78.6	83.3	81.9	91.9	83.3
Ac	77.4	80.0	83.3	80.6	80.0	83.3	80.6	83.3	83.3

The testing was carried out on Caucasian (31), Mongolian (32) and Negroid (30) image samples. In figure 9, the confusion matrix represents 25 were classified correctly for 31 Caucasians. The success rate for Mongolian was 25 out of 31 and for Negroid, 25 out of 30. The overall classification accuracy is 82.4% and misclassification was found to be 17.6%.

Figure 9. Confusion Matrix for facial features (Masood, 2017)

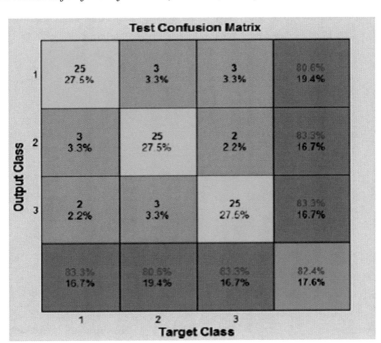

In this study, ANN classification algorithm is built on the feature extraction of the images taken as numerical values for the inputs. The best performance was obtained at 17 HLs with learning rate of 0.17. The confusion matrix shows the total classification and misclassification reading; describe the classification model performance. It is also referred as error matrix.

CONVOLUTIONAL NEURAL NETWORKS

CNN is a class of DL, applied to analyzing visual imagery. A CNN learns from the filters and thus requires little preprocessing. The term "fully-connectedness" of the networks makes them prone to overfitting data. CNN have wide applications in image and video recognition, image classification and medical image analysis.

Two Main Applications of CNN

The key applications of CNN are Image classification and segmentation.

Image Classification

In classification, a given image is shrunk and made to go through all the convolution layers to fully connected layers, which generates one outcome, a predicted label for the input image. Here, the given image is downsized only to get a single predicted outcome.

Image Segmentation

A digital image is divided into multiple parts is referred as segmentation; the aim of segmenting any image is to extract more deep weighted information. A label is allocated to each pixel on segmentation so that every pixel of similar characteristics holds same characteristics.

Figure 10. The schematic view of CNN

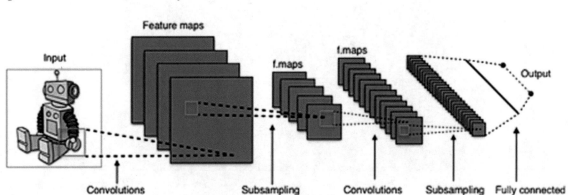

THE STRUCTURE OF CNN

ConvNet is built with one input and output layer; along with many HLs in between. HLs have a sequence of conv layers that convolve with a multiplication or dot product. RELU is used as the activation function, followed by add-ons conv layers namely pooling, fully connected and normalization, referred to as hidden layers because their inputs and outputs are masked by the activation function and final conv. This final conv involves back propagation to get more accurate weights at the end product in figure 10.

An image is considered as an array of pixel values, each value lies amid the range value of 0-255 shown in figure 11.

Figure 11. Image is read as matrix values in pixels

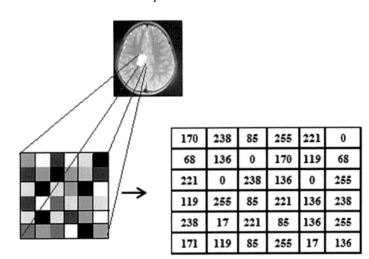

Figure 12. 32 x 32 x 3 image

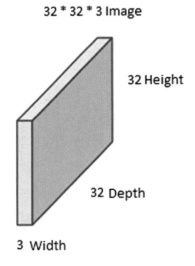

Convolution (Conv) is a process of merging two functions by multiplying them. To convolve means to move together. Basically, a Conv nets accomplishes a kind of search. Image detection uses Conv nets and takes many search-rounds over a single image in all directions -horizontal lines, diagonal ones, vertically several times as a visual element is required. A kernel is well-defined matrix of numbers which is used in image conv. Kernels of various sized comprises different patterns of numbers provide various outcomes and the kernel size is arbitrary but 3 x 3 is used maximum.

In figure 12, the INPUT image [32 x 32 x 3] will have grip of the pixel values of an image with width as 32, height as 32, and red, blue and green as three different color channels. The dot product output between the filter and small 5 x 5 x 3 chunk of the image (3 x 3 x 3 = 75 - dimension dot product + bias) is shown in figure 13. Whenever the filter slides or convolves upon the input image, the product of the filter values with the original image pixel values, i.e. element-wise multiplications is the net result.

A Conv Net has four principal layers:

Figure 13. Applying 5 x 5 filters

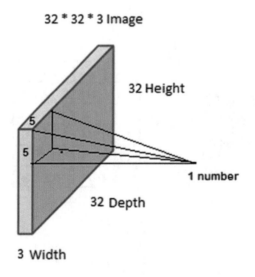

32 * 32 * 3 Image

32 Height

5

5

1 number

32 Depth

3 Width

Figure 14. Filter applied on image

Image (x)

8	2	4
6	8	3
4	8	0

Filter (A) B=0

5	-3	6
-4	2	6
-2	6	-5

Result

(8*5)+(2*-3)+(4*6)+
(6*-4)+(8*2)+(3*6)+
(4*-2)+(8*6)+(0*5)=

$\sum x_n \cdot A_n + B$

= 108

1. CONV
2. Sub-Sampling is referred as Pooling (figure 14).
3. ReLU (Rectified Linear Unit) in figure 15.
4. Dense.

Each neuron outputs are associated to the inputs and are computed by CONV layer, each dot product is computed by their weights and smaller areas connected to the feed-in volume.

The element-wise activation function is applied to ReLU layer, such as the max(0, *x*). When the thresholding is at zero, the size is left unchanged [32 x 32 x 12]. Down-sampling operation is performed

Figure 15. Conv–ReLU layers

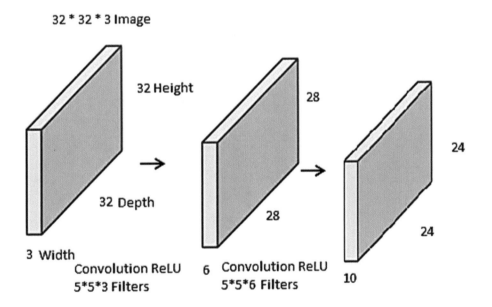

Figure 16. Maxpooling

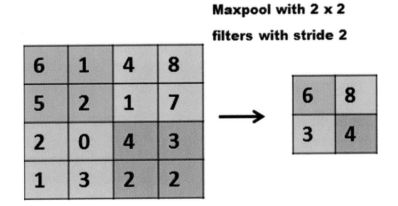

by pooling layer along the spatial dimensions (width, height), resulting in volume such as [16 x 16 x 12] and used widely in CNN. ReLU's does not activate all neurons at the same time rather converts all negative inputs to zero from the image and neurons do not get activated. This makes it very efficient in computational as few neurons are activated per time. ReLU converges six times faster than TanH and sigmoid activation functions and does not saturate at the positive region.

The process of subsampling is known as Maxpooling, here the input representation i.e image or matrix from hidden layer output is down-sampled by dropping its network parameters that helps in shrinking overfitting. The peak pixel value from a region depends on its size is considered by maxpooling. The pool size is 2 x 2 pixels as in figure 16 (Chen & Seuret, 2017).

Neurons at FC, fully-connected layer are found at the rear of CNN generating the resulting size [1 x 1 x 10] and in output layer, MLP uses a softmax activation function (Prisilla & Iyyanki, 2019). The preceding layer's neurons of the fully connected layers are associated with each neuron in the next layer. Finally, the segmented images get the appropriate labels after retrieving all of the features from each image as shown in figure 17.

Figure 17. Fully convolutional neural network (Prisilla & Iyyanki, 2019)

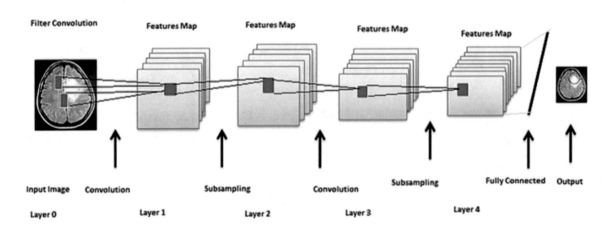

Conv Layer - Feature Maps

In CNN, an image is taken as input and processes through Conv kernel (layers). Each kernel output gives a feature map (FM) shown in figure 18; the first FMs capture high-level features. Moving deeper in the network, produces smaller FMs because of the pooling layers. Each of them represents a different kind of feature that it captures. Hence, FMs are present in between the input first layer and final outcome as shown in figure 17.

Let's visualize all given filters in one image that can be practically implemented on digital brain tumor image. The activation maps (AM) are also known as FMs that captures the results of filter's applied to any input image or another FM. The key of feature map visualization for any specific image is to understand the features of the input image that are detected or preserved in the FMs. FM provide data augmentation of one image which can be in different order seen in figure 19. FMs closer to the input

Figure 18. FMs obtained from conv layers

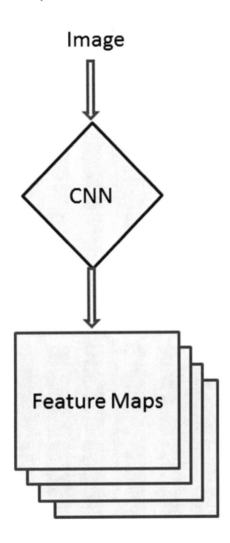

model captures more fine details in the image and as it moves deeper into the model, the FMs show less and lesser information (Brownlee, 2019). FM helps one to understand the various factors of the image. One can notice that as the receptive field is moved from activation to activation by a pixel, then the field will overlap with the previous activation by (field width - 1) input values.

The visualization of FMs from BT images is carried out using python on Spyder for data augmentation and to understand how FMs varies with the change in filters applied and with increase or decrease in conv layers and the squares is shown in the figures 19 and 20.

The figure 19 is achieved with input_shape = (256, 256, 3) and 3 x 3 filter in Con2D(512, (3,3)) and 1 x 1filter in Con2D(1024, (1,1)) that produced for only conv layers as output in table 6.

Figure 20 is achieved with input_shape = (256, 256, 3) and 3 x 3 filter in Con2D(512, (3,3)) and 1 x 1filter in Con2D(1024, (1,1)) that produced for only conv layers as in table 7.

The feature maps are the output of conv layer, pooling, ReLU and dense layer. They depend on the number of layers in between the first and fully connected layer that numbers of FMs are obtained and they give information of the image features. The nearer FM to the image provides more information about the image. Deep inside or away from the input image means FM does not provide the clarity.

Figures 19. FMs extracted from the CNN Model with 5 conv in 4 blocks and 8 squares

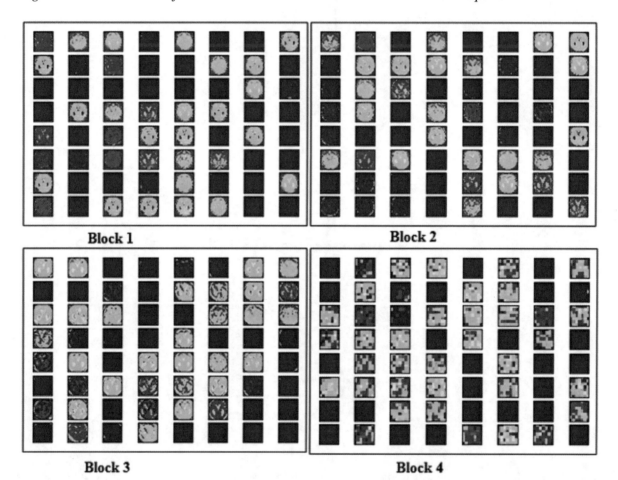

Block 1   Block 2

Block 3   Block 4

Table 6. Five conv layers of CNN model in four blocks

Layers	Output shape	Parameters
con2d_1 (Con2D)	(None, 256, 256, 512)	14336
con2d_2 (Con2D)	(None, 256, 256, 1024)	525312
con2d_3 (Con2D)	(None, 83,83,256)	2359552
con2d_4 (Con2D)	(None, 25,25,256)	590080
con2d_5 (Con2D)	(None, 7,7,64)	65600

Figure 20. FMs extracted from the CNN Model with 3 conv layers in 3 blocks and 16 squares

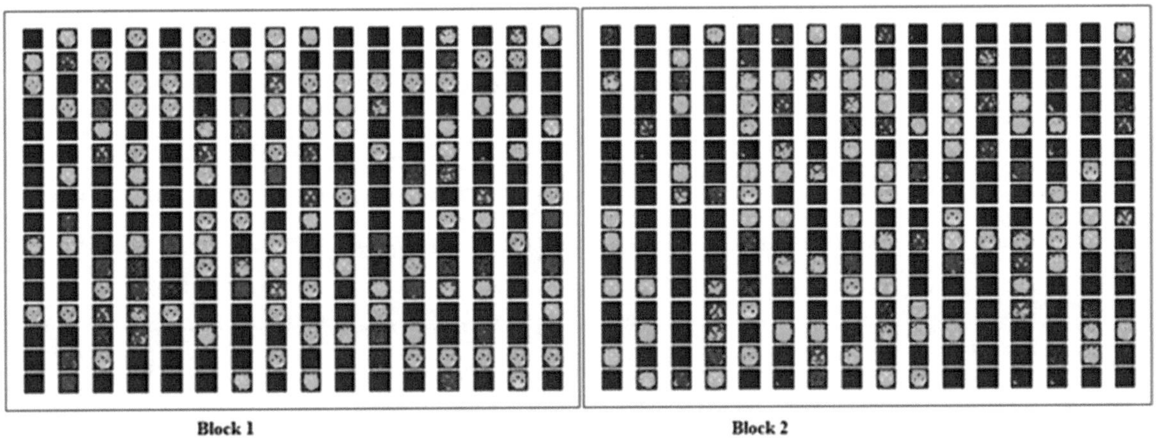

Block 1   Block 2

Table 7. Three conv layers of CNN model in three blocks

Layers	Output shape	Parameters
con2d_1 (Con2D)	(None, 256, 256, 512)	14336
con2d_2 (Con2D)	(None, 256, 256, 1024)	525312
con2d_3 (Con2D)	(None, 83,83,256)	2359552
con2d_4 (Con2D)	(None, 25,25,256)	590080

Accuracy and Loss Graph: CNN Model

Using the computation of the loss and accuracy curves on training dataset of brain tumors indicates the

Figure 21. Image training for 20 epochs

```
Found 5000 image belonging to 2 class.
Found 2000 image belonging to 2 class.
E 1/20
2000/2000 [==================] - 629s 315ms/step - lo: 0.4127 - ac: 0.8289
E 2/20
2000/2000 [==================] - 699s 350ms/step - lo: 0.2865 - ac: 0.8820
E 3/20
2000/2000 [==================] - 691s 346ms/step - lo: 0.1750 - ac: 0.9316
E 4/20
2000/2000 [==================] - 715s 357ms/step - lo: 0.0970 - ac: 0.9641
.....
E 18/20
2000/2000 [==================] - 560s 280ms/step - lo: 0.0135 - ac: 0.9957
E 19/20
2000/2000 [==================] - 559s 280ms/step - lo: 0.0129 - ac: 0.9959
E 20/20
2000/2000 [==================] - 574s 287ms/step - lo: 0.0136 - ac: 0.9958
```

Figure 22. Image training for 50 epochs

```
Found 5000 image belonging to 2 class.
Found 2000 image belonging to 2 class.
E 1/50
8000/8000 [==============================] - 2410s 301ms/step - lo: 0.2438 - ac: 0.8997
E 2/50
8000/8000 [==============================] - 2335s 292ms/step - lo: 0.0470 - ac: 0.9833
E 3/50
8000/8000 [==============================] - 2291s 286ms/step - lo: 0.0258 - ac: 0.9912
E 4/50
8000/8000 [==============================] - 2282s 285ms/step - lo: 0.0188 - ac: 0.9938
.......
E 47/50
8000/8000 [==============================] - 2317s 290ms/step - lo: 0.0033 - ac: 0.9993
E 48/50
8000/8000 [==============================] - 2391s 299ms/step - lo: 0.0035 - ac: 0.9992
E 49/50
8000/8000 [==============================] - 2369s 296ms/step - lo: 0.0035 - ac: 0.9992
E 50/50
8000/8000 [==============================] - 2335s 292ms/step - lo: 0.0032 - ac: 0.9993
```

errors in the loss of the image obtained as result. Loss is known as the summation of the errors produced

Figure 23. The graphical representation for 20 epochs

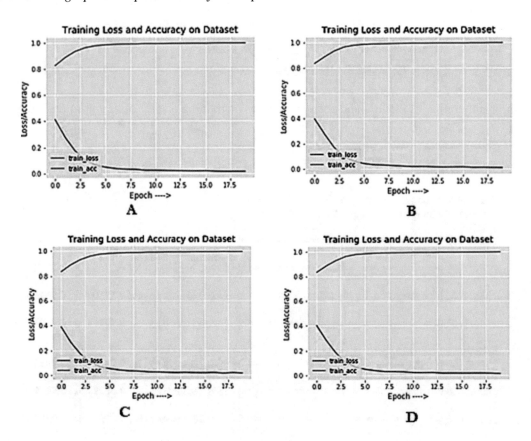

from each image during training or validation. And when the accuracy reaches closer to 1, then the model is learning well and is referred as overfitting model. Lesser the loss, the minimum is the error. The graph for epochs of 20 and 50 are shown in figure 21, and 22 respectively. By analyzing the calculations and graphs in the figures 23 and 24. It was observed that accuracy and minimal loss achieved are better when the image was trained for 50 epochs. Comparing with figure 21, loss initiates at 0.4 and drops near 0.01 and accuracy begins at 0.8 and reaches 0.9 for 20 epochs and E and l represents epochs and loss respectively. Whereas in figure 22, the loss initiates at 0.24 and declines at 0.0032 and accuracy begins at 0.89 and rises to 0.9. Least loss obtained infers that the model is performing well.

The accuracy and loss graph help us to understand that the minimum loss is the better model. The loss gives the information about the minimal error in the model producing the output.

Figure 24. The graphical representation for 50 epochs

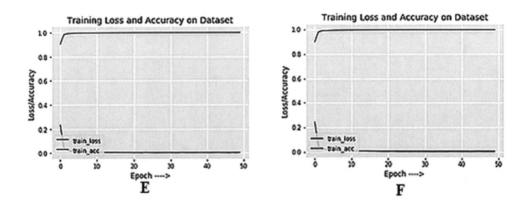

Detection of Brain Tumors Using CNN

In a case study of detection, CNN algorithm was implemented to detect the tumors. This was carried out by a python code by capturing an image as input and labeling the outcome is known as image detection. The propound CNN was experimented on 100 brain tumor patient's x-ray images; on the average the CNN takes about 1s processing time for each image. Hence, the results obtained by CNN is more accurate in object detection that helps in analyzing the tumor part in the images of the brain thus helping the surgeon to diagnose it faster and further analysis.

The Calculations for Parameters in CNN

The filter is slided / convolved across the input image, and a dot product is computed to give an activation map. The output size of activation map is computed by (N + 2P - F) / stride + 1, where N is the input size, P is padding, F is the filter size and the stride applied on the image. The output obtained can be overfitting, fit or under-fitting, where fit is assumed to be the better result. Figure 25 illustrates the way the filters are applied to the CNN model to create a feature map.

Figure 25. Filter applied to 2D CNN to create a feature map

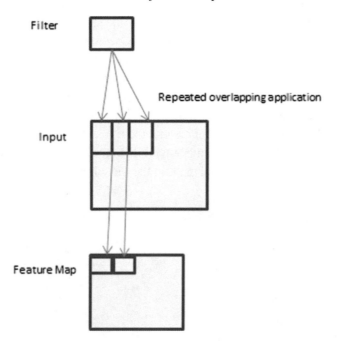

For further understanding, the feature map and activation map does almost the same operations. But, AM does mapping corresponding to the activation of different parts of the image, whereas a feature map maps to certain kind of features found in the image. High activation means a certain feature was found.

Zero padding is used when the strides are applied on input size 7 x 7, the output size shrinks and maintains the size. The next layer parameters are reduced by pooling; max-pooling and average pooling is used. No parameters are present at pooling and dense layer. Coming to the fully connected (FC) layer is at the end of CNN, and all the neurons in FC must have full connections to all activations in the previous layer. The process is trained for number of epochs mentioned.

In this study, for object detection, one needs to identify region of interest which is the brightest area that is obtained by applying Guassian Blur on the image. This will circle the area where the object is bright and compares the area to the other background area as shown in figure 26. In figure 27, the CNN architecture is shown with the different layers and number of parameters.

In this study, the tumors are detected based on the CNN algorithm and application of GuassianBlur on the image with consideration of region of interest. CNN has more than two layers which include conv layer, followed by subsampling or pooling layers where maxpooling or average pooling takes place based on the requirement. Next is the ReLU where activation function occurs –sigmoid function. Flatten and dense layers are followed by fully connected layer where the output is obtained.

THRESHOLDING: TECHNIQUE OF IMAGE SEGMENTATION - CNN

Image segmentation is the process of segmenting the image for deeper image analyzing. The practical applications are content-based image retrieval, object recognition, image compression, and image editing.

Figure 26. Tumor part detected in the brain tumor image using CNN model (Prisilla & Iyyanki, 2019)

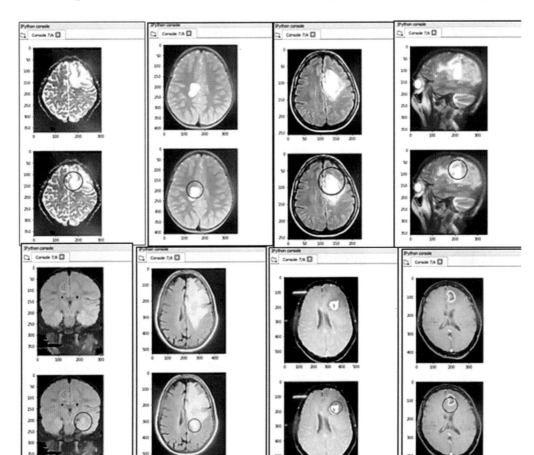

Figure 27. CNN with all the layers and number of parametersin its output (Prisilla & Iyyanki, 2019)

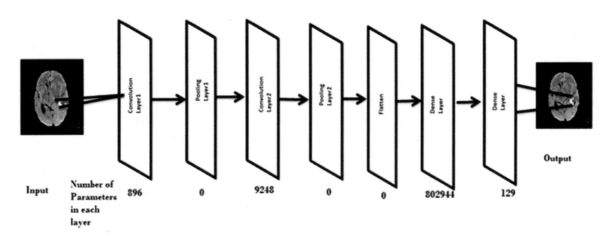

The contextual and non-contextual are two segmentation techniques. A method of non-contextual image segmentation is a non-linear operation that transforms a gray-scale into a binary image is thresholding. A binary image reduces the data complexity and simplifies the process of recognition and classification that is obtained by non-contextual image.

Contextual Segmentation

This involves separating individual objects as it is intended for pixels closeness that belongs to an individual object. Image Segmentation emphasizes on discontinuity of signal and similarity. Identifying the boundaries by encircling within the uniform regions and visualizing the abrupt signal changes through each boundary are discontinuity-based technique. Similarity-based is the region created uniformly by grouping together all similar pixels that satisfies definite criteria. Two methods reflect each other and holds responsible for a complete one region boundary splits into two parts.

Non-Contextual Segmentation

Non-contextual segmentation group pixels based on the definite attribute with no regard to the relative locations of the image feature.

Thresholding

The simpler technique of non-contextual segmentation is where transformation of grey-scale image to binary and is known as binary region map (BRM). The two disjoint regions of BRM are: one that has input data pixel values lesser than threshold; and other the pixels values with the same value of threshold or greater. Thresholding is a non-linear operation that transforms a gray-scale image into a binary image. The two levels assigned to pixels are below or above the specified threshold value i.e if a pixel value is above threshold value, it is assigned one value for white, and other is assigned black. The grey-level pixel thresholding is given as: $p(a,b) = 0$ if $q(a,b) < Th$ and $p(a,b) = 1$ if $q(a,b) >= Th$, where Th represents the threshold. Two thresholds, $Th_1 < Th_2$, grey level region1 range is given as $p(a,b) = 0$ if $q(a,b) < Th_1$ or $q(a,b) > Th_2$ and $p(a,b) = 1$ if $Th_1 <= q(a,b) <= Th_2$.

Color Thresholding

Color segmentation gives more accurate information at the pixel level when compared with grey images. HSL (hue, saturation, and luminance), HSV, HSI, or other models are used in computer vision and image analysis for feature detection or image segmentation. The robust interrelated color components are present in Red-Green-Blue (RGB) standard color, while other colors (HSI) reduces redundancy, and decide the actual object/ background colors regardless of illumination, and achieve better stable segmentation. The color space partitioning in RGB or HSI is performed out by division of color images. The method is established on a leading color (R_0, G_0, B_0) and Cartesian distances threshold from each pixel color $f(a,b) = (R(a,b), G(a,b), B(a,b))$:

$$g(a,b) = \begin{cases} 1 \text{ if } d(a,b) <= d_{max} \\ 0 \text{ if } d(a,b) > d_{max} \end{cases} \quad d(a,b) = \sqrt{(R(a,b)-R_0)^2 + (G(a,b)-G_0)^2 + (B(a,b)-B_0)^2}$$

(5)

where *g(a,b)* represents the BMP after the process of thresholding. The principles of thresholding defines sphere in RGB space as focused on the reference color. If the pixels are placed inner of the sphere then those are referred to region 1 and region 0 has other pixels.

Thresholding is process of segmenting foreground with background regions where binary value is considered. And whole image takes only gray and white color.

Brain Tumor Segmentation Using HSV Color Space

In a case study of brain tumor segmentation using HSV color space was carried using python code where image blurring is convolved with filter and this eliminates the unnecessary noises in the images. Later guassian blur is applied to the image to obtain the tumor segment in the image. The three components of HSV color space- hue / dominant wavelength, saturation and value. A lower threshold and higher threshold mask for these modules are considered. Any pixel within these thresholds will be set to 1 and the remaining pixels will be zero (Efford, 2000). The extracted image in green color can be seen in the Lt bottom most of each tumor image with the black background.

Figure 28. Result of extracted tumor images

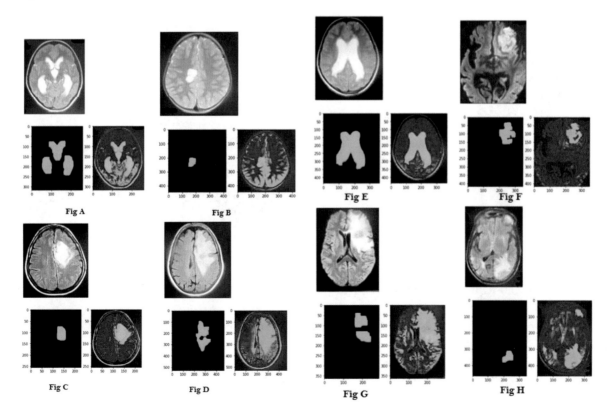

Results of the extracted image segmentation using thresholding and HSV color space are shown in the figure 28.

Segmentation Using Watershed Segmentation and Morphological Operation

Yet other case study of image segmentation of brain tumor, the significant task is segmenting the image accurately for exact diagnosis. In this approach, it includes to examine MRI of brain image for processing, image filtering, skull stripping, and segmentation using OTSU and watershed, morphological operation, calculation of the tumor area and determination of the tumor location (Maha Lakshmi, 2016).

Cancer Segmentation Using Fuzzy Entropy Level Set

In other case study of tumors, (Maolood, 2018) suggested an approach for cancer segmentation on fuzzy entropy with a level set (FELs) thresholding. The approach was successful in segmenting cancer images and efficiently performed the segmentation of test ultrasound image, brain MRI, and dermoscopy image. The results in figure 29 gave an excellent performance by using the method in detecting cancer image segmentation in terms of accuracy, precision, specificity, and sensitivity measures.

Figure 29. Fuzzy Entropy Level extracting cancer part through image segmentation (Maolood, 2018)

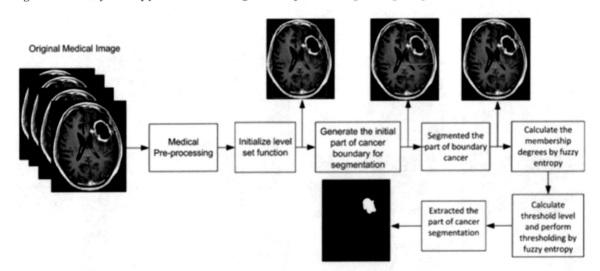

DEEP NEURAL NETWORKS

DNN can accomplish modeling of complex non-linear relationships. DNN is a feedforward network, where data from the input layer flows to the output layer without looping back. Such a network with only one hidden layer would be a non-deep (or shallow) feedforward NN. But in a DNN, the number of HLs are more as in figure 30, let's say, 100. But a network if it has layers greater than 3 are considered as DNN.

Figure 30. Deep Networks with many HLs

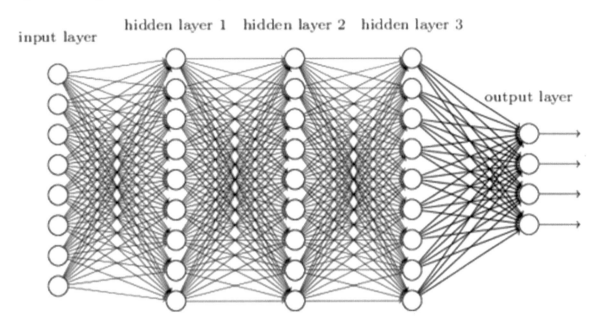

In this study, implementation of DNN with semantic segmentation (SS) was analyzed for image segmentations. The indispensable task for analysis of the image is termed segmentation; it involves understanding the images comprehensively by dividing it whereas the SS describes each image pixel with a class label. The image processing uses filters, gradient and color information and classification of pixel is one of the pre-defined class labels for each pixel. The input image is segmented into the regions corresponding to the objects of any image scene.

Two recognition tasks are the object class detection and semantic segmentation. Object detection addresses localization of objects with the classes. Minimum bounding rectangles (MBRs) of the objects are the ideal output. The approach involves sliding window concept with varying size and classifies sub-images defined by the window. Object detection reduces to semantic easily; Deep CNN results may be blurred and inaccurate; adding CRF to it gives accurate results.

DNN are of various models where the base is inherited from CNN but with many additional layers in it makes it complex and faster. This model nets are used for classification, object recognition and detection with masking and bounding boxes in the results. This models run on GPU to give more accurate and faster results than on CPU.

Semantic Segmentation: FCN-8 - FCN-16

In a case study, FCN-8 and FCN-16 was implemented using keras library. DL models can be built using keras that provides high-level building blocks for developing a model. ImageNet images are acquired and the corresponding weights are used from pascal-fcn8s-dag.mat and pascal-fcn16s-dag.mat. The images undergo preprocessing, labeling and are further classified into the different groups. For training, PASCAL VOC 2012 segmentation class images are used.

Next, training an end-to-end FCNs includes two steps (1) pixel-wise prediction and (2) supervised pre-training. Prediction of dense outputs is carried for fully conv of existing networks from arbitrary-sized inputs. The pixel-wise prediction is carried out by up-sampling layers and subsampled pooling for learning nets.

The model is trained on the ImageNet prior for classification. Dense prediction is implemented by the VGG and its last fully connected layer is replaced with conv layer with addition of 1x1 conv with channel dimension 21. For predicting scores for each of the PASCAL classes includes background at each of the coarse result locations; followed by a transpose conv layer with the stride 16 for bilinear up-sampling the coarse results to pixel-dense outputs.

Low Level Layers Merged With the Features of Higher Ones

The blend of layers in the hierarchy features by the new processing FCN for segmentation refines the spatial precision of the outcome. The final prediction layer is pooled with lower layer of the finest strides by adding skips connections; a direct acyclic graph with edges is curved with a line topology that skip ahead from lower layers to higher ones. The finest scale prediction associates fine and coarse layers that allow the model from local prediction.

The output stride is split into two halves by predicting from a 16 pixel stride layer. Then 1x1 conv is added on the top of pool 4 to yield surplus class predictions. The predictions output compute on the top of conv 7 (conv fc7) at stride 32 by adding a 2 x up-sampling layer and summing both the predictions. Lastly, the stride 16 predictions are up-sampled back to the image; this net is known as FCN-16s (Shelhamer, 2016).

Post Processing

Predicting labels are performed by CRF, and a discriminative model is used for predicting labels. It uses contextual information from previous labels that increases the amount of information to make a good prediction that corresponds to inputs. The post processing phase refines the segmentation outcome and enhances the ability to capture fine-grained details to use for CRF. Merging the image information at low level between pixels with the outcome of multi-class inference systems generates per pixel class scores. When CNN fails to capture the long range dependencies, the merging method supports and fine local details are also taken into account (Garcia, 2017).

While predicting FCN uses labels for each pixel independently including its surrounding labels, this may result in coarse segmentation. Two inputs considered by CRF are one for the original image and the second is for the predicted probabilities for each pixel. Highly efficient inference algorithm uses CRF for fully connected CRF models in which the pair wise edge potentials are defined through a linear combination of Gaussian kernels in an arbitrary feature space. The trained parameter outputs uses both FCN, given by the number of conv layers where as deconv size is 528 for FCN-16 given image size is 512 X 512 and deconv size is 2 for FCN-8 for the same image size (Prisilla, Iyyanki & Aruna, 2018).

During training fully conv, patch-wise training is carried out and it lacks efficiency. Implementing sampling corrects the imbalance class and mitigates the spatial correlation of dense patches. In this training, class balance can be achieved by the weight loss, and spatial correlation is addressed by loss sampling (Shelhamer, 2016).

Figure 31. Implementation of CRF for the outcome of FCN- 8 and FCN-16 (Prisilla, Iyyanki & Aruna, 2018)

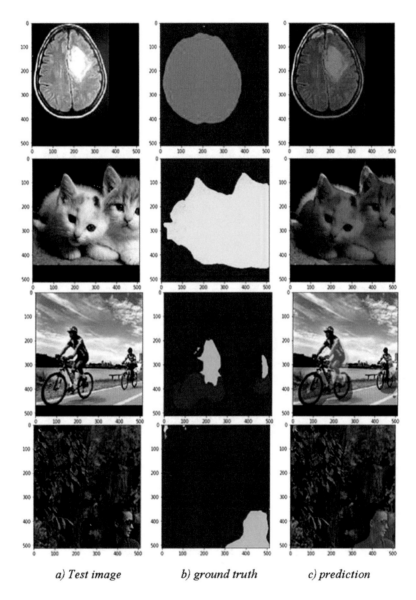

a) Test image   *b) ground truth*   *c) prediction*

Pooling (Prisilla, Iyyanki & Aruna, 2018) helps in classifying networks and decreases the resolution with loss of information; hence the skip connections are implemented. The output of FCN-8 and FCN-32 model is depicted in the Appendix -2, and the results of the images passed in this model are shown in figure 31.

The transfer learning paradigm helps in saving lot of computational cost, time and increases the efficiency. Hence fully conv nets segments image at pixel level, this promotes the use of end-to-end conv network for semantic segmentation.

CONCLUSION

- Deep Learning techniques explore in image segmentation as one of the major concern in the medical imaging.
- ANN is used on numerical datasets of the patient's health details and validates the best performance by training datasets by varying the HL.
- The study was carried out using brain tumors numerical and image datasets.
- Using CNN, brain tumors images were processed to detect the tumor in the images.
- The feature maps provide data augmentation of one image in CNN and gives very fine details of the image.
- HSV color space technique is applied on the tumor images where color and image blurring plays a significant role in the segmentation of tumors.
- FCN-8 and FCN-16 nets produced a semantic segmentation on the various images.
- In simple terms, these techniques of DL proved in detecting the tumors and in validating the best performance of the model.

To summarize,

- ANN has the learning ability and models complex non-linear relationships.
- ANN generalizes and can predict on unseen data.
- ANN has the ability to process large amount of data.
- Thresholding segregates an image into a forepart and back part. In this type of segmentation it isolates objects by transforms grayscale images into binary.
- Thresholding divides an image into a foreground and background. In this image segmentation it isolates objects by converting grayscale images into binary images.
- HSV Color space is used for image segmentation, processing and image analysis.
- Finally, the features of semantic segmentation are used in image classification, image detection and image segmentation. It gives ground truth and masking of the objects detected.

REFERENCES

Abiodun, O. I., Jantan, A., Omolara, A. E., Dada, K. V., Mohamed, N. A., & Arshad, H. (2018). State-of-the-art in artificial neural network applications: A survey. *Heliyon*, *4*(11), e00938. doi:10.1016/j.heliyon.2018.e00938 PMID:30519653

Agatonovic-Kustrin, S., & Beresford, R. (2000). Basic concepts of Artificial Neural Network (ANN) modeling and its application in pharmaceutical research. *Journal of Pharmaceutical and Biomedical Analysis*, *22*(5), 717–727. doi:10.1016/S0731-7085(99)00272-1 PMID:10815714

Amato, F., López, A., Peña-Méndez, E. M., Hamp, P. A., & Havel, J. (2013). Artificial Neural Networks in medical diagnosis, *Journal of Applied Biomedicine. Appl. Biomed*, *11*(2), 47–58. doi:10.2478/v10136-012-0031-x

Bishop, C. M. (1995). *Neural Networks for Pattern Recognition.* Oxford University Press.

Brownlee, J. (2019). Applications of Deep Learning for Computer Vision. *Machine Learning Mastery.*

Brownlee, J. (2019). How to Visualize Filters and Feature Maps in Convolutional Neural Networks. *Deep Learning for Computer Vision.* Retrieved from https://machinelearningmastery.com/how-to-visualize-filters-and-feature-maps-in-convolutional-neural-networks/

Chen, K., & Seuret, M. (2017). *Convolutional Neural Networks for page segmentation of historical document images.* Academic Press.

Efford, N. (2000). *Segmentation. Digital Image Processing: A Practical Introduction Using Java*™. Pearson Education.

Ferguson, J. (2019). Neural Networks in Healthcare. *Journal of Healthcare.* Retrieved from https://royaljay.com/healthcare/neural-networks-in-healthcare/

Garcia-Garcia, A., Orts-Escolano, S., Oprea, S. O., Villena-Martinez, V., & Garcia-Rodriguez, J. (2017). *A Review on Deep Learning Techniques Applied to Semantic Segmentation,* arXiv: 1704.06857

Gharehchopogh, F. S., Maryam, M., & Freshte, D. M. (2013). Using Artificial Neural Network in Diagnosis of Thyroid Disease: A Case Study. *International Journal on Computational Sciences and Applications, 3.*

Hinkelmann, K. (2016). *Neural Networks.* University of Applied Sciences Northwestern Switzerland.

Lavrac, N. (1999). Selected techniques for data mining in medicine. *Journal of Artificial Intelligence in Medicine, 16*(1), 3–23. doi:10.1016/S0933-3657(98)00062-1 PMID:10225344

Maha Lakshmi, T., Swathi, R., & Srinivas, A. (2016). MATLAB Implementation of an Efficient Technique for Detection of Brain Tumor by using Watershed Segmentation and Morphological Operation. *GRD Journals- Global Research and Development Journal for Engineering, 1*(4).

Maolood, I. Y., Abdulridha Al-Salhi, Y. E., & Lu, S. (2018). Thresholding for medical image segmentation for cancer using fuzzy entropy with level set algorithm. *De Gruyter. Open Medicine: a Peer-Reviewed, Independent, Open-Access Journal, 13*(1), 374–383. doi:10.1515/med-2018-0056 PMID:30211320

Masood, S., Gupta, S., Wajid, A., Gupta, S., & Ahmed, M. (2017). Prediction of human ethnicity from facial images using neural networks. *Data Engineering and Intelligent Computing. Advances in Intelligent Systems and Computing, 542,* 217–226. doi:10.1007/978-981-10-3223-3_20

Moghimi, F. H., & Wickramasinghe, N. (2014). *Artificial Neural Network Excellence to Facilitate Lean Thinking Adoption in Healthcare Contexts.* Springer Science + Business Media. DOI doi:10.1007/978-1-4614-8036-5_2

Prisilla, J., & Iyyanki, M. K. (2019). Convolution Neural Networks: A case study on brain tumor segmentation in medical care. *Proc. International Conference on ISMAC-CVB. Springer Nature Switzerland AG.* 10.1007/978-3-030-00665-5_98

Prisilla, J., Iyyanki, M.K., & Aruna, M. (2018). Semantic Segmentation Using Fully Convolutional Net: A Review. *IOSR Journal of Engineering, 8*(9), 62-68.

Ravi, D., Wong, C., Deligianni, F., & Berthelot, M. Andreu- Perez, J., Lo, B., & Yang, G. (2016). Deep Learning for Health Informatics. *IEEE Journal of Biomedical and Health Informatics*. Advance online publication. doi:10.1109/JBHI.2016.2636665 PMID:28055930

Sarasvathi, V., & Santhakumaran, A. (2011). Towards Artificial Neural Network Model to Diagnose Thyroid Problems, *Global. Journal of Computer Science and Technology*, *11*, 53–55.

Saravanan, K., & Sasithra, S. (2014). Review on Classification Based on Artificial Neural Networks. *International Journal of Ambient Systems and Applications*, *2*(4), 11–18. doi:10.5121/ijasa.2014.2402

Sharma, P. (2019). *Image Classification vs. Object Detection vs. Image Segmentation*. Analytics Vidhya.

Shelhamer, E., Long, J., & Darrell, T. (2016). *Fully Convolutional Networks for Semantic Segmentation*, arXiv: 1605.06211

Shi, X., Schillings, P., & Boyd, D. (2004). Applying Artificial Neural Networks and Virtual experimental design to quality improvement of two industrial processes. *International Journal of Production Research*, *42*(1), 101–118. doi:10.1080/00207540310001602937

Teodorrescu, T., Kandel, A., & Jain, L. C. (1998). *Fuzzy and Neuro-fuzzy systems in medicine*. CRC Press.

Zhao, Z., Zheng, P., Xu, S., & Wu, X., (2017). Object Detection with Deep Learning: A Review. *IEEE Transactions on Neural Networks and Learning Systems for Publication*.

KEY TERMS AND DEFINITIONS

Algorithm: A step-by-step approach for solving a problem.

Bias: An additional parameter used to tune the output along with the weighted sum of the inputs to the neuron.

Binary Image: It has two colors usually black and white; they are given in two levels - 0 or 1.

Greyscale: The image has the value of each pixel with the amount of light that carries only intensity information.

Image Processing: It uses the computer algorithms to perform image processing on digital images.

Image Classification: It groups items based on the categories.

Image Segmentation: It is the process of dividing an image into multiple segments. The aim is to get more meaningful information and easier for analyzing.

Neural Network: It is a circuit of neurons and with layers of a modern sense.

Neurons: It is a cell and specialized to transmit information to the entire system.

Object Detection: It deals with detection of semantic objects of a certain class (such as humans, buildings, or cars) in digital images and videos.

Perceptron: A perceptron is a single-layer neural network. It includes input values, weights and bias, net sum, and an activation function.

Thresholding: The method is based on a threshold value to turn a gray-scale image into a binary image.

Weights: They are numerical parameters in an ANN that converts an input to control the output.

APPENDIX 1

The following algorithm is used for extracting a color object from the given image, i.e. tumor chunk from the X-ray or MRI of brain tumor. The following steps are:

1. Read the input image in gray model.
2. Convert the gray image to RGB or BGR.
3. Now convert the RGB or BGR to HSV image.
4. Apply image blurring, this is achieved by convolving the image with a low-pass filter kernel on the HSV image.
5. Apply erosion for eroding the foreground object borders.
6. Compare the extracted results with RGB or BGR or gray model.

APPENDIX 2

Generation of FCN-8 and FCN-16 Models using python code.

Figure 32.

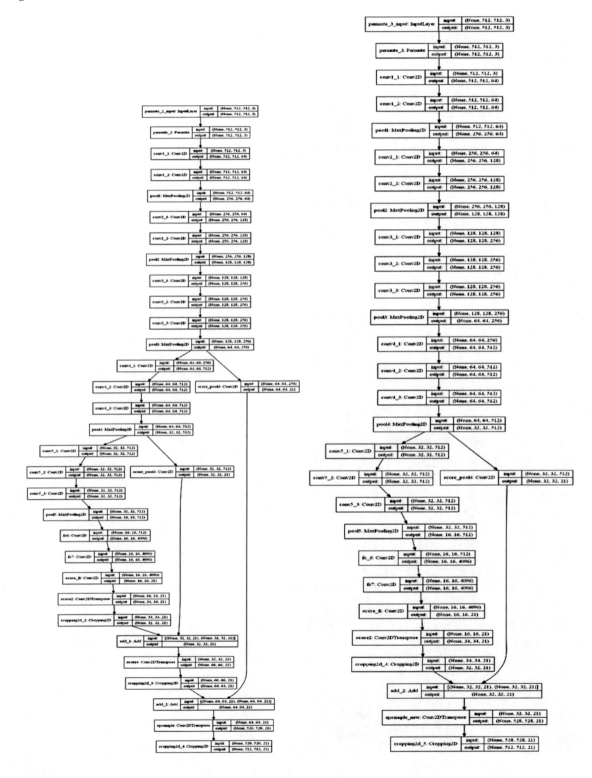

Chapter 10
Demystification of Deep Learning–Driven Medical Image Processing and Its Impact on Future Biomedical Applications

R. Udendhran

ⓘ https://orcid.org/0000-0002-4044-0663

Department of Computer Science and Engineering, Bharathidasan University, India

Balamurugan M.

Department of Computer Science and Engineering, Bharathidasan University, India

ABSTRACT

The recent growth of big data has ushered in a new era of deep learning algorithms in every sphere of technological advance, including medicine, as well as in medical imaging, particularly radiology. However, the recent achievements of deep learning, in particular biomedical applications, have, to some extent, masked decades-long developments in computational technology for medical image analysis. The methods of multi-modality medical imaging have been implemented in clinical as well as research studies. Due to the reason that multi-modal image analysis and deep learning algorithms have seen fast development and provide certain benefits to biomedical applications, this chapter presents the importance of deep learning-driven medical imaging applications, future advancements, and techniques to enhance biomedical applications by employing deep learning.

INTRODUCTION

Researchers calm that deep learning, Quantum Computing and Internet of Things will revolutionize the world similar the way electricity did a century ago. This chapter presents the important opportunities as well as challenges experienced in medical image applications. Generally, biomedical imaging and

DOI: 10.4018/978-1-7998-3591-2.ch010

healthcare industry works under the rule of doctor-patient confidentiality, however, this becomes a challenge for biomedical industry with the integration of deep Learning, for instance:

- Will the data be safe after entering into the system.
- What will happen to the patients' profile and data?
- What factors contributes to the accountability and integrity of automated decision making of deep learning driven image interpretation and the machines utilization of data?

Information about the same patient will be generated and entered into the database through various tools in own proprietary data formats are employed. In this case, if any flawed user input are entered and imperfect database design leads to data inconsistency and redundancy (Würfl et al (2018)).

Figure 1. An example of Multi-modality for 3-D image of the brain which presents multi-information

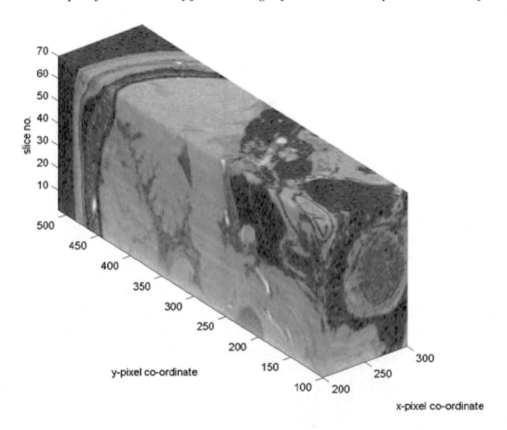

In order to preclude data redundancy, inconsistency problems and less expenditure to create effective database structures before they are deployed. But in some cases, database which does not have effective data structures that can suffer from these problems, a process of database normalization should be implemented. The purpose of database normalization is to re-modify tables such a way that the relations among them are logical, so that database is scalable without any anomalies and avoid data redundancy, inconsistency problems and less expenditure. It is recommendable to design the database in the OLTP format which are highly normalized which avoids data duplication errors. In such cases, healthcare server and computers can employ Linux operating system which is a powerful multi-user operating system which allows several users to access it simultaneously. Linux precludes any changes when employed in the mainframes as well as servers, however, it needs to address the need for security since an attacker can corrupt confidential data, in order to solve this challenge Linux has classified the authorization into ownership and permissions. The reasons for the network failure and other problems in healthcare information systems maybe due to the IP Address and Network Card issues since more than one computers allocated to the same IP address in the data center could cause these kinds of network problems, generally, network card links computers and problems may arise from network card, another reason can be weak radio signals in parts of locations. Drop-in Internet connections can be considered as another potential problem for network issues and the firewall settings must be checked. Before you start the troubleshooting, we need to check if all the hardware are switched on and working well. And make sure the router is not switched off and all switches are in correct positions.By proper power cycling the modems, routers and systems can provide a solution for solving simple network problems. For example, if the administrator encounters network failure in healthcare database server, type "ipconfig" in the command prompt in the terminal. Now check if computer's IP address starts with 169, this means the system will not receive a valid IP address. You can solve this problem by typing "ipconfig /release" followed by "ipconfig /renew" to get a new valid IP address. By using the command "nslookup" we can perform a DNS check. Review database logs. Check all the database logs are working properly sometimes the database maybe full or malfunctioning, it could be reason for the problems that flow on and affect your network performance.

In health and insurance,a central semantic store approach can be deployed which concentrates on logging as well as storing all the rules employed by the database integration process in a single centralized repository. The reason for this approach is that data sources are updated and new ones which that included do not fall outside data integration rules as shown in figure 2.

Deep Learning for Multi-Modal Medical Images

Traditional medical images such as computed tomography (CT) as well as magnetic resonance imaging often require scanning protocols in its toolbox and these protocols provide various property of the underlying tissue, for instance to determine the ischemic stroke lesion, three modalities are required, namely, cerebral blood volume (CBV), cerebral blood flow (CBF), and time to peak of the residue function (Tmax) by employing three modalities the perfusion imaging is extracted and these modal images may differ in interpretation (Yinan, L. Jiajin, Z. Wenxue, and L. Chao (2016)). In order to utilize the multi-modal medical data effectively, early fusion and late fusion methods are employed, early fusion works by stacking several modalities as various input channels and late fusion consists of outputs of networks from various modalities(Christopher Krauss, Xuan Anh Do, Nicolas Huck, (2017)). These methods have gained major importance in medical images Both fusion methods do not employ complementary information from several input modalities to its full potential for instance if three modalities, namely, CBV, CBF and Tmax. are determined, it is observed that from these modalities that lesion area present in the CBV as well as CBF are darker when compared to the normal area and lighter in the modalities of MTT and Tmax. In this case, if we employ early-fusion method, the complementary information can be wrongly fused and if we employ late fusion method which uses encoder-decoder for each modality, then full network becomes harder to converge as well as increases computational expensive. Tensors are

Figure 2. Implementation of Multimodal based-MRI

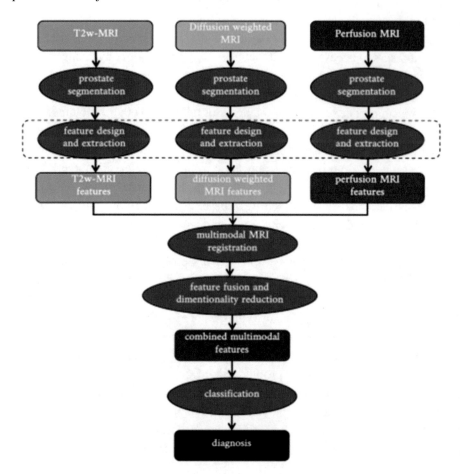

three-dimensional arrays for deep learning, they are generalization of scalars, vectors, arrays. Tensors are most useful for deep learning since you can represent more physical entities into your deep learning, tensors are like data structures for deep learning and gears of neural networks.

LeNet and AlexNet are recognised nets which takes fixed-sized inputs and generate non-spatial outputs. The fully connected layers of these nets consist of fixed dimensions and preclude spatial coordinates. These fully connected layers are considered as convolutions with kernels which takes the full input regions and transforming into fully convolutional networks which can take any input size as well as produce output classification maps as shown in figure 3.

Convolution makes the output size much smaller whereas deconvolution is a process where the output size will be large by using upsampling and it referred as up convolution, and transposed convolution in which deep features can be extracted but spatial location information will be missed in this process. That means output from shallower layers have more location information. However, by integrating both, the result can be enhanced.

Image detection and recognition refers to the problem of identifying a particular element in a medical image as shown in figure 4 and 5.

Figure 3. An Example of Convolution Neural Network Process

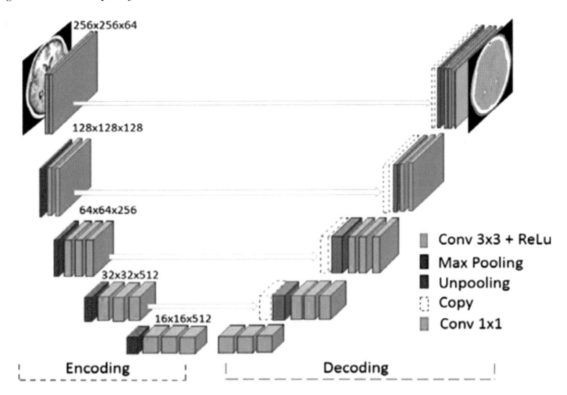

Further advancements include detecting anatomical landmark in 2D X ray projection images as proposed by Bier et al. The authors have employed deep network and trained projection-invariant feature descriptors present in 3-D annotated landmarks. Another alternative method is to use region proposal convolutional neural networks proposed by (Akselrod-Ballin A et al, (2016)) in which they were able to identify tumours in mammographic images. Recognition and detection are often integrated by several other modalities. But the chapters focus on cell detection and classification and the first developments were led by Aubreville et al (2017). employing guided spatial transformer networks that enable refinement of the identification which takes place before the classification is carried out. The perfect example of this technique is mitosis classification which makes use of this technique to its fullest potential.

In the process of image segmentation, detection of an organ as well as anatomical structure is the outcome of image segmentation as shown in figure 6. And convolutional neural networks is widely employed for this purpose such as Holger Roth's deeporgan et al (2016) employed CNN for brain MR segmentation, (Moeskops et al (2017)) developed a fully convolutional multi-energy 3-D U-net and later enhanced by (Chen et al. (2018)) and a U-net-based stent segmentation in X-ray projection domain was developed by (Breininger et al. (2017)). Segmentation employing deep convolutional networks was implemented in 2-D and presented by (Nirschl et al (2017)) for histopathologic images.

(Ali and K. A. Smith (2016)) implemented a project by mapping Frangi's vesselness with a neural network and adjusted the convolution kernels so that it can be effective in solving the specific task of vessel segmentation in ophthalmic fundus imaging. Another interesting group of segmentation algorithm are recurrent networks used in medical image segmentation. It was further explored by Ahmed et al

Figure 4. (a) An simulated image, (b) three-dimensional plot (c) two-dimensional signal and the box filter function, and (d) low frequency image data in time domain obtained from the inverse DFT. Scalar product of the filter and signal in frequency domain stand for convolution in time domain

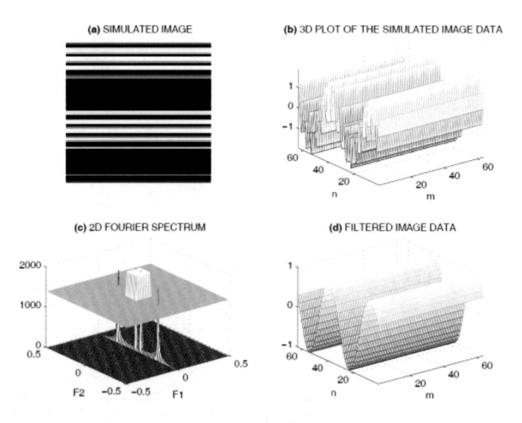

(2002). who presented the full potential of recurrent fully convolutional neural network and deployed it on multi-slice MRI cardiac data, and Aubreville M et al ((2017)) deployed it on brain segmentation and presented the effectiveness of GRUs.

(Suresh, A., Udendhran, R. & Balamurgan, M (2019) enhanced imaging technique by integrating registration metric which acted as a loss function for learning optimal feature representations. In-order to registration problems in 2-D/3-D registration, it is necessary to determine the 3-D position directly from the 2-D point features. Deep learning base algorithms are extensively employed in full volumetric registration, for instance quicksilver algorithm can determine a deformable registration as well as employ patch-wise prediction that can be performed directly from the image appearance and another method is to consider registration problem as a control problem by employing agent and reinforcement learning. Based on this method, a rigid registration which can predict the next effective movement so that it can be align both the volumes and later this method was deployed to non-rigid registration. Therefore, it is possible to employ agent-based techniques for solving point-based registration problems (Hammernik et al. (2018)) . Based on agent-based technique (Zhong et al (2018)) present and employed it for intra-operative brain shift employing imitation learning.

Figure 5. Wavelet functions

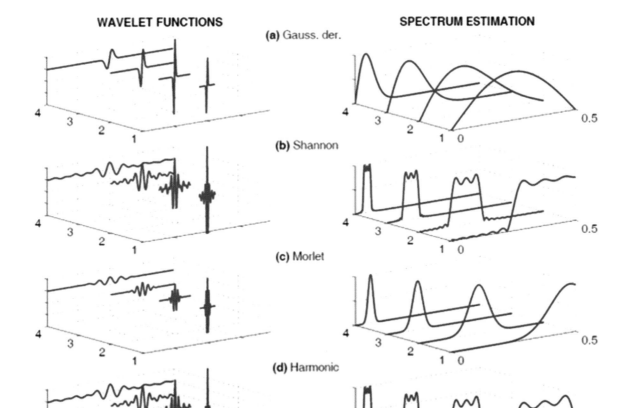

Computer-aided diagnosis is considered as one of the most difficult challenges in the field of medical image processing as shown in figure 7 and 8. Therefore, human error must be avoided and a solution was provided by (Diamant et al. (2018)) employing transfer learning methods. And similar methods were deployed by ophthalmologists in the reading of volumetric optical coherence tomography data. In the latest developments, Google's deep Mind has employed referral decision support.

Physical simulation can regarded as a emerging application of deep learning and it is widely employed in the gaming applications to create realistically appearing physics engines and smoke simulation in real-time. Many researchers proposed deep learning networks for bio-medical applications.

Deep learning has provided many benefits in the field of medical image reconstruction((Diamant et al. (2018))).

Figure 6. Process of Image Segmentation

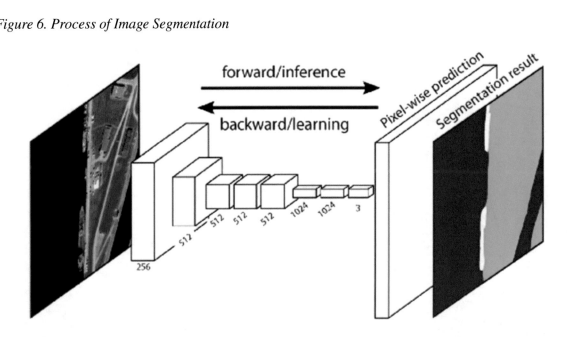

Figure 7. Original medical image data

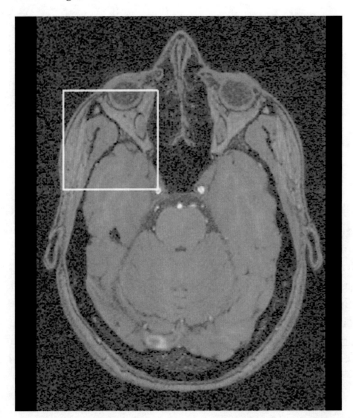

Figure 8. Image reconstruction

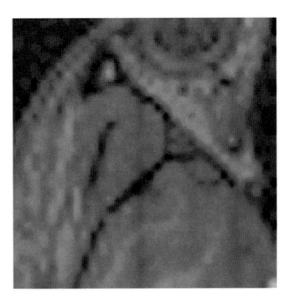

FUTURE RESEARCH DIRECTIONS

In this chapter, we have discussed the advance developments in deep learning employed for medical imaging focusing on detection, interpretation, and segmentation tasks. Deep learning has been discussed which can be applied in medical images and it can be trained as a ConvNet which acts as binary classifier. In latest developments,there are many ways in which medical image can be represented for instance medical image which consists of 224 x 224 can generate important characteristics which can led to enhanced image resolution. However, creating a deep neural network requires careful consideration, for instance, VGG16 (using 16 layers). In-order to avoid over-fitting, the size of the training set is small when compared to number of parameters which consists of 1000 positive + 1000 negative instances. Keras is employed for creating the neural network in which its consists of convolution layers, the amount and size of filters in each convolution layer as well as amount and size of maxpooling layers, amount of fully connected layers, amount of neurons in fully connected layers, activation functions following convolution layers and its activation functions in fully connected layers. Permutation have been employed to generate hyper-parameters and tested by optimizer and binary cross entropy as well as mean squared error with its learning and its batch sizes. Permutation have been employed to generate hyper-parameters and tested by optimizer and binary cross entropy as well as mean squared error with its learning and its batch sizes. Several epochs are trained with least average loss on the validation set.

Microsoft Machine learning server is an important development and one of the most popular solution for healthcare industries as shown in figure 9. Microsoft Machine learning server main objective is to be an enterprise software for data science, providing R and Python interpreters and deployment of manifest files which are packaged together as a container image. The term Cognitive Service containers refers to faster starting, stateless instances in the OS kernel leading to less usage of resources since the Cognitive Service containers does not depend on the OS. This advantage has gained major importance for deployments of faster work-load projects which employs cloud-native application as well as micro

Table 1. Widely used techniques for medical image tasks

Approach	Advantages	Disadvantages
Convolution Neural Network for medical image tasks	Accuracy in image recognition. Higher performance. Enhanced accuracy when trained with lots of data . Need less time for classification tasks Hyper-parameter tuning is non-trivial	High computational cost Slow to train for complex tasks. Need a lot of training data.
KNN for medical image classification	Fresh data can be includes seamlessly Simple classifier that works well on fundamental recognition problems.	Lazy Learner and Instance based learning which means no training period is needed. Need to employ ClassificationData data structure to train the KNN classifier It is well suited for large medical dataset since large medical datasets will results in higher cost of calculating the distance between the new point and each existing points and lessens the performance of the algorithm. It is not suited for high dimensional data since the large number of dimensions will make the KNN to calculate the distance in each dimension. Need feature scaling Not suited to noisy data, missing values and outliers
Fuzzy C-mean for medical image clustering	It is considered as a an unsupervised algorithm which has been performed successfully and employed for clustering purpose on medical images. It is well suited for overlapped data set and comparatively effective than k-means algorithm. FCM is the most important suited clustering techniques for medical image segmentation	Need Apriori specification of the number of clusters Improved results are obtained with lower value of β but needs more number of iteration. Overlapping generally presents in several grey-scale medical images for various tissues Euclidean distance measures can unequally weight underlying factors
Expectation Maximization (EM) algorithm	It is an unsupervised algorithms employed for density estimation of medical data pixels. It is widely employed for fitting mixture distributions. It is simple to implement. It is the most general inference algorithms.	Slow convergence.

services-based components. Machine learning server Server is most compared with Ubuntu Core which is also a lightweight OS. There are many ways in which the Machine learning server Server varies from the Server Core. Machine learning server is primarily built for 64-bit applications as well as tools. Machine learning server Server lacks GUI features, Internet Information Services, Domain Name System as well as local log-on. The admins employ Windows PowerShell, Windows Remote Management along with Windows Management Instrumentation for specific state configuration for the container's settings. Unlike the Server Core, the Machine learning server Server lacks Active Directory Domain Controller, group policy, network interface card teaming, proxy server access and important tool, namely, System Center Configuration Manager and Data Center protection Manager. After the advent of Cognitive Service containers by Microsoft, container has been the new buzzword for the companies and the main

Figure 9. Microsoft Machine Learning Server for predictive service which can be used for Healthcare

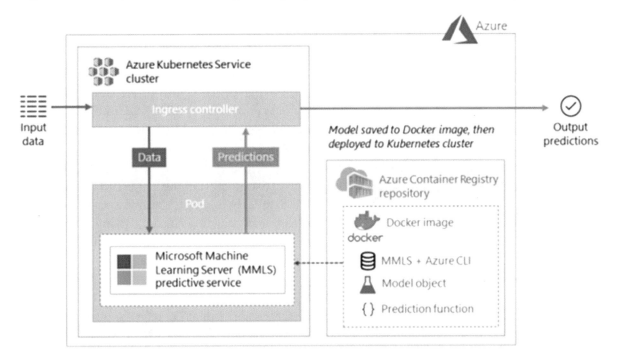

reason is that Cognitive Service containers offers greater agility, since they load the right amount of kernel resources and run-time code to execute an application in the container, thus leading to flexibility among the operating environments. Cognitive Service containers are widely employed in public cloud, for example in Red hat's OpenShift PaaS. The Cognitive Service containers are still evolving technology and it still needs major development in the field of security, load balancing as well as trusted connections.

The evolution of virtualization is going to continue for several years since it has found its application in almost every field of technology and most important field is cloud computing and mobile devices, its evolution has led to the new use-cases for smart phones. As mentioned by Gartner, virtualization in smart phones will increase by 60% in upcoming years. Since virtualization can only solve the need for higher storage space in smart phones. Most advance operating systems contains high level of functionality, thus leading to poor security mess and virtualization solves this challenge, for instance, the virtualization provides better security by partitions. There are several challenges encountered in cloud computing and distributed applications, few common risks involve virus and malware attacks, intentional attacks, man in the middle attack. The most important challenges would be:

- Data Breach: Since large amount of medical data is stored in the cloud, hackers will try to make the most of confidential data. This could be resolved by implementing security controls and deploy encryption to secure confidential data.
- Network security: Sometime the leakage of confidential medical data may occur during interconnection of devices in a network, to avoid this, reliable transmission protocol such as TCP (Transmission Control Protocol) and effective network encryption for secure traffic network.

- Denial of Service attacks: It overpower the resources in your cloud services so that users in your cloud services cannot to connect to applications as well as services. This type of attack is more common in in a distributed computing condition.

In-order to prevent this attack, one can deploy honeypot tools which monitors and alerts if DoS attacks occurs. Maintaining better security policies which provides better security of data in a network. The importance of providing Multi-factor authentication should emphasized and aggrandized, for instance, OTP and mobile code can secure data from unauthorized access. Encryption of confidential data as well cloud backups of data is encouraged. Software-defined WAN has gained major importance since it is a game changer for healthcare industry. The main benefits include easy-to-deploy architecture enabling healthcare organisations to deploy over-the-top of any existing networking infrastructure regardless of location. Integrating SD-WAN enhances any clinic to quickly as well as seamlessly gain access to those applications. However, SD-WAN needs a thorough understanding since it dynamically changes often, for instance, several benefits could be attained by integrating SD-WAN with the medical sector that drives the technology. Though there are certain difficulties, certain developing countries like India, Malaysia, Ghana and South Africa witnessed to improve the healthcare with the SD-WAN technology. To enhance the outcomes of the patients, SD-WAN embraces with the healthcare that forms a great answer to all healthcare organisations. The interaction among the patients, doctors through the internet to the cloud. This process starts with the patients and their visit to the doctor who makes a verification of the patient's details in the e-Health (Cloud) system. The doctor analysis the patient's health with the details provided and offers certain medicine for their health with the possible medication. Through this system the interactions could be very soon, and the details of the patients could once again be updated in the SD-WAN. In the world of healthcare SD-WAN is said to be an emerging technology and seems to be very effective when compared to the current health systems.

Google has developed several applications for healthcare and brining all health-related information into their search engine and similar to their mission which is to organize the world's important information such a way that it can be accessed for useful purposes. Google make use of these channels to gather image data:

- Google Web browser Search
- Web crawling
- Website analytics

In a recent development of Google's announcement in smart display for healthcare applications, a disputable feature which watches the users continuously, this feature is known as Face Match brought by Google Nest Hub Max which employs smart display front-facing camera for providing enhanced security during real-time video calls. This Face Match collects photos, texts as well as calendar details, the reason for gathering these data so that it provides faster recognition. However, this image collection may sound normal for consumers but in the aspect of privacy, Google collects, stores, analyses these gathered image and confidential information present in the images are prone to certain hacking. The Google Nest Hub Max assists several user profiles through fingerprint log-in. Google's Face Match enables to create face framework and it's faster than manual log-in. Google provided a statement in defence to this controversial feature, stating that face image data are stored as well as processed on Nest Hub Max and sometimes retrieves facial data image to aggrandise product experience. Google stated that facial image data stored in cloud storage will be deleted after processing. Other companies consist of privacy policies to certain degree, however, if a user allows to collect and process then it will be stored in cloud in one capacity or another. For effective data transfer in healthcare, among the various existing routing protocols, Destination-Sequenced Distance Vector was introduced which prevents loop free by employing the concept of sequence numbers. The concept of sequence numbers is assigned to each route along with the recent sequence number. This protocol was introduced for MANETs with extra features which lessen traffic minimisation. The forwarding tables consists of entries for all the nodes which are reachable in a system. Ad-hoc On-Demand Distance Vector Routing is the enhanced version of Destination-Sequenced Distance Vector which allows forwarding tables to reach as well as users for discovery of fresh active destination users. Another algorithm is loop-free path-finding which maintains loop-free routing tables and updates routing tables frequently based on the topological as well as link-cost. However, employing techniques such as maximum hop count, spilt horizon, route poisoning, Hold-down Timers can aggrandise routing protocols such a way making the protocols loop-free and effective.

For enhancing security for healthcare related images and sensitive information, HIPAA encryption requirements must be followed, hence GPG encryption is a well-suited encryption for this purpose. The GPG produces random numbers and GPG needs entropy to do this, therefore the system must be kept busy while generating a keypair during keypair generation. The system will be busy and providing random number generator enough entropy to produce a random number. If your system display error stating the need for more random bytes, it means that your system needs more entropy to produce the keypair, therefore open a new shell session and install rng-utils. Generally, the source for the entropy could be hardware RNG device and the entropy sink could be a kernel entropy pool which is employed for generation of keypair. GPG can support these asymmetric encryption algorithms: RSA, ELGAMAL as well as Digital Signature Algorithm. For RSA it needs to be between 1024 to 4096 bits, ELGAMAL needs key length from 1024 to 4096 bits while digital Signature Algorithm needs 1024, 2048 as well as 3072-bit. It employs SHA1, SHA256, and SHA512.

When a healthcare organization expands, the network must support the healthcare organization, therefore the original network plan should contain techniques from upgrading the original network. In certain case, the healthcare organization may need to include several network hardware devices, of different quality, from different manufacturers, as well as different network connection technologies, to connect fresh patients. Whenever, a fresh component is included in a network, it may degrade the quality of the network too, therefore the network should be designed with utmost care. It best recommended to include managed service provider when designing a network. Before designing a network, the on-site technician should perform a site survey to determine the where network equipment can be installed in the network.

The number of clients as well as how many network users, printers, and company servers will the network types support should be determined, Generally, we should obtain how many users will be included in the next 12 months as well as how many network servers the network has to contain. For wired plus Wireless Local Area Network, a wireless-only network (802.11x) can be employed. A high-speed Internet connection DSL router should be employed for faster internet connectivity. The need for specific applications such as voice as well as video need extra network device configuration as well as fresh ISP services.

As soon as the network is connected to the Internet, it opens physical links which consists of more than 50,000 unknown networks as well as all their unknown users. Therefore, Integrated Services Routers (ISR) should include firewall and other capabilities.

In artificial intelligence, the evolutionary algorithms by Darwin which are dependent on survival of the fittest principle can be considered as optimization techniques. The Darwin's theory states that the fittest individuals in nature have more chance to survive. Similarly, in evolutionary algorithms, there are various disciplines present like Evolutionary programming, Evolution strategies, Genetic programming, Different evolution and Genetic algorithms. All the techniques share alike characteristics like variation and selection operators for solving problems but have been independently developed. The evolutionary algorithms are divided based on the solution representation. Swarm intelligence algorithms is another technique which is gaining more popularity in multi-modality. In optimization theory, No Free Lunch theorem was developed and widely employed and well-developed theorem. According to this theorem, every algorithm is equal over a set of possible functions. None of the algorithms could be considered as superior according to this theorem. This theorem is applied only to the finite search spaces however it is not known that the theorem could be applied to the infinite search spaces. The search algorithms could be applied to all finite search spaces with the computer implementations. Swarm Intelligence (SI) is concentrated on the insect behaviour that develops meta-heuristics with the finding abilities of the insects. This interaction among insects is in the form of the collective Intelligence among the insect colonies. This communication system among the insects could be a solution for many scientific problems. Among the global Optimization problems there are many SI algorithms in the history. Function optimization introduced by (Damodaram, R. and Valarmathi, M. L. (2012)) and many other applications such as determining the optimal routes (De Leon, D. (2012). scheduling Deb, K. (2014), data processing in the field of biomedical and classification (Deb, K., Pratap, A., Agarwal, S., and Meyarivan, T. (2002)), image and data analysis (Deb, K., Pratap, A., and Meyarivan, T. (2001)), and many more utilizes the swarm optimization computational modeling inspired by the natural behaviour of swarms. Several other more applications developed have received its inspiration from the natural behaviour of various species. Many biologists have received their inspiration from these ant colonies, and it had motivated the researchers to develop the algorithm based on the ant colony named as the ant colony inspired algorithm. Several algorithms such as clustering and continuous optimization as well as the static and dynamic combinatorial optimization have been developed by the ant metaphor. These developed algorithm shows several similarities based on the ant colony in the nature and they have several advantages during their mechanism by allowing the entire colonies to survive effectively during the evolutionary process (Eberhart, R. C. and Shi, Y. (2001).). The collective behaviour of the insects works with the two effective examples such as the Cemetery formation and brood sorting. Hunting the prey, predator and prey interaction etc, has been the other behaviour of ants observed. The nature could be better understood, and it could be better by finding the underlying mechanism with the replication of behaviour of the insects. Moreover, we could determine the effective techniques if these natural insect behaviours are applied in the field of computer science. Searching, data mining and experimental analysis in data could be best achieved by implementing the clustering and sorting model of insects in the computer science.

CONCLUSION

The advancements in healthcare are not implemented in developing countries because of the poor healthcare infrastructure. The solution for this problem is integrate health sensing devices with portable devices such as smart phones and deploy them in the cloud. By following this technique, poor people can make use of healthcare by employing smart phones which are cheap these days. Deep learning can be explored more in image registration, which can yield interesting results. There are few interesting works which focus on predicting results from the image and other approaches deploy reinforcement learning-based methods for image registration as on optimal control problem. Several advantages can be derived by integrating deep networks for learning representations with registration approaches.

Recent deep learning publications focus on computer-aided diagnosis which is gaining more attention. We predict that simple deep learning tasks that work effectively in a high workload for healthcare doctors will be given priority. However, for advance diagnoses, current deep learning methods which predict results are not suited for this purpose.

REFERENCES

Ahmed, M., Yamany, S., Mohamed, N., Farag, A. A., & Moriarty, T. (2002). A modified fuzzy c-means algorithm for bias field estimation and segmentation of MRI data. *IEEE Transactions on Medical Imaging*, *21*(3), 1939. doi:10.1109/42.996338 PMID:11989844

Akselrod-Ballin, A., Karlinsky, L., Alpert, S., Hasoul, S., Ben-Ari, R., & Barkan, E. (2016). A region based convolutional network for tumor detection and classification in breast mammography. In *Deep learning and data labeling for medical applications* (pp. 197–205). Springer. doi:10.1007/978-3-319-46976-8_21

Al-Fuqaha, Guizani, Mohammadi, Aledhari, & Ayyash. (2015). Internet of Things: A Survey on Enabling Technologies, Protocols, and Applications. IEEE Communication Surveys & Tutorials, 17(4).

Ali & Smith. (2016). On Learning Algorithm Selection for Classification. *Journal on Applied Soft Computing*, *6*(2), 119–138.

Alpaydin, E. (2010). *Introduction to machine learning* (2nd ed.). MIT Press.

Aubreville, M., Knipfer, C., Oetter, N., Jaremenko, C., Rodner, E., Denzler, J., Bohr, C., Neumann, H., Stelzle, F., & Maier, A. (2017). Automatic classification of cancerous tissue in laser endomicroscopy images of the oral cavity using deep learning. *Scientific Reports*, *7*(1), 41598–017. doi:10.103841598-017-12320-8 PMID:28931888

Aubreville, M., Krappmann, M., Bertram, C., Klopfleisch, R., & Maier, A. (2017). A guided spatial transformer network for histology cell differenti- ation. In T. E. Association (Ed.), Eurographics workshop on visual computing for biology and medicine (pp. 21–25). Academic Press.

Aubreville, M., Stöve, M., Oetter, N., de Jesus Goncalves, M., Knipfer, C., Neumann, H., & ... . (2018). Deep learning-based detection of motion artifacts in probe-based confocal laser endomicroscopy images. *International Journal of Computer Assisted Radiology and Surgery*. Advance online publication. doi:10.100711548-018-1836-1 PMID:30078151

Chen, S., Zhong, X., Hu, S., Dorn, S., Kachelriess, M., & Lell, M. (2018). Automatic multi-organ segmentation in dual energy CT using 3D fully convolutional network. In B. van Ginneken & M. Welling (Eds.), MIDL. Academic Press.

Damodaram, R., & Valarmathi, M. L. (2012). *Experimental study on meta heuristic optimization algorithms*. Academic Press.

Davenport, T. H., & Lucker, J. (2015). Running on data. *Deloitte Review*, (16), 5–15.

De Leon, D. (2012). Unconstrained optimization with several variables, in math 232- mathematical models with technology, spring 2012 lecture notes. Department of Mathematics, California State University, Fresno.

Deb, K. (2014). Optimization for engineering design. Algorithms and examples. PHI Learning Private Limited.

Deb, K., Pratap, A., Agarwal, S., & Meyarivan, T. (2002). A fast and elitist multiobjective genetic algorithm: Nsga-ii. *IEEE Transactions on Evolutionary Computation, 6*(2), 182–197. doi:10.1109/4235.996017

Deb, K., Pratap, A., & Meyarivan, T. (2001). Constrained test problems for multi-objective evolutionary optimization. In *Evolutionary Multi-Criterion Optimization* (pp. 284–298). Springer. doi:10.1007/3-540-44719-9_20

Diamant, I., Bar, Y., Geva, O., Wolf, L., Zimmerman, G., Lieberman, S., & ... . (2017). Chest radiograph pathology categorization via transfer learning. In *Deep learning for medical image analysis* (pp. 299–320). Elsevier. doi:10.1016/B978-0-12-810408-8.00018-3

Eberhart, R. C., & Shi, Y. (2001). Particle swarm optimization: developments, applications and resources. In *Evolutionary Computation (CEC'01), Proceedings of the 2001 Congress on*, (vol. 1, pp. 81–86). 10.1109/CEC.2001.934374

European Commission. (2016). *Protection of personal data*. https://ec.europa.eu/justice/data-protection/

Hammernik, K., Klatzer, T., Kobler, E., Recht, M. P., Sodickson, D. K., Pock, T., & Knoll, F. (2018). Learning a variational network for reconstruction of accelerated mri data. *Magnetic Resonance in Medicine, 79*(6), 3055–3071. doi:10.1002/mrm.26977 PMID:29115689

He, D., & Zeadally, S. (2015). *An Analysis of RFID Authentication Schemes for Internet of Things in Healthcare Environment Using Elliptic Curve Cryptography*. doi:10.1109/JIOT.2014.2360121

IBM. (2015). *Watson for a Smarter Planet: Healthcare*. Available at: https://www-03.ibm.com/innovation/us/watson/

Krauss, Do, & Huck. (2017). Deep neural networks,gradient boosted trees, random forests: Statistical arbitrage on the S&P 500. *European Journal of Operational Research*.

Moeskops, P., Viergever, M. A., Mendrik, A. M., de Vries, L. S., Benders, M. J., & Isgum, I. (2016). I'sgum I (2016). Automatic segmentation of MR brain images with a convolutional neural network. *IEEE Transactions on Medical Imaging, 35*(5), 1252–1261. doi:10.1109/TMI.2016.2548501 PMID:27046893

Nirschl, J. J., Janowczyk, A., Peyster, E. G., Frank, R., Margulies, K. B., Feldman, M. D., & ... . (2017). Deep learning tissue segmentation in cardiac histopathology images. In *Deep learning for medical image analysis* (pp. 179–195). Elsevier. doi:10.1016/B978-0-12-810408-8.00011-0

Roth, H. R., Lu, L., Farag, A., Shin, H.-C., Liu, J., & Turkbey, E. B. (2015). DeepOrgan: multi-level deep convolutional networks for automated pancreas segmentation. In *International conference on medical image computing, computer-assisted intervention*. Springer. 10.1007/978-3-319-24553-9_68

Suresh, A., Udendhran, R., & Balamurgan, M. (2019). A Novel Internet of Things Framework Integrated with Real Time Monitoring for Intelligent Healthcare Environment. *Springer-Journal of Medical System*, *43*, 165. doi:10.100710916-019-1302-9

Suresh, A., Udendhran, R. & Balamurgan, M. (2019). Hybridized neural network and decision tree based classifier for prognostic decision making in breast cancers. *Springer - Journal of Soft Computing*. doi:10.100700500-019-04066-4

Udendhran, R. (2017). A Hybrid Approach to Enhance Data Security in Cloud Storage. *ICC '17 Proceedings of the Second International Conference on Internet of things and Cloud Computing at Cambridge University, United Kingdom*. Doi:10.1145/3018896.3025138

Würfl, T., Hoffmann, M., Christlein, V., Breininger, K., Huang, Y., Unberath, M., & Maier, A. K. (2018). Deep learning computed tomography: Learning projection-domain weights from image domain in limited angle prob- lems. *IEEE Transactions on Medical Imaging*, *37*(6), 1454–1463. doi:10.1109/TMI.2018.2833499 PMID:29870373

Yinan, Y., Jiajin, L., Wenxue, Z., & Chao, L. (2016). Target classification and pattern recognition using micro-Doppler radar signatures. *Software Engineering, Artificial Intelligence, Networking, and Parallel/Distributed Computing, Seventh ACIS International Conference on*, 213-217.

Zhong, X., Bayer, S., Ravikumar, N., Strobel, N., Birkhold, A., & Kowarschik, M. (2018). Resolve intraoperative brain shift as imitation game. MIC- CAI Challenge 2018 for Correction of Brainshift with Intra-Operative Ultrasound (CuRIOUS 2018).

ADDITIONAL READING

Zhang, Y., & Yu, H. (2018). Convolutional neural network based metal artifact reduction in X-ray computed tomography. *IEEE Transactions on Medical Imaging*, *37*(6), 1370–1381. doi:10.1109/TMI.2018.2823083 PMID:29870366

Chapter 11
Artificial Intelligence and Reliability Metrics in Medical Image Analysis

Yamini G.
Bharathidasan University, India

Gopinath Ganapathy
Bharathidasan University, India

ABSTRACT

Artificial intelligence (AI) in medical imaging is one of the most innovative healthcare applications. The work is mainly concentrated on certain regions of the human body that include neuroradiology, cardiovascular, abdomen, lung/thorax, breast, musculoskeletal injuries, etc. A perspective skill could be obtained from the increased amount of data and a range of possible options could be obtained from the AI though they are difficult to detect with the human eye. Experts, who occupy as a spearhead in the field of medicine in the digital era, could gather the information of the AI into healthcare. But the field of radiology includes many considerations such as diagnostic communication, medical judgment, policymaking, quality assurance, considering patient desire and values, etc. Through AI, doctors could easily gain the multidisciplinary clinical platform with more efficiency and execute the value-added task.

INTRODUCTION

AI plays a vital role in the field of medical imaging systems that deal with the interpretation, image processing, data mining, data storage, diagnosis, image acquisition and many more applications. Several techniques are involved in the field of AI, which refers to the field of computer science that makes the system to perform the task without the necessity of human intelligence by the integration of several technologies. A sub-branch of AI is said to be the Machine Learning (ML) technology, which involves the system to learn from the occurred data that does not need any explicit programming applied in the

DOI: 10.4018/978-1-7998-3591-2.ch011

field of imaging. An example of AI stages undergone throughout the entire medical imaging system is shown in figure1.

Figure 1. Stages underwent through the medical imaging process through Artificial Intelligence

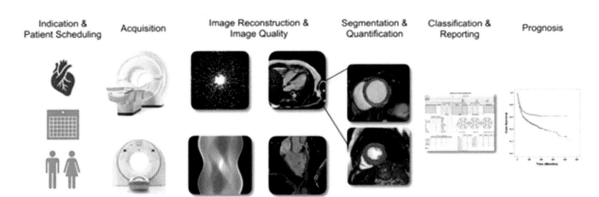

Certain industries should learn from certain platforms that follow their rule in order to maintain their person's safety (e.g. Aviation). Therefore, continuous improvements should be met by the organizations, which are found to be the key for maintaining the reliability in healthcare. This roadmap is followed and being experimented with the Stanford University Hospital, Thedacare and Stanford Children's. This manages and leads several ways by applying several principles and tools, which were operated in the field of healthcare manufacturing process. This relies on dealing with everyday problem by relying the development of frontline staff and then the frontline worker gets connected with their purpose of the organization(Sharma et al. 2018).

Highly reliable playbook had been designed as a principle that becomes an outcome for the field of Indian health care organizations with the quality and the cost problems in the clinic and in the hospitals. At last, the cost effective with the safest system could be the result for every person. The question arises whether the health care system could be more reliable than the airline industry? In recent years, stakeholders, payers, providers and healthcare consumers have demanded better business outcomes and patients care by achieving better reliable performances and organization status. This industry should catch with other consumer-based industries and invest with several efforts and resources that could maintain a solid track to get operated with the reliable organization model to enhance the patient's care outcomes and business result performance as shown in table 1.

For the practical application, deep learning methodologies are not gullible for the practical platforms but theoretically, they have higher performance criteria of the simple machine learning networks. Computer-Aided Diagnosis and detection (CAD) are found to be very familiar among the radiologists from the year 1960 that could be used in the examination of chest x-ray and mammography applications. At a higher-end level with superior decision making and functionality, AI could be used at a wide range to make effective usage of the computational resources. An overview of AI, ML and DL are shown in figure 2. This chapter provides a basic use case for artificial intelligence in the imaging world, and how can AI tools amend workflows to advance the detection and diagnosis of potentially fatal conditions?

Table 1. High Reliability principles for Healthcare

Important principles	Care takers behaviour	Examples
Failed preoccupations	Attitude	Physicians and other health care professionals marking the correct surgical site
Operational sensitivity	System based value practices	Maintaining a good record of the team with the incoming and outgoing information and their present situation status to enhance the accuracy of the team
Simplification disinclination	Meta-cognitive skills	Patients admitted at the critical situation and the fellow residents should know their roles and responsibilities
Determination for the organization	Emotional Intelligence and assertion	Nurses have the right to promote their advice to the physician regarding their allergies or any other physical information, which they might have known before.
Respect the relevant and the qualified experts	Competency skills and leadership	Patient's healthcare should be monitored at the regular basis by the nurse who had been promoted to monitor their regular activities.

WHY DEVELOP USE CASES?

(Wolterink et al. 2016). Another question arises as a fact of finding all those pathological situations and conditions, "Is it necessary to deal with that particular case and forms as a worthy solution while it is has been solved?"

If the AI algorithm is just used to train a computer in detecting the fingers of a hand from the radiograph, then it tends to become unworthy (Vinge, 1993). An alternative for this system would be to use the narrow AI system, which deals in identifying every minute scaphoid bone from the hand, thereby helping the radiologist in determining any sort of miniature fractures. The classification and segmenta-

Figure 2. Overview of Artificial Intelligence, Machine Learning and Deep learning

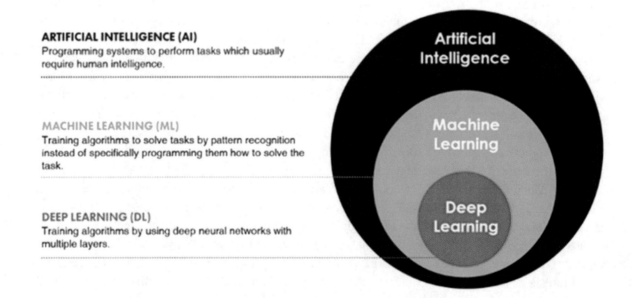

ARTIFICIAL INTELLIGENCE (AI)
Programming systems to perform tasks which usually require human intelligence.

MACHINE LEARNING (ML)
Training algorithms to solve tasks by pattern recognition instead of specifically programming them how to solve the task.

DEEP LEARNING (DL)
Training algorithms by using deep neural networks with multiple layers.

tion process should be integrated with the original part of the system as shown in figure 3, so that the initial workflow of the radiologists' armamentarium will not be disturbed. The algorithm developed will be clinically accepted and integrated with the standard workflow only it promotes added value to the patient's care.

Figure 3. Schematic representation of the AI imaging value chain based on the imaging value chain

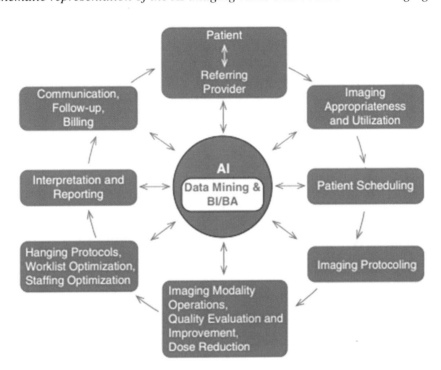

USE CASES

Detection of Cardiovascular Anomalies

Detection of disease related to the cardiovascular system could be easily identified by observing various structures of heart, which can be treated with the normal medicines or through surgery (Chopra and White, 2011). Taking x-rays is the most commonly preferred imaging test, which helps the radiologist or the doctors to make the decision quickly with less diagnosis error at certain times. According to American College of Radiology Data Science Institute ACR DSI, "the first imaging system preferred for the patients coming with the shortness and difficulties in breathing will be suggested to examine with the chest radiograph". Cardiomegaly is considered to be the medical situation in which the heart is inflated or enlarged. This condition may vary from person to person. The situation could also lead to high blood pressure. Chest radiography is used as an initial tool in determining this situation and also it is accepted to be a quick visual assessment done by the radiologist. But in certain cases, this technique is found to be inaccurate.

AI plays a role in obtaining left atrial enlargement that refers to the left atrium dilation, which occurs due to multiple disease states and also increases the pressure of the left atrium desperately. These include valvular heart disease, cardiomyopathies, congestive cardiac failure and congenital cardiac faults. AI focuses to obtain the enlargement region from the chest x-rays thereby promoting relevant treatment for the patients suffering from cardiac and certain other pulmonary problems (Turing, 1950). Applying AI to imaging data helps in detecting carina angle measurement, aortic valve analysis and pulmonary artery diameter automatically. AI tools could also be used to determine the muscle thickening in the left ventricle wall and it also monitors the blood flow rate from the expansion of synergetic arteries.

This automated process could also avoid certain diagnosis errors, save the interpretation time of the doctors and provide a structured quantitative data through the automated flow quantization of pulmonary arteries that could be further used for the future studies and risk lamination schemes (Giliker, 2011). Data are produced in the form of computerized reports that could be saved and later on considered for the verification process thereby saving the human efforts and time.

Figure 4. Role of AI in the cardiovascular medicine (Johnson et al. 2018)

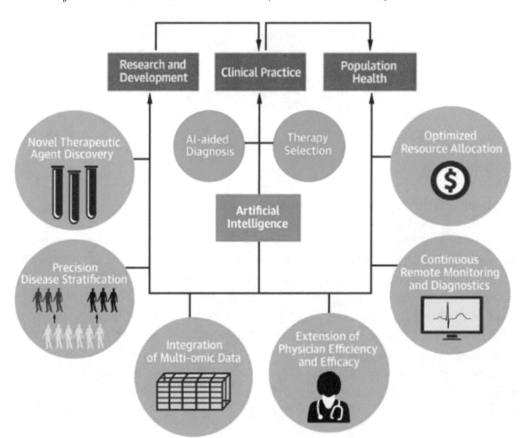

Use of AI in Several Cardiovascular Imaging Modalities

Echocardiography is the highly portable and affordable that is said to be a necessary tool in the clinical cardiovascular imaging toolbox (Krittanawong, Zhang, Wang, Aydar and Kitai, 2017). However, the system is completely dependent on the operator and the process from data acquisition till the complete workflow is an elongated process as shown in figure 4. Machine learning techniques could speed up the reporting rate thereby enhancing the accuracy and also interprets digital 'assistance' for promoting echo images. In the year 1970, patients suffering from 'idiopathic hypertrophic sub aortic stenosis', mitral valve prolapse and mitral stenosis had undergone M-mode image investigation with Fourier analysis that automates the image classification process from the normal one. Currently, the advancement in echocardiogram with the diagnosis and segmentation had enhance the reproducibility and feasibility of real-time full-volume 3D transthoracic echocardiography to automatically measure the volumes and function of both atrial fibrillation and sinus rhythm. The above method also includes the automatic segmentation of left and right atrial and ventricular parameters that is said to be hasty and simple and also adapted by various vendors (King, 2018).

Computed Tomography (CT) is used in the detection of coronary artery disease (CAD) and also helps in analyzing the entire anatomy of heart. Coronary atherosclerosis is a condition that affects the arteries, which supply blood to the heart. CAD could be determined with certain techniques that include coronary CT angiography (CCTA that uses an iodine-containing contrast material and CT scan to determine the narrowing of arteries) and Coronary artery calcium scoring (CACS that evaluates the amount of calcium deposit in the valves of the artery muscles). According to Wolterink et al. the dose of the radiation usage could be highly reduced for the CACS process if convolutional neural network (CNN) has been trained thereby reducing the noise factor. Machine Learning also advances the appraisal of coronary calcium and atherosclerotic plaque (plaque deposit is said to be made of cholesterol that builds up in the wall of arteries). Manual analysis of such disease is said to be a burden and time-consuming and hence it is not practical for the clinical process. With the supervised learning process, CAC could be directly identified and the recurrent CNN could be trained to perform the entire automated analysis with high accuracy when compared to the manual segmentations. Once the plaques are detected then machine learning process could be used for detailed analysis with the help of radiomic techniques (Maes, Collignon, Vandermeulen, Marchal and Suetens, 1997). Radiomics deals with the process of supervised machine learning, which includes the extraction of radiology image features in a large amount and classifies them subsequently in order to obtain predictive performance and diagnosis. Therefore, radiomics is also termed as texture analysis (TA) that leverage interpixel relationship among the radiographical images by quantifying them.

Machine learning algorithm could also be used for the detection of stenosis, which is considered to be hemodynamically significant (internal carotid artery peak systolic velocity is greater than 125 cm/sec). This is due to the fact that the specificity of CCTA is small by using visual examination. Deep Learning techniques could also be used with the CCTA scan images for analyzing the left ventricle (LV) myocardium for the patients suffering from coronary artery stenosis (Zreik et al.). With the multiscale convolutional neural network (CNN), segmentation process is carried out in LV myocardium by subsequently encoding it with the unsupervised convolutional auto encoder (CAE). Since LV myocardium region is divided in the form of spatially connected clusters and computed statistical encoding, support vector machine (SVM) classifier is used for classifying the patients with the extracted features.

Computational fluid dynamics (CFD) method could also be used for the amendment of significant CAD, which is said to be the workstation-based algorithm and it takes around 30-60 minutes for the manual extraction of coronary tree. The accuracy of this approached had been improved with the trained method that uses 12,000 synthetic 3D coronary models of several anatomies and CAD degrees that computes the CT-FFR values based on CFD. Machine learning based approach had been used in training the coronary models with the CT-FFR application (Schier, 2018). This CT-FFR model is again trained with the Deep learning model that incorporates the extracted features from the coronary tree geometry.

AI has the capability to shift the Magnetic Resonance Imaging (MRI) to the next level. A novel method based on the cascade CNN for dynamic MRI reconstruction had been developed (Schlemper et al.). Cardiac MRI depends on the under-sampling scan time, however de-aliasing requires complex algorithm implementation for obtaining the resultant image. Current prevailing methods depend on the sparsity requirement and incoherence among the sparsity and sampling domains, which are known to have intense computation (Pluim, Maintz and Viergever, 2003). De-aliasing the images with the generic strategy could be done by deep learning methodology easily without any formulation rules.

Practically, AI methodology greatly adapts in dealing with the classification and segmentation process in the medical imaging that has been widely adapted by the radiologist. Cardiac MRI had been greatly used and it also varies from patient to patient with the scanning time and also depends on several parameters, which enforce several challenges in the interpretation and also make necessity for the pre-processing (Chen et al. 2018). Normalization, orientation alignment and bias correction are the common factors required by the AI algorithm before feeding the data into the tool. Quantitative measures are mainly evolved in retrieving the present diagnosis criteria, which gives a normal indication and variation among the normal and pathologic process obtained from several degrees of the body. These parameters are determined from the manual operators usually from the radiologist. Difficulties in the process automation arises due to the inconsistency of cardiac shape, basal and apical slices, inconsistency among several scanners and imbricate among the heart and the background with the noise and fuzzy margins.

Cardiac structures could be obtained by nuclear imaging techniques that includes single-photon emission computed tomography (SPECT) and positron emission tomography (PET) in case of CT imaging modalities that needs the ionizing radiation to obtain the structure of the heart and other surrounding membrane (Prior, Brunsden and Hildebolt, 2009). The evaluation of the diagnosis should be done with the lowest possible radiation so that it does not cause any harm to the human body. The proposed CNN techniques had not been used widely in the field of medical imaging, but it can work with the image that had been obtained with the low dose rate, which could be noisy and blurry at certain times. The diagnostic accuracy of SPECT myocardial perfusion imaging (MPI) could be enhanced with the machine learning approach that is still under study.

Detection of Musculoskeletal Injuries

Patients with any miniature fractures and musculoskeletal injuries may suffer from chronic pain if they are untreated and not quickly determined. The constrained mobility and hospitalized elderly persons will face a major problem of hip fractures, which are sustained for a long term and also very difficult to treat if they persist. Fractures, soft tissue injuries and certain dislocation are difficult to detect but with the AI this becomes simple to detect that helps the caretakers and surgeons to have more confident with their treatment prospects.

Fractures can sometimes be left and heralded when medical diagnosis is carried out for trauma related problem, while it's imaging is considered to be a primary clinical concern. Example of trauma related image diagnosis problem with AI technology includes the analysis of cervical spine fractures if the patient had been reported for neck and head trauma. Very delicate and exquisite variations in the human body abnormal could also be detected with the AI imaging tools, which is very difficult for the normal diagnostic tools to detect (Lee et al. 2018). This helps the doctor to make variations among the condition that requires surgery in case of perpetual instability. AI could also help provide safety measures for the patient suffering with hip joint pain in providing sufficient surgery treatment for the replacement (Tang et al. 2018). More that 4,00,000 total hip arthoplasties (THAs) having performed in a year and patience to arrive for the musculoskeletal radiology have to be there for about 100 examination per day for every annual follow-up exams. If the device used for the joint replacement has become stackened or the device does not react to tissues then the patients might need an extra orbitant and invasive alterations. However, detecting abnormalities around the site could be quite challenging task. Detecting the abnormal trauma condition with the X-ray examination is not sufficient and it requires comparison with certain other progressive examinations and the delay. These examinations could also cause delay in the treatment of diseases.

Radiologist could also gain advantage in reducing the patient's risk, false negative rate and also the medical legal risk with the involvement of AI tools. Patients under high risk situation could be segregated for high serum Cobalt levels and they are moved to MRI imaging for other supplementary analysis.

Neurological disorder disease identification and diagnosis: Amyotrophic lateral sclerosis (ALS) is said to be a serious nervous system disease that tends to weaken the muscle and also impact the physical condition. ALS is considered to be a death-dealing neurological situation for a patient and there is no cure for the disease to be treated (Rueckert et al. 1996). When such situation is identified at the earlier cases, then treatment could be provided for the disease to get controlled promoting long term tolerance.

It is necessary to define the situation among PLS and ALS. Primary lateral sclerosis (PLS) falls under the category of motor neuron disease, which makes the nerves among the brain to break down slowly. PLS also causes a heavy fragility among the voluntary muscles that helps in functioning of legs, hands and tongue. Through the proper analysis condition the depth of the disease could be identified. It is the work of the radiologist to identify whether the obtained lesions are pertinent or it simply imitates any other disease structure that leads to the false positive value, which is assumed to be the most happening cases in the field of radiology. In this chapter, the use cases of AI in several medical imaging modalities had been explained. Current articles promote the insight of AI integrated with the radiology and more than 50% of the articles deals with the data collected from the Computed Tomography (CT) and Magnetic Resonance Imaging (MRI). The application mainly concentrated on the cardiac, thoracic, trauma, muscle injuries and fractures and finally with the detection of breast cancer. Advances in technology arise day by day. An advancement from AI is said to be Machine Learning and other evolution deals with the Deep learning is the other evolving technology that is known to be a subset of machine learning in AI, which is composed of several networks capable of learning from the obtained data that may be unlabeled or unstructured (unsupervised).

Detecting new biomarkers is the main focus in current research area by enhancing the accuracy as well as the seed of the diagnosis. Recently valuation in motor cortex that includes Quantitative Susceptibility mapping (QSM) and manual segmentation consumes more time and also seems to be very challenging. Propitious biomarker in the imaging field could be developed by the automation of the above methodology with the machine learning techniques. The image relevant to ALS or PLS could be obtained by the machine learning algorithm very accurately with the clear indication of susceptible risk area provided in the form of a clear image. Efforts of the radiologist will also be reduced by automatic report generation thereby reducing the workflow.

Indication of Thoracic Abnormalities and Condition

Quick actions are required for the patients suffering from pneumonia and pneumothorax, both are said to be the dangerous situation that may lead to death of the suffering. AI algorithm also helps in dealing with the prime target of certain diseases. If untreated, pneumonia inflames the air sacs in both the lungs with the infection caused with the filling of fluids, which can cause issues with the breathing and also leads to threaten the life (Holzinger, 2016). The condition of the lungs could be diagnosed for pneumonia with the radiology images and helps in differentiating the condition with the bronchitis. If the patient is already suffering from cystic fibrosis or malignancies then it will cause difficulty for the radiologist to detect the disease by reading the images. Therefore, conventional imaging methods tend to cause difficulties in analyzing pneumonia if the patient is already suffering from any sort of lung problems.

If little amount of pneumonia exceeds the dome of the chest below the diaphragm, then it will be necessary for the CT scan to be taken that could be difficult for the normal imaging techniques to identify, which could be avoided with the use of AI tools and algorithms (Ashburner, 2007). For effective and quickest treatment of diseases, AI algorithm promotes speed access to the x-ray images for evidence of opacities with the effective diagnosis of pneumonia (Mazura et al. 2012).

Pneumothorax is a condition that occurs when air leaks into the space among the chest walls and the lungs. The disease is caused mainly due to the penetrating chest injury (blunt) or certain lung infections that causes high risk to the patients, which can be identified with AI when the patients are suspected for the disease. This disease introduces air pockets among the chest and the lungs that could also lead to trauma or invasive interruption. This condition could be treated at the earlier stage but when left untreated could also lead to dangerous situation. AI helps in prioritizing the severity and the type of pneumothroaces that might require immediate treatment (Viceconti, Henney, Morley-Fletcher, 2016). The patients could also be monitored with the AI tools over time. The main advantage of AI in detecting pneumothoraces is that it helps in differentiating the size of the disease from the previous state. According to ACR DSI, "though several use cases are required for the Artificial Intelligence to be tested for its efficiency, it appears to be a confident tool in the field of medical imaging". Disease that might to be difficult for the human eye to visualize could be easy for the AI tools in diagnosing and making decisions, leading to the betterment of treatment with injuries, diseases and certain other situations that may be cured when identified at earlier cases.

Detection of Breast Cancer

Medical imaging system could also be used to detect the colon and breast cancer so that it might be helpful for the radiologist to diagnose the disease at the earlier stage. Micro-calcification is a term that refers to the tiny deposits of calcium salt in the tissues that could make the imaging methodologies difficult for the radiologist to detect the tumor cells as malignant or benign. Sometimes the false negative results which meant that the disease is wrongly indicated or some missed malignancies would results in unnecessary invasive testing or worse diagnosis (Shin, Roth and Gao, 2011). According to ACR DSI, "Certain variation could occur due the interpretation of micro-calcifications done by the radiologists during the diagnosis period". The most accurate method which uses quantitative imaging features and identifying the micro-calcification by its level could be AI. Thus, AI could decrease the rate of unnecessary non-cancerous biopsies.

Promoting risk scores based on the spreading of tumor cells under the particular region of interest could promote informed decisions to get in progress with the treatment or testing. During the routine check for colorectal cancer, if polyps are identified (a small clump of cells that surrounds the lining of rectum or colon) during diagnosis, and then the patients could have a productive conversion with the doctors regarding the condition, as they might lead to cancer if left undetected.

CT colonography (CTC) promotes invasive structural examination at a minimum rate in the rectum and colon in order to detect the indication of polyps, high examination time rate could be taken by the inexperienced radiologist in detecting the traces of polyps, which might also left unidentified or missed traces. The accuracy of detection could be enhanced by the AI that reduces the medical legal risk and also the false positive rate for the radiologists. Malignant tumor could be identified easily with the AI technology, which could be treated for them from spreading to other cells. Extranodal extension (ECE) of cancers could be detected only during the surgery that has poor prognosis.

ECE could be identified with the potential algorithm at the earlier cases that does not lead to the surgery, thereby promoting betterment in the treatment lamination. Automatic detection and classification of ECE could improve the radiotherapy targeting of nodal basins and also optimizes the treatment for post-operative imaging-detected nodal disease (Kohli, Summers and Geis, 2017). AI also helps in detecting cancer from several parts that include colorectal, head, neck, cervical and abdomen. The cancer outcomes with the minimized morbidity could be promoted by the AI approach, though the approach has not yet proven with the real-time outcomes.

The above mentioned use cases requires high understanding and study, in order to test the images with the AI tools though the medical images could be readily applied with the Artificial Intelligence. Decision making and diagnosis supplementary with AI offers the patients as well as the radiologist in having the clear visual insights of the injuries, diseases and conditions, which may be difficult to detect with the conventional tools.

Hidden Risks and Dangers

Development of AI products and their use in radiological practice is still in developing stage. In usage of machine learning based algorithms in medical imaging, the known and unknown risks are more. It is difficult to identify the problems in existing methods compared to the earlier methods with lack of basic knowledge in machine learning techniques (Clark et al. 2013). To understand the process and to conclude the process whether it is correct or wrong, one should have the basic knowledge on computer science, statistics and deep learning. This can also be compared with the basic medical physics knowledge of radiologists. For example, in MRI machines and MRI related studies, radiologist's training is associated with physical principles so that the machine learning principles should also be considered in training.

Quality and Validation of Data

In training and testing of data in deep learning, there will be more amount of data needed. To order those data in an order, it consumes more computer power and it is labor oriented. Restoring and detecting the right data is not easy if it is not managed in good manner. Additionally, it also focuses on volume and technical quality of data which is carried out from earlier stage. While training the datasets, some abnormalities should be avoided which includes motion artifacts, image noise, beam hardening, partial voluming. For example, the algorithm will underachieve if it does not include the contrast phases in detecting hepatocellular carcinoma in CT or MRI. The validation of supervised learning process will be based on gold standard (usually a diagnostic test or benchmark that is best available under reasonable condition). The algorithm depends on mainly on quality of the gold standard test but there is no complete standard for obtaining such gold standard.

Data Security and Privacy

Some privacy, security issues regarding management and interchange of patient-related data should not be rejected. Securing patient data is tedious but it should be highly confidential since more smart algorithms are there for mining patient information in database and used in mishandling for other mischievous purposes or commercial earnings. The clients have the right to own their data and also have the power to make certain modifications, which is considered to be "privacy by design" law. Without an AI provider, hospital cannot share data spontaneously by giving patient ownership and control over data (Van Leemput, Maes, Vandermeulen and Suetens, 1999). The term confidentiality and privacy are related to one and another but they cannot be interchangeable. The right of individuals to have their control on their health data and information is referred to be 'data privacy'. The accountability of the trusted data in maintaining privacy is considered to be 'confidentiality'.

Ethics and AI

Still machine learning approach in medical imaging field is in the developing stage and it requires the complete knowledge of the radiologists and physicians regarding AI tools in the upcoming generation so that it will avoid the ethical challenges and also promotes awareness in implementing AI with health care. When the decisions made by the doctors are mistaken then there arises a question regarding, "Who will be responsible for dealing with the fault analysis?" "What if the doctor accepts the diagnosis report of a patient that has been produced as a false analysis?" "What if the machine makes life-threatening errors due to its malfunctions and causes the death of a patient?" With the help of autonomous AI-based systems, transformation could be made in the decision-making field that impacts both the doctors and the patients in knowing the disease earlier and also gaining sufficient treatment in the earlier stage (especially in the cancer diagnosis). The intelligent autonomous based system should be ahead above the conventional radiologist system that advances the suitable treatment offered to the patients. Three main categories dealing with the machine learning system based on the ethical issues:

- Ethics related to data
- Ethics related to algorithm
- Ethics related to practice

CONCLUSION

The question arises regarding, "Why the radiologist must consider the use cases of AI?" Generally, a computer science expert gains the basic knowledge in operating a machine and provides proper training for the computers to process images but still they do not have significant knowledge in selecting the relevant part for which the study is to be carried out. This knowledge will be completely obtained by the radiologist. Without guidance it is difficult for the experts from the non-medical background to obtain the clinically relevant terms, which requires the detection of specified image findings, exact quantification of imaging biomarkers, determination and analysis of pathological situation or obtaining the changes occurred from the previous study.

REFERENCES

Ashburner, J. (2007). A fast diffeomorphic image registration algorithm. *NeuroImage, 38*(1), 95–113. doi:10.1016/j.neuroimage.2007.07.007 PMID:17761438

Boland, G. W., Duszak, R. Jr, McGinty, G., & Allen, B. Jr. (2014). Delivery of appropriateness, quality, safety, efficiency and patient satisfaction. *Journal of the American College of Radiology, 11*(1), 7–11. doi:10.1016/j.jacr.2013.07.016 PMID:24387963

Chen, L., Jones, A. L., Mair, G., Patel, R., Gontsarova, A., & Ganesalingam, J. (2018). Rapid automated quantification of cerebral leukoaraiosis on CT images: A multicenter validation study. *Radiology, 288*(2), 573–581. doi:10.1148/radiol.2018171567 PMID:29762091

Chopra, S., & White, L. F. (2011). *A legal theory for autonomous artificial agents*. The University of Michigan Press. doi:10.3998/mpub.356801

Clark, K., Vendt, B., Smith, K., Freymann, J., Kirby, J., Koppel, P., Moore, S., Phillips, S., Maffitt, D., Pringle, M., Tarbox, L., & Prior, F. (2013). The Cancer Imaging Archive (TCIA): Maintaining and operating a public information repository. *Journal of Digital Imaging, 26*(6), 1045–1057. doi:10.100710278-013-9622-7 PMID:23884657

Coenen, A., Kim, Y. H., Kruk, M., Tesche, C., De Geer, J., Kurata, A., Lubbers, M. L., Daemen, J., Itu, L., & Rapaka Giliker, P. (2011). Vicarious liability or liability for the acts of others in tort: A comparative perspective. *J Eur Tort Law., 2*(1), 31–56.

Holzinger, A. (2016). Project iML – proof of concept interactive machine learning. *Brain Informatics, 3*(2), 119–131. doi:10.100740708-016-0042-6 PMID:27747607

Johnson, K. W., Soto, J. T., & Benjamin, S. (2018). Artificial Intelligence in Cardiology. *Journal of the American College of Cardiology, 71*(23), 2668–2679. doi:10.1016/j.jacc.2018.03.521 PMID:29880128

King, B. F. Jr. (2018). Artificial intelligence and radiology: What will the future hold? *Journal of the American College of Radiology, 15*(3), 1–3. doi:10.1016/j.jacr.2017.11.017 PMID:29371088

Kohli, M., Summers, R., & Geis, R. (2017). Medical image data and datasets in the era of machine learning— Whitepaper from the 2016 C-MIMI meeting dataset session. *Journal of Digital Imaging, 30*(4), 392–399. doi:10.100710278-017-9976-3 PMID:28516233

Krittanawong, C., Zhang, H., Wang, Z., Aydar, M., & Kitai, T. (2017). Artificial intelligence in precision cardiovascular medicine. *Journal of the American College of Cardiology, 69*(21), 2657–2664. doi:10.1016/j.jacc.2017.03.571 PMID:28545640

Lee, J.-G., Jun, S., Cho, Y.-W., Lee, H., Kim, G. B., Seo, J. B., & Kim, N. (2017). Deep learning in medical imaging: General overview. *Korean Journal of Radiology*, *18*(4), 570–584. doi:10.3348/kjr.2017.18.4.570 PMID:28670152

Maes, F., Collignon, A., Vandermeulen, D., Marchal, G., & Suetens, P. (1997). Multimodality image registration by maximization of mutual information. *IEEE Transactions on Medical Imaging*, *16*(2), 187–198. doi:10.1109/42.563664 PMID:9101328

Mazura, J. C., Juluru, K., Chen, J. J., Morgan, T. A., John, M., & Siegel, E. L. (2012). Facial recognition software success rate for the identification of 3D surface reconstructed facial images: Implications for patient privacy and security. *Journal of Digital Imaging*, *25*(3), 347–351. doi:10.100710278-011-9429-3 PMID:22065158

Pluim, J. P. W., Maintz, J. B. A., & Viergever, M. A. (2003). Mutualinformation-based registration of medical images: A survey. *IEEE Transactions on Medical Imaging*, *22*(8), 986–1004. doi:10.1109/TMI.2003.815867 PMID:12906253

Prior, F. W., Brunsden, B., Hildebolt, C., Nolan, T. S., Pringle, M., Vaishnavi, S. N., & Larson-Prior, L. J. (2009). Facial recognition from volume rendered magnetic resonance imaging data. *IEEE Transactions on Information Technology in Biomedicine*, *13*(1), 5–9. doi:10.1109/TITB.2008.2003335 PMID:19129018

Rueckert, D., Sonoda, L. I., Hayes, C., Hill, D. L., Leach, M. O., & Hawkes, D. J. (1996). Nonrigid registration using freeform deformations: Application to breast MR images. *IEEE Transactions on Medical Imaging*, *18*(8), 712–721. doi:10.1109/42.796284 PMID:10534053

Schier, R. (2018). Artificial intelligence and the practice of radiology: An alternative view. *Journal of the American College of Radiology*, *15*(7), 1004–1007. doi:10.1016/j.jacr.2018.03.046 PMID:29759528

Sharma, S., Schwemmer, P., Persson, C., Schoepf, A., Kepka, U. J., Hyun, C., Yang, D., & Nieman, K. (2018). Diagnostic accuracy of a machine-learning approach to coronary computed tomographic angiography-based fractional flow reserve: Result from the MACHINE consortium. *Circulation: Cardiovascular Imaging*, *11*, e007217. PMID:29914866

Shin, H. C., Roth, H. R., Gao, M., Lu, L., Xu, Z., Nogues, I., Yao, J., Mollura, D., & Summers, R. M. (2011). Deep convolutional neural networks for computer-aided detection: CNN architectures, dataset characteristics and transfer learning. *IEEE Transactions on Medical Imaging*, *35*(5), 1285–1298. doi:10.1109/TMI.2016.2528162 PMID:26886976

Tang, A., Tam, R., Cadrin-Chênevert, A., Guest, W., Chong, J., Barfett, J., Chepelev, L., Cairns, R., Mitchell, J. R., Cicero, M. D., Poudrette, M. G., Jaremko, J. L., Reinhold, C., Gallix, B., Gray, B., Geis, R., O'Connell, T., Babyn, P., Koff, D., ... Shabana, W. (2018). Canadian association of radiologists white paper on artificial intelligence in radiology. *Canadian Association of Radiologists Journal*, *69*(2), 120–135. doi:10.1016/j.carj.2018.02.002 PMID:29655580

Turing, A. M. (1950). Computing machinery and intelligence. Mind. *New Series.*, *59*(236), 433–460. doi:10.1093/mind/LIX.236.433

Van Leemput, K., Maes, F., Vandermeulen, D., & Suetens, P. (1999). Automated model-based tissue classification of MR images of the brain. *IEEE Transactions on Medical Imaging, 18*(10), 897–908. doi:10.1109/42.811270 PMID:10628949

Viceconti, M., Henney, A., & Morley-Fletcher, E. (2016). *In silico clinical trials: how computer simulation will transform the biomedical industry*. Avicenna Consortium.

Vinge, V. (1993). The coming technological singularity: How to survive in the post-human era. NASA, Lewis Research Center. *Vision (Basel), 21*, 11–22.

Wolterink, J. M., Leiner, T., de Vos, B. D., Coatrieux, J. L., Kelm, B. M., Kondo, S., Salgado, R. A., Shahzad, R., Shu, H., Snoeren, M., Takx, R. A., van Vliet, L. J., van Walsum, T., Willems, T. P., Yang, G., Zheng, Y., Viergever, M. A., & Išgum, I. (2016). An evaluation of automatic coronary artery calcium scoring methods with cardiac CT using the orCaScore framework. *Medical Physics, 43*(5), 2361–2373. doi:10.1118/1.4945696 PMID:27147348

Zreik, M., Lessmann, N., van Hamersvelt, R. W., Wolterink, J. M., Voskuil, M., Viergever, M. A., Leiner, T., & Išgum, I. (2018). Deep learning analysis of the myocardium in coronary CT angiography for identification of patients with functionally significant coronary artery stenosis. *Medical Image Analysis, 44*, 72–85. doi:10.1016/j.media.2017.11.008 PMID:29197253

Zreik, M., van Hamersvelt, R. W., Wolterink, J. M., Leiner, T., Viergever, M. A., & Išgum, I. (2018). *Automatic detection and characterization of coronary artery plaque and stenosis using a recurrent convolutional neural network in coronary CT angiography*. Academic Press.

KEY TERMS AND DEFINITIONS

Ambient Assisted Living: AAL provides an effective IoT platform governed by artificial intelligence algorithms, thereby satisfying the reliability metric in monitoring patient's health in their place of living in a safe manner. The AAL system includes activity monitoring of patients which is important for patient suffering from Alzheimer's disease, bedsore, diabetes, and osteoarthritis.

Intelligent Information System: An intelligent information system is said to be the set of software and hardware that involves the skilled people for the process of decision making and co-ordination among the organization.

Machine Learning: A sub-branch of AI is said to be the Machine Learning (ML) technology, which involves the system to learn from the occurred data that does not need any explicit programming applied in the field of imaging.

Rational Decision Making: Rational decision making is said to be a crucial process unlike several other sectors, which offers certain provisions to the community.

Chapter 12
Transforming Biomedical Applications Through Smart Sensing and Artificial Intelligence

Harini Akshaya T. J.
National Engineering College, Kovilpatti, India

Suresh V.
National Engineering College, Kovilpatti, India

Carmel Sobia M.
National Engineering College, Kovilpatti, India

ABSTRACT

Electronic health records (EHR) have been adopted in many countries as they tend to play a major role in the healthcare systems. This is due to the fact that the high quality of data could be achieved at a very low cost. EHR is a platform where the data are stored digitally and the users could access and exchange data in a secured manner. The main objective of this chapter is to summarize recent development in wearable sensors integrated with the internet of things (IoT) system and their application to monitor patients with chronic disease and older people in their homes and community. The records are transmitted digitally through wireless communication devices through gateways and stored in the cloud computing environment.

DOI: 10.4018/978-1-7998-3591-2.ch012

INTRODUCTION

Healthcare system faces tremendous challenges but with the developments made in the healthcare systems, many residents are living a healthy and safe life with the normal health condition in many industrialized countries. Several people around the world face severe health condition, among them 285 million visually-impaired persons are also living worldwide, says the World Health Organization (WHO). Though the healthcare reform with their monitoring facility hasn't yet reached many people, within a few upcoming years 32 million newly insured people who face severe trauma also tend to use this facility and they could also gain their maximum benefits. According to the information of the World report on disability, more than one billion people are found to have a certain disability with 150 million schools that are operated to train the disabled children (Renier and De Volder, 2010).

Body-wearable sensor network has increased due to the development made in the wearable embedded system technologies with the remote monitoring of human activities and health (Perkins, Da and Royer 2000). This system improves the quality of life by promoting the promising solution that smooth the progress of independent living and even save people from sudden risk (Intanagonwiwa, Govindan, Estrin, Heidemann and Silva 2003). Daily physical activities and exercises are all monitored with the simple smart phones fixed with the motion sensors or with the help of pedometers. Numerous benefits are gained by the use of wearable devices within the home as well as the community. The long term monitoring of patient health with the gathering of physiological data could lead to the improvement in diagnosing and treating the heart patients (Manjeshwar and Agarwal, 2001). Through this technology, the drawbacks of the available technology could be neglected that includes the recording of rare occurring instances in the diagnosis. Patients suffering from motion disorder also have a huge impact on clinical management in-home monitoring. For instance, the severity of rigidity, bradykinesia, akinesia, and dyskinesia could be avoided by monitoring the symptoms of Parkinsonian disease and thereby facilitating medication titration (Gietzelt, Wolf, Haux, 2011). It is necessary to provide field monitoring to the patients suffering from Parkinson disease. But certain technical limitations tend to stop the methodologies adopted by the clinical applications. From the year of 1990, researchers tend to work hard in implementing a technology adapted towards the growth of in-home monitoring of patients through wireless technology and e-textile methods (Malhotra, 2005). For these technologies, small sensors with the operating systems dedicated to the miniature-sized devices available with low-power radios had been developed with the sensor network. So many miniature-sized devices are used today and they also create a huge impact with the growing technologies. Generally, the people with disabilities are categorized with these groups: 1) Mobility impairments - persons, who do not have to ability to interact with the physical world, manipulate the object and they cannot move freely. 2) Visual impairments – People with low vision ability or they could not view anything (Sánchez and Elías, 2007). 3) Hearing impairments – People who face difficulty in detecting the sound and they could adapt only sign languages. 4) Cognitive impairments: People suffering from dyslexia, dementia, autism etc. Such people face difficulty in learning, understanding and gaining knowledge and also lacks with their perception. Many changes had been determined in the past decades for helping the disabled people of our society. Moreover, these standards and legislations bought had made many changes with the discrimination made against the disabled people to participate entirely with their education, community and employment (Heinzelman, Chandrakasan and Balakrishnan, 2000). Several sensor-based devices have been developed to help disabled people, thereby making their lives easier. This could be termed as the assistive technology, which guides the disable people in meeting their social aspects. This technology does not tend to operate by its own; rather it requires the consideration of a human. The impact or the outcomes of an assistive device should allow the measurement of making a betterment quality of life with the patient and make an impact on the technology used among the community of disabled people (Suresh, Udendhran and Balamurgan, 2019). Clinical processes are categorized with a high degree of communication among the caretakers and the particular patient. This could be highly possible with the integration of the information system that could have a rapid information flow from the patient locality to the doctors or the caretakers (Tapu, Mocanu, Tapu, 2014). Yet there is a lag in these kinds of information system and the requirement to develop them is blooming in these areas. This chapter deals with several sensors based technologies involved in helping the disabled people and also for monitoring the patient's healthcare. Moreover, machine learning techniques that are involved in dealing with progress could also be discussed. The main objective of this chapter is to promote an idea regarding the sensor-based health monitoring devices with the technologies adopted by the system and the way these devices are integrated into an IoT environment with the cloud-based storage (Hoang and Chen, 2010).

BACKGROUND

Wearable system with the miniature-sized sensor with the microcontroller-based electronic device plays a significant role in today's technology. The problem arises with the sensing technology is due to the data transmission and gathering by the small-sized sensor and an electronic circuit that causes difficulty for adapting to a long-term monitoring environment. Recent advancement had made a way, by adding up with amplification circuits, radio transmissions and advanced in microcontroller functions.

The circuit allows the system to gather the physiological data and transmit them wirelessly with a low-power radio to the data logger system. Micro Electro Mechanical System (MEMS) is the advancement in developing miniature-sized inertial sensors, which had been widely adopted in health status monitoring and motor activity systems. The reduction in the size of the sensor could be achieved by the batch fabrication process and techniques. Microelectronics tends to achieve an integrated System-on-Chip facility by adding microprocessor and other radio communication devices under a single circuit (Sichitiu, 2004).

E-textile is another advancement that had been bought with the sensing systems and circuits into garments. The electrodes could be weaved into the garments of the users at the appropriate positions (electrocardiograph and electromyograph) and gather data by printing conductive elastomeric-based components on the fabric and sense the resistance to stretch associated by the fabric while the subject is making some actions (Akyildiz, Su, Senkorasubramanioam and Cayirci, 2002). Advancement in weaving garments with electrodes could improve the technology to print the entire circuit board onto a fabric. Sensor networks integrated with the wireless communication protocols could be limited body-worn or integrated body-worn with ambient sensors. This body-worn sensor system is also known as a body sensor network previously, was composed of several wires for transmitting data that should be necessarily placed into the pocket of the user (Antiochia, Lavagnini, and Magno, 2004). Moreover, this system made a huge disadvantage for not making a long-term monitoring system. This had been further advanced with the advancement of Bluetooth, IEEE 802.15.4/ZigBee and other tethering systems. Standard IEEE 802.15.4/ZigBee based on the Ultra-wide-band (UWB) impulse radio is said to be a high data rate sensor network, which consumes less power and also low in cost. It also produces high accuracy with the location estimation. Many monitoring application requires the gathered data to be sent to the hospital server of the nearby locality for analysis. Smart phones and personal computer forms an information gateway serves this purpose. Many developed countries have already achieved universal broadband connectivity. The monitored data could be communicated from the personal computer or smart phones through the internet (Asada et al. 2003). The patient could also be monitored outside the home environment with the mobile communication standards (4G and 5G) that promotes pervasive and continuous healthcare monitoring facility. Monitoring through smart phones has a huge impact due to the growth of market value and it seems to grow annually with 220 million units shipped by the year 2010 (Rao, Sundararaman, Parthasarathi and Dhatri, 2010). Due to their "ready to use" platform, smart phones are generally preferred when compared to the traditional data loggers. Mobiles could also be used as an information processing units that are integrated with the GPS tracking system that determines the location of the patients in case of emergency situations. The storage could be done in the cloud-based platform (Udendhran, 2017).

Sensium is another low power sensor interface that could also be used for monitoring the health of the patients. Several parameters such as heart rate, skin moisture, temperature, regular energy disbursement etc, could be monitored by using smart bands fitted with the appropriate sensors. This usage of wearable devices helps in enhancing the diversity of the system. Moreover, this system should also be sufficient and suitable for the users. Monitoring physical body is necessary and this also envelops the use of wearable devices by storing and processing the data collected by those devices. The connection is also made among the mobile and the home appliance devices that could disable the TV remote if sufficient user does not perform any activity for a period of time (Li et al. 2005). This activity performance is monitored by accelerometers and through this progress; the health of the human body could be increased. Photoplethysmographic sensor implemented in the form of a ring could be capable of providing the data related to the heartbeat rate and also the oxygen content in the blood and the data is transmitted to the base station wirelessly (Miao, Djouani, Kurien and Noel, 2012). Moreover, this sensor is also fitted as the earpiece and positioned around the outer ear that consists of light-emitting diodes. In-ear sensors are capable of transmitting the heart rate data that could be enhanced in the future for monitoring the oxygen content in the blood.

Figure 1. Block diagram of biological sensor system

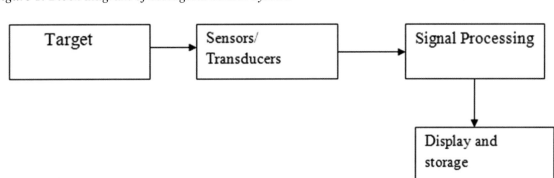

HUMAN HEALTHCARE MONITORING SYSTEM

A lot of research activities are going on in the field of human health monitoring facilities and the globally considered parts are as follows: 1) type of sensor used 2) suitable wireless protocol to be employed for the transmission 3) considered health parameter to be monitored 4) Methodology to determine and extract the important features 5) development of light-weight, powerful and low-cost sensor modes 6) ability to get connected with the mobile devices 7) does not provide any difficulties for the system configuration and implementing them with the human body. A simple block diagram of biological system is shown in figure 1 and sensor-based data sources and their properties are shown in Figure 2.

Figure 2. Sensor-based data sources and properties (Marschollek et al. 2012)

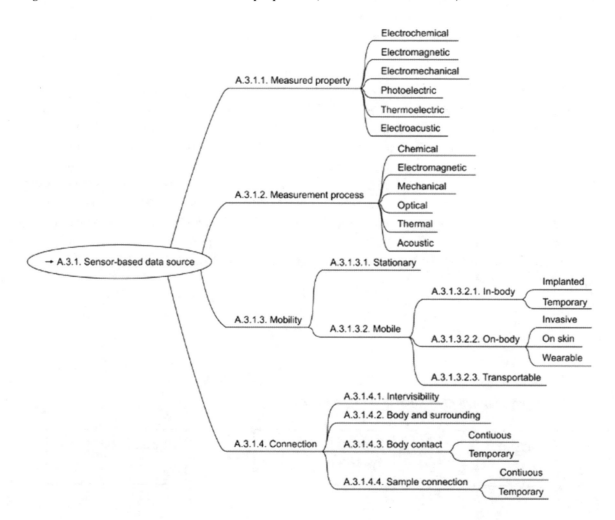

The processor collects the raw data from the sensor and then they are displayed and stored based on the users' decision. Normal people use these wearable devices during their regular activities while running, walking and jogging and the measured parametric values are displayed from the sensors. With the help of the wireless data transmitting facility, the data could be transmitted to the central station with the help of the transceiver (Karl and Willig, 2003). Most of the data retrieved from the sensor could be processed and displayed in the graphical format and based on the complexity the results could be accessible from the remote places through the possible websites. Certain physiological parameters such as the heart rate, body temperature, sugar level, blood pressure etc, could be monitored with sufficient sensing technologies. Several sensors that are used to determine the human body parameters are as follows:

Accelerometers

Several motion sensors deployed with smart phones and pedometers are the most widely used devices to continuously monitor the activities of the patients during their regular physical workout. Sensium is said to be the ultra low power sensor that is incorporated with the smart band aims to work well with the medical health monitoring systems. Several metrics that includes skin moisture, energy expenditure, body temperature, heart rate and other vital human signs could be measured by the smart band-aids. Fitbit and BodyBugg are the other technologies that come to wearable devices. But due to the discomforts and misplacement of sensors, these technologies find limitations in the pervasive computing society. For the optimal validity of data, work is required to overcome several challenges. Ignoring these challenges could lead to false physical activity recording and interfaces.

Considering these limitations, accelerometers finds a scope that consumes less power and could be implemented easily and also present in several devices. They consume very low power and yield excellent performances. Accelerometers had already been implemented with animals such as rats (Chang, Hightower and Kveton, 2009) and cheetahs (Link, 2010) to study their movements and also their behavioral patterns. There are 4 major reasons for the implementation of these devices and determining the position of the human body.

Adaptability and Usability- Based on the lifestyle, clothing and activities undertaken by the users, the pervasive monitoring systems must have the capacity to adapt to the varying location. This system promotes a pathway for the software developers for designing smart devices that can adapt to the surrounding environment. Automatic device localization permits smart phones to automatically adapt to the user's requirements. Based on the configuration and their usage the system should adapt to that particular circumstance. For instance, if the smart phone is mounted by the user to their left arm position, then it could activate the cardiac workout mode.

Accurateness- The patient's conditional that include ultraviolet exposure, skin moisture and temperature are all determined by the Wearable non-intrusive embedded systems and their functioning greatly depends on the accurate placement of the position of the sensor. If the temperature of the body is determined then each body parts shows a varying temperature such as oral (33°C - 38°C), Rectal (34.5°C - 38°C) and Ear (35.5°C - 38°C). For the accurate hypothesis, the sensor should be placed on the particular region of the body as it may vary based on the different body location. This could misguide the caretakers or the clinicians due to the variation in data.

Communication Optimization- Another wireless communication technology includes Intrabody communication (IBC) that utilizes the human body as a transmission medium (Naik, Singh and Le, 2010). There are two different categories under this technology: electric-field IBC (Miluzzo, Zheng, Fodor and

Campbell, 2008) and wave-guide IBC (Shah et al. 2008). If the body-wearable sensor networks rely on the human body as the transmission medium then it finds a great advantage among the research field, but environmental factors and its effect has huge impalpable than the effect it has with the human body. Moreover, the human body also has certain deviation with the performance of the radio transceivers. This effect may also cause a heavy deterioration with the on-body and off-body communication systems performance (Wegmueller, 2007) (Xu, Zhu and Yuan, 2011). According to the quality of the channel, the communication power of the wireless device will be controlled by the automatic body localization that could have compensation with the negative impact of the body.

Accuracy- The correct orientation and the placement of the motion sensors is highly required for the optimal operation of motion and energy expenditure that includes gyros and accelerometers (Misra, Dias Thomasinous, 2010). The relation among the device placement and the accuracy could be determined by the position of pedometer that could be fitted to any parts of the human body other than the waist, as the movement of the users will be projected with variations on the accelerometer. Table 1 shows the location of the pedometer having a greater impact with the step count and error estimation.

Table 1. Pedometer location vs. Step count (Vahdatpour et al. 2011)

Location	thigh	forearm	waist	chest	Upper arm
Step count	148	181	230	0	177
Error $\frac{(real - estimated)}{real}$	32%	17%	4%	100%	19%

It is observed that a simple dynamic threshold control technique could have a greater impact on the accuracy of the pedometer. Energy expenditure estimation through motion sensing is the other impact where the produced accuracy depends on the proper classification of the movement types done by the patients and the accelerometer position on the human body. Accuracy of the activity detection highly reduces when the patient wears the accelerometer in the foot instead of waist. This is due to the tuning of the device based on the waist motions. This could lead to overestimation or under-estimation of patient's calorie value.

One of the most commonly used motion sensors is the accelerometer that could be used in several applications such as game controlling, activity discovery, fall detection, device orientation etc. This system could be integrated with the machine learning algorithm that could be comprised of 2 phases: unsupervised activity discovery and supervised location estimation. According to (Vahdatpour, 2011), the unsupervised technique discovers the activity based on the frequency and consistency during the long time interval. Once the activity had been discovered, then supervised classification technique could be used for analyzation. In [9], support vector machine (SVM) is used to detect the location of the device based on the time interval and frequency properties. The most frequent activity discovery had developed an interest in the research community. Two steps are generally considered in the discovery of most frequent activity. First, frequent patterns are determined by analyzing the accelerometer data separately for obtaining each individual dimension. Binning approach could also be used to determine the similar patterns in the single-dimensional accelerometer data. Second, clustering approach is used for grouping the classified patterns to form the activity data. SVM approach is used to detect the location of the accelerometer on the human body by analyzing the walking sequence. The impact on each step is evaluated by the frequency domain features due to the energy distribution among the foot and the floor, which is shared among the entire body. The closer the sensor location from the foot (origin of strike for every step) the stronger will be the energy sensed by the accelerometer. The range of motion in all three directions will be analyzed and evaluated by time-domain features. Certain limbs move freely than the other, which is necessarily considered. SVM algorithm first approximates the sensor orientation at every time interval of 10s through the DC value calculation by Fourier transformation. By considering the consecutive approximations, the total amount of orientation variables is estimated for the long duration time series.

Through this technique, the least classification technique could also be determined since on the upper extremities it has the capability to mimic the motion characteristics. In the most research study, accelerometers had been used to make a gender differentiation by detecting the hip and shoulder movements. Females walk with more pelvic movement and least step width and on the contrary, males move their shoulders very often. The gait analysis is less invasive for deploying biometric authentication systems, conflicting to other techniques such as fingerprint, voice and iris movement analysis.

Kinect Sensors

Several advanced sensing hardware is incorporated with the Kinect sensors that contain a color camera, depth sensor and microphone array, which promotes facial recognition, 3D motion capture of an entire body and voice identification capacities. The inside view of the Kinect sensor is composed of an IR camera and an IR projector (depth sensor consist of these elements). The depth sensor is of monochrome complementary metal-oxide semiconductor (CMOS) and the IR laser acts as a projector. The Kinect sensor has an extended usage apart from the gaming technology, which could be used as a sensing technology from helping the autism children to guiding the doctors in the operation. This sensor works in the structured light principle. The IR laser that acts as a projector is passed through the diffraction grating system and it turns into a set of IR dots. This IR dots could be viewed by the IR camera. With the triangulation, the 3D image could be constructed if the dot observed in the image could be matched with the dot observed in the projector pattern. Normalized cross-correlation technique could also be used in comparing the small neighborhoods from the IR image and the projector pattern as the produced dot pattern is random. The depth values are encoded with the grey values; if the point is closer to the camera, then the pixel will be dark. If the points are too far then the depth value could not be compared and computed accurately and if the points are too close then a blind region will be developed indicating a necessity for a field of view among the camera and the projector.

At certain cases, if the calibrated value of the IR camera and the projector are not valid, then this may lead to inaccurate depth values. This is due to the vibration or heat produced by IR projector during any sort of drift or transportation. This could be well answered by recalibrating the Kinect sensors through the camera calibration techniques. The depth value depends on the following equation:

$$Z_{measured} = \alpha Zt_{ime} + \beta \tag{1}$$

Here, α and β are the values determined for the recalibration. Rather than this technique, complex distortion model could also be used that follows the same procedure. The feature points of the 3-D coordinate on the calibration card is determined by the recalibration technique with the help of RGB camera that is said to be known as the true values. Meanwhile, Kinect sensors also promote feature points of the 3-D coordinates in the IR camera's coordinate system. The values of α and β are determined by minimizing the distance among the two set points thereby promoting a rigid transformation among IR and RGB cameras. A human body is represented with the number of joints from the neck, head, shoulders and arms in the skeletal tracking. The joints are represented in the form of 3D coordinates. The main aim of this system is to determine the 3D coordinates of a body with effective usage of resources and permits fluent interactivity. As an intermediate process, per-pixel body part recognition has been proposed (Thevenot, Toth, Durs and Wilson, 1999). This process could avoid combinatorial search in various joints of the human body. Over fitting could be avoided by using deep randomized decision forest classifier while classifying numerous images. The classifier could also run separately on each pixel with the Graphics Processing Unit (GPU) that could speed up the entire process. Finally, by computing spatial modes of the inferred per-pixel distributions with the mean shift could result in the 3D joint proposals. The schematic representation of Kinect sensor with the skeletal tracking of various body parts and per-pixel body part recognition has been shown in figure 3-5.

Figure 3. Schematic representation of Kinect sensors (Thevenot, 2011) Source (Thevenot, 2011)

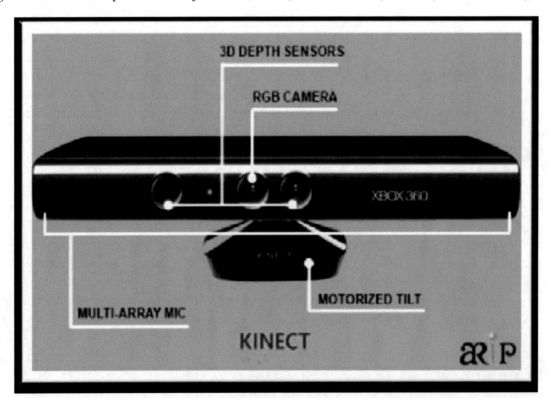

Biosensor and Its Application

According to IUPAC, "A biosensor is an integrated self-sufficient device that is capable of promoting quantitative or semi-quantitative analytical information with the transducer element, which is directly

Figure 4. 3D representation of skeletal body parts (Thevenot, 2011) Source (Thevenot, 2011)

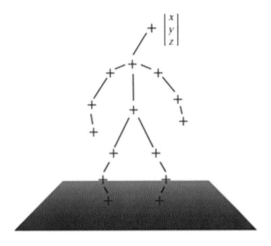

Figure 5. Intermediate stage of per-pixel body part recognition (Thevenot, 2011) Source: (Thevenot, 2011)

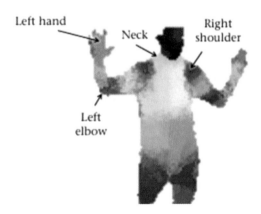

cooperated spatially with the biological recognition element. A biosensor varies from the bio-analytical system that requires reagent for processing. Moreover, a bio probe is different as it should be discarded after a single measurement". A biosensor is composed of 2 main elements:

1) A bio-receptor for determining the analyte such as antigen, enzyme and complementary DNA, which is said to have immobilized sensitive biological element (antibody, probe or enzyme) (D'Souza, 2001).
2) A transducer element that is capable of converting a chemical signal into an electrical parameter based on the interaction with the bio-receptor. The generated signal intensity will be directly or inversely proportional to the analyte concentration.

Based on the signal transduction, the biosensor may be classified as optical, electrochemical (Riv and Pedano, 2006), piezoelectric and thermal sensors. The conductometric and potentiometric sensor comes under the category of electrochemical biosensors (Bilitewski and Turner, 2000). Biosensors are widely used in many fields that include medical applications, diagnostic, bioreactors, process control, nano technology (Ye and Ju, 2005) (Cai et al. 2003), agriculture, quality control (Bergveld, 1996), bacterial and viral diagnostic etc. Some of the advantages of bio-sensor are as follows(Matrubutham, 1998):

- Extremely practical system
- A short period of response time
- Non-polar molecules could be measured, which is very difficult to be measured with other devices
- Control is said to be continuous and rapid with the biosensors (Wang and Liu, 2011).

A few disadvantages are also possible with the use of biosensors that are as follows:

- Due to the denaturalization, heat sterilization is not possible since the biological materials present in the biosensors functions with their natural properties that may be disturbed with the ions, pH and temperature.
- At certain cases, certain other molecules could intoxicate the biosensors cell, which has the capability to diffuse through the membrane.

Cancer diagnosis and treatment had made a huge interest among the researchers due to the disease identification, repetition after treatment and death rate. Cancer can occur in 200 evident forms in various parts of the human body that includes lungs, stomach, throat, mouth, skin, brain, ovarian, breast, leukemia, intestine and colon (Brett and Brett, 1994). The occurrence of cancer may be due to the environmental factors and genetic impacts. Some types of cancer could also lead to viral and bacterial infections. In medical science, biosensor has the capability to determine the blood glucose level, determine pathogens and also helps in detecting and monitoring cancer. The patients' quality of life could be enhanced with the early detection of cancer cells and promoting appropriate treatment at the early stage since when it becomes too late in detecting the disease then it may lead to difficulty in treatment that may also lead to death. This is possible by measuring the level of protein secreted by the tumor cells and biosensors could be able to detect the stages of the tumor (benign or cancerous) (Lambrianou, Demin and Hall, 2008).

Smart Cane

The Central Michigan University's students developed a device known as Smart Cane, which is portable equipment employed with the sensor system that is comprised of an ultrasonic sensor, vibrator, water detector, buzzer and finally a microcontroller element that was basically designed to help the visually-impaired persons. The obstacles present on the path could be determined by the sensor, servo motors and fuzzy controller, which could be indicated in the form of vibration or a voice message. The feedback of the obstacle's position is indicated by the servo motors. The ultrasonic sensor will detect the obstacles and finally, the decisions will be made by the fuzzy controller to navigate the user (Tiponut, Ianchis, Bash and Haraszy, 2011). If the person does not have the hearing ability then a vibrator with gloves will provide sufficient vibration as a message. The performance of the vibrator could also be enhanced by fixing the motors to each finger, with each one having a specific meaning (Tiponut, V.; Popescu, Bogdanov and Caleanu, 2008). The water sensor will not work unless the water level is 0.5 cm apart and the buzzer will also not stop is producing the sound unless the water is wiped and cleaned. Buzzer timing could also be adjusted in order to solve certain issues.

Tongue-Placed Electro-Tactile Device (TED)

This device along with the tiny dipole antenna could be used in assisting the partially and completely blind persons. Wireless transmission is provided among the TED device and the electrodes through the antenna (Dakopoulos and Bourbakis, 2010). The system is mainly composed of three important elements: Tongue-placed electro-tactile device (TED), detective cameras attached to the sunglasses and finally a host computer. The sensor will be placed through the tongue and the other blocks fitted externally include transmission block, central processing unit (CPU) and finally a battery. The sensor placed through the tongue is composed of electrode arrange in 3×3 matrix, which produces a distribution of 8 pulses (Liu et al. 2010). Every pulse specifies to a direction and the other parts are integrated into the form of a circuit.

A Low Cost Navigation Assistive System (LowCost Nav)

This device helps the visually-impaired people in navigation, which is composed of Raspberry Pi, GPS receiver and 3 main buttons to run the system that is located generally in the pocket of the user. For a better experience, the device is promoted with the speech and voice recognition device. Magnetic compass and gyroscope are used in the system that calculates the distance among the user and the object. The entire navigation process is controlled with the help of Raspberry Pi. Both the LowCost Nav and Geo-Coder-US modules are used for generating a pedestrian route. The user could use the microphone to feed the address into the system. If the address is already stored then the button should be pressed. Once the address is fed to the system, the middle button should be pressed to ensure whether the given address is correct or wrong with the help of the database that had been already stored in the system. The system has 6 important modules: 1) Loader- forms as the controller of the system 2) Initializer - verify with the existing libraries and data 3) User interface - receives the necessary address from the user 4) Address query - translates the given address to the geographic coordinates 5) Route query - gets the current location of the user from GPS and finally 6) Route transversal - audible instructions are provided to the user to reach the desired destination.

Benefits of IoT in Healthcare

The rise of chronic diseases with the increase in the global population has made healthcare services expansive as it is in the state of great despair. Health service accessibility is said to be very low with the growing population and among the rural areas. Technologies cannot stop the ageing of people but it could stop the spreading or increase in the chronic disease. Hospital bills are at a high cost with the medical diagnosis. This could be made easier by having access to healthcare in the patient's home. An accurate diagnosis could also abate the need for the hospitalization. The home-centric application of healthcare could be highly possible with the Internet of Things (IoT) paradigm (Miao, Djouani, Kurien and Noel, 2012). This helps the patient in achieving better healthcare treatment and it forms as a mutual hope by making the medical centers to exercise more adeptly.

Cotemporary Monitoring and Reporting

Real-time monitoring for the patients suffering from asthma, heart attacks, diabetes etc, is extremely necessary as it could save lives in case of any emergency medical situations. Smart medical devices connected along with the smartphone application could monitor patients at a real-time and with the connected devices, the data are collected and the sufficient information is sent to the physician locality. According to the study conducted by the Center of Connected Health Policy says that there was a 50% reduction in 30-day readmission rate due to the real-time monitoring of patients suffering from heart failures (Naik, Singh and Le, 2010). The data collected by the IoT devices are as follows: oxygen, blood pressure, weight, sugar level and ECG rate. The data obtained could be stored in the cloud and it can be transmitted to the authorized person who could be an external consultant, a health firm, physician or the patient's insurance company, which provides permission to have a look at the occurred data despite their time, place and system.

End-To-End Affordability and Connectivity

The patients' care workflow could be automated by IoT with the healthcare mobility solution and certain other technologies with next-generation healthcare amenities. With the help of IoT, several services that include machine-to-machine communication, interoperability, information and data exchange could be possible that makes the services to be more efficient and productive. Several protocols used for connectivity includes Wi-Fi, Bluetooth LE, ZigBee, Z-wave and other modem protocols that could function based on the ailments and illness in patients as well as finds innovation in treating the patients. Moreover, unnecessary travels visits can be avoided thereby decreasing the travelling cost and also enhancing the planning and allocation with better resource quality.

Diversity in Data with Analysis

If there is no accessibility to the cloud platform, then a large amount of healthcare data obtained in a very short duration span becomes hard in managing and accessing them. The data occurring from several healthcare devices becomes difficult for the analyzers as well as for the providers. IoT devices can report, collect and analyze the real-time data and also negociate the necessity for storing the raw data. This could all be possible by cloud computing as this tends to display the final data in the form of graphs to the providers. Additionally, the decision-making process becomes easy with the error-less insights for the organization to achieve vital healthcare data analytics.

Tracking and Open-Eyed

In the event of life-threatening circumstances, on-time alerts are very crucial. Data are gathered with the IoT medical devices and those data are transmitted to the doctors for the real-time tracking and the people could receive notifications regarding the body parts in a critical situation through their mobile apps and certain other associated devices. Through this facility on-time treatment with the appropriate decisions could be provided. Therefore, with the help of IoT real-time tracking and monitoring with alerts could be promoted thus enabling hands-on treatment with better accuracy and the patient will also find it simple and updated regarding their health situation with the better delivery result.

Remote Medical Assistance

Healthcare promotes mobility solutions through which the medics can abruptly check the condition of the patients and detect the malaise on the go. Through the linked devices, numerous health delivery chains are functioning in delivering the medicine to the patient's locality with the prescription promoted by the doctors or with the analysis of malaise-related data.

Research Purposes

Data collected through the IoT devices tend to be very simple when compared to the data when collected manually might take make years for analyzation. These data collected could be used as a statistical study for the researchers in the medical field. This could not only save the analyzing time but also promoted cost-benefits thereby promoting better and bigger medical treatments with the advanced technological systems. Moreover, patients could also be satisfied with quality healthcare services. Embedded chips are used widely for updating the old and existing devices with IoT based smart devices. The care and the assistance for the patients could be higher with the use of embedded chips.

Challenges in IoT and Cloud

Data Privacy and Security

The work of IoT is to capture and transmit data at real-time and store them in the cloud. But, still, many IoT devices lack in the data standards and protocols. The data ownership regulation also lacks in ambiguity and all these factors make the IoT devices prone to cybercriminals who could easily hack the system and retrieve the Personal Health Information (PHI) from both the doctors and the patients. This creates a lack of data privacy as well as security since these cybercriminals could misuse these data by creating fake IDs to purchase medical drugs and types of equipment. The worst case of this situation may lead to a false policy claim in the name of the unknown patients.

Multiple Devices and Protocols - Integration

Integration of IoT with the healthcare could become difficult if there is a purpose of integrating multiple devices thereby causing hindrance. This is caused due to the device manufacturers if they hadn't reached a consensus regarding the communication standards and protocols. Although varies devices are connected with the IoT, the variation in the communication protocol causes complication in the data aggregation process. The whole process becomes slow by reducing the scope of scalability among the IoT healthcare devices with the non-uniformity of the connected devices.

Accuracy and Data Overload

As stated earlier, data aggregation becomes complicated if varying communication standards and protocols are used. But IoT devices can record billions of data thereby providing vital insights. Due to the tremendous amount of data and insights from them, doctors find it difficult to choose the appropriate content and also make it challenging for them to make decisions. The complication could increase with

Table 2. Types of sensors with its functionality

Types of sensors	Functionality
Accelerometer based motion sensors	Could be used to detect the position of sensor and movement of the human body during activities and exercises, which consume only low power and yield excellent functionality. It could be used in several other applications such as game controlling, activity discovery, fall detection, device orientation etc.
Kinect Sensors	It promotes 3D projection of a human body or an object, which could be used as a sensing technology from helping the autism children to guiding the doctors in the operation. The main aim of this system is to determine the 3D coordinates of a body with effective usage of resources and permits fluent interactivity. Ease of introducing reminder system.
Biosensors	It is an integrated self-sufficient device that is capable of promoting quantitative or semi-quantitative analytical information with the transducer element, which is directly cooperated spatially with the biological recognition element. It has a bio-receptor and a transducer element. Based on the signal transduction, the biosensor may be classified as optical, electrochemical, piezoelectric and thermal sensors.
Smart Cane	A portable device that is composed of ultrasonic sensor, vibrator, water detector, buzzer and finally a microcontroller element, which helps the visually impaired persons in detecting the obstacles
Tongue-placed Electro-tactile device (TED)	The sensor placed through the tongue is composed of electrode arrange in 3×3 matrix, which produces a distribution of 8 pulses. Every pulse specifies to a direction and assist the blind people.
A Low Cost Navigation Assistive System (LowCost Nav)	This device helps the visually-impaired people in navigation, which is composed of Raspberry Pi, GPS receiver and 3 main buttons to run the system that is located generally in the pocket of the user

the increased amount of devices associated with the system. This could also increase the cost and this should be considered in the developing countries. IoT in healthcare is known to have a promising and fascinating idea. IoT should tend to cut down the cost to one-tenth that facilitates the developing nation to increase the accessibility of this facility during critical situations.

CONCLUSION

In this chapter, sensor-based IoT devices with their benefits and the way they could be integrated with the IoT environment are studied in detail. The real-world variables are captured by the smart sensors, which are the part of IoT components and these data are in the form of digital streams for transmission to a gateway. The built-in microprocessor unit performs the application process with certain algorithms that tends to perform filtering, compensation and certain other signal processing and conditioning tasks. The microprocessor also reduces the load of data that flows towards IoT and it could also spot production parameters that go beyond the accepted values by generating alarms accordingly. This enables the operator to take preventive action before heavy damages occur. In certain cases, the data could be transmitted by the sensor if they exceed the previous sample values and this facility is determined by "report by exception" mode. This reduces the power requirements of the sensors and also reduces the loads on the central computing resources. Sufficient battery sources should also be promoted to the IoT system to continue its function in case of power failure. Additionally, a probe can also contain 2 sensors, which work together for improved monitoring feedback. These devices could be helpful in assisting the visually-impaired people and the patients who are suffering from chronic disease. Early detection of certain diseases like cancer could also be done with the help of this integrated environment.

REFERENCES

Akyildiz, I. F., Su, W., Senkorasubramanioam, Y., & Cayirci, E. (2002). Wireless Sensor Networks a Survey. *Computer Networks*, 28(4), 383–422.

Antiochia, R., Lavagnini, I., & Magno, F. (2004). Amperometric Mediated Carbon Nanotube Paste Biosensor for Fructose Determinatio. *Analytical Letters*, 37(8), 1657–1669. doi:10.1081/AL-120037594

Asada, H. H., Shaltis, P., Reisner, A., Rhee, S., & Hutchinson, R. C. (2003). Mobile monitoring with wearable photoplethysmographic biosensors. IEEE Engineering in Medicine and Biology Magazine, 2003(22), 28–40. doi:10.1109/MEMB.2003.1213624 PubMed

Bergveld, P. (1996). The Future of Biosensors. *Sensors and Actuators. A, Physical*, 56, 65–73.

Bilitewski, U., & Turner, A. P. F. (Eds.). (2000). *Biosensors for Environmental Monitoring*. Harwood Academic.

BodyBugg. (n.d.). http://www.bodybugg.com/

Brett, C. M. A., & Brett, A. M. O. (1994). *Electrochemistry Principles, Methods, And Applications*. Oxford University Press.

Cai, H., Cao, X., Jiang, Y., He, P., & Fang, Y. (2003). Carbon Nanotube- Enhanced Electrochemical DNA Biosensor for DNA Hybridization Detection. *Analytical and Bioanalytical Chemistry*, 375(2), 287–293. doi:10.1007/s00216-002-1652-9 PubMed

Cai, H., Xu, Y., He, P., & Fang, Y. (2003). Indicator Free DNA Hybridization Detection by Impedance Measurement Based on the DNA-Doped Conducting Polymer Film Formed on the Carbon Nanotube. Modified Electrode Electroanalysis, 9(15), 1864–1870. doi:10.1002/elan.200302755

Chang, K., Hightower, J., & Kveton, B. (2009). Inferring identity using accelerometers in television remote controls. PerCom '09.

D'Souza, S. F. (1999). Immobilized enzymes in bioprocess. *Current Science*, 77, 69–79.

D'Souza, S. F. (2001). Microbial biosensors. *Biosensors & Bioelectronics*, 16(6), 337–353. doi:10.1016/S0956-5663(01)00125-7 PubMed

Dakopoulos, D., & Bourbakis, N. G. (2010). Wearable obstacle avoidance electronic travel aids for blind: A survey. *IEEE Transactions on Systems, Man, and Cybernetics*, 40(1), 25–35. doi:10.1109/TSMCC.2009.2021255

FitBit. (n.d.). http://www.tbit.com/

Gietzelt, M., Wolf, K. H., & Haux, R. (2011). A nomenclature for the analysis of continuous sensor and other data in the context of health-enabling technologies. *Studies in Health Technology and Informatics*, *169*, 460–464. PubMed

Heinzelman, W., Chandrakasan, A., & Balakrishnan, H. (2000). Energy-Efficient Communication Protocol for Wireless Microsensor Networks. Proceedings of the Hawaii Conference on System Sciences, 1-10. doi:10.1109/HICSS.2000.926982

Hoang, D. B., & Chen, L. (2010). Mobile Cloud for Assistive Healthcare (MoCAsH). Services Computing Conference (APSCC), 2010 IEEE Asia-Pacific, 325-332.

Karl, H., & Willig, A. (2003). *A Short Survey of Wireless Sensor Networks*. Technical University Berlin, Telecommunication Networks Group.

Lambrianou, A., Demin, S., & Hall, E. A. (2008). Protein Engineering and Electrochemical Biosensors. *Advances in Biochemical Engineering/Biotechnology*, *109*, 65–96. doi:10.1007/10_2007_080 PubMed

Li, G., Liao, J. M., Hu, G. Q., Ma, N. Z., & Wu, P. J. (2005). Study of Carbon Nanotube Modified Biosensor Formonitoring Total Cholesterol in Blood. *Biosensors & Bioelectronics*, *20*(10), 2140–2144. doi:10.1016/j.bios.2004.09.005 PubMed

Link, J. A. B., Fabritius, G., Alizai, M. H., & Wehrle, K. B. (2010). Seeing the world through the eyes of rats. The 2nd IEEE International Workshop on Information Quality and Quality of Service for Pervasive Computing.

Liu, J., Liu, J., Xu, L., & Jin, W. (2010). Electronic travel aids for the blind based on sensory substitution. Proceedings of the 2010 5th International Conference on Computer Science and Education (ICCSE). doi:10.1109/ICCSE.2010.5593738

Malhotra, B. D., Singhal, R., Chaubey, A., Sharma, S. K., & Kumar, A. (2005). Recent Trends in Biosensors. *Current Applied Physics*, *5*(2), 92–97. doi:10.1016/j.cap.2004.06.021

Manjeshwar, A., & Agarwal, D. P. (2001). Teen: a Routing Protocol for Enhanced Efficiency in Wireless Sensor Networks. Proc. 15th Int. Parallel and Distributed Processing Symp., 2009–2015. doi:10.1109/IPDPS.2001.925197

Marschollek, M., Gietzelt, M., Schulze, M., Kohlmann, M., Song, B., & Wolf, K. H. (2012). Wearable Sensors in Healthcare and Sensor-Enhanced Health Information Systems: All Our Tomorrows? *Healthcare Informatics Research*, *18*(2), 97–104. doi:10.4258/hir.2012.18.2.97 PubMed

Matrubutham, U., & Sayler, G. S. (1998). *Enzyme and Microbial Biosensors: Techniques and Protocols*. Humana Press.

Miao, L., Djouani, K., Kurien, A., & Noel, G. (2012). Network Coding and Competitive Approach for Gradient Based Routing in Wireless Sensor Networks. *Ad Hoc Networks*, *10*(6), 990–1008. Advance online publication. doi:10.1016/j.adhoc.2012.01.001

Miao, L., Djouani, K., Kurien, A., & Noel, G. (2012). Network Coding and Competitive Approach for Gradient Based Routing in Wireless Sensor Networks. *Ad Hoc Networks*, *10*(6), 990–1008. Advance online publication. doi:10.1016/j.adhoc.2012.01.001

Miluzzo, E., Zheng, X., Fodor, K., & Campbell, A. T. (2008). Radio characterization of 802.15.4 and its impact on the design of mobile sensor networks. Proc 5th European Conf on Wireless Sensor Networks (EWSN 08, 171–188. doi:10.1007/978-3-540-77690-1_11

Misra, S., & Dias Thomasinous, P. (2010). A Simple, Least-Time and Energy-Efficient Routing protocol with One-Level Data Aggregation for Wireless Sensor Networks. *Journal of Systems and Software*, *83*(5), 852–860. doi:10.1016/j.jss.2009.12.021

Naik, R., Singh, J., & Le, H. P. (2010). Intelligent Communication Module for Wireless Biosensor Networks. In P. A. Serra (Ed.), *Biosensors* (pp. 225–240). InTech., doi:10.5772/7212.

Perkins, C., Das, S. R., & Royer, E. M. (2000). Performance Comparison of Two On-demand Routing Protocols for Ad Hoc Networks. *IEEE Infocom*, *2000*, 3–12.

Rao, G. S. V. R. K., & Sundararaman, K. (2010). A Pervasive Cloud initiative for primary healthcare services. Intelligence in Next Generation Networks (ICIN), 2010 14th International Conference on, 1-6.

Renier, L., & De Volder, A. G. (2010). Vision substitution and depth perception: Early blind subjects experience visual perspective through their ears. *Disability and Rehabilitation. Assistive Technology*, *5*(3), 175–183. doi:10.3109/17483100903253936 PubMed

Rivas, G. A., & Pedano, M. L. (2006). Electrochemical DNA Biosensors. In *Craig Encyclopedia of Sensors*. American Scientific Publishers.

Sánchez, J., & Elías, M. (2007). Guidelines for designing mobility and orientation software for blind children. Proceedings of the IFIP Conference on Human-Computer Interaction, 10–14. doi:10.1007/978-3-540-74796-3_35

Schurgers, C., & Srivastava, M. B. (2001). Energy Efficient Routing in Wireless Sensor Networks. Proc. Communications for Network-Centric Operations: Creating the Information Force. IEEE Military Communications Conf. MILCOM, 2001(1), 357–361.

Shah, R. C. (2008). On the performance of bluetooth and ieee 802.15.4 radios in a body area network. *BODYNETS*, *08*, 1–9.

Sichitiu, M. L. (2004). *Cross Layer Scheduling for Power Efficiency in Wireless Sensor Networks*. Academic Press.

Suresh, A., Udendhran, R. & Balamurgan, M. (2019). Hybridized neural network and decision tree based classifier for prognostic decision making in breast cancers. Springer - Journal of Soft Computing. doi:10.100700500-019-04066-4

Suresh, A., Udendhran, R., Balamurgan, M., & Varatharajan, R. (2019). A Novel Internet of Things Framework Integrated with Real Time Monitoring for Intelligent Healthcare Environment. Springer-Journal of Medical System, 43(6), 165. doi:10.1007/s10916-019-1302-9 PubMed

Tapu, R., Mocanu, B., & Tapu, E. (2014) A survey on wearable devices used to assist the visual impaired user navigation in outdoor environments. Proceedings of the 2014 11th International Symposium on Electronics and Telecommunications (ISETC). doi:10.1109/ISETC.2014.7010793

Tashev. (n.d.). Recent Advances in Human-Machine Interfaces for Gaming and Entertainment. Int'l J. Information Technology and Security, 3(3), 6976.

Thevenot, D. R., Toth, K., Durst, R. A., & Wilson, G. S. (1999). Electrochemical Biosensors: Recommended Definitions and Classification. *Pure and Applied Chemistry*, 7, 2333–2348.

Tiponut, V., Ianchis, D., Bash, M., & Haraszy, Z. (2011). Work Directions and New Results in Electronic Travel Aids for Blind and Visually Impaired People. *Latest Trends Syst.*, 2, 347–353.

Tiponut, V., Popescu, S., Bogdanov, I., & Caleanu, C. (2008). Obstacles Detection System for Visually-impaired Guidance. New Aspects of system. Proceedings of the 12th WSEAS International Conference on SYSTEMS, 350–356.

Udendhran, R. (2017). A Hybrid Approach to Enhance Data Security in Cloud Storage, ICC '17 Proceedings of the Second International Conference on Internet of things and Cloud Computing at Cambridge University, United Kingdom. 10.1145/3018896.3025138

Vahdatpour, A., Amini, N., Xu, W., & Sarrafzadeh, M. (2011). Accelerometer-based on-body sensor localization for health and medical monitoring applications. *Pervasive and Mobile Computing*, 7(6), 746–760. doi:10.1016/j.pmcj.2011.09.002 PubMed

Wang, P., & Liu, Q. (2011). Biomedical Sensors and Measurement. Zheijang University. doi:10.1007/978-3-642-19525-9

Wegmueller, M. S., Kuhn, J. F. A., Oberle, M., Felber, N., Kuster, N., & Fichtner, W. (2007). An attempt to model the human body as a communication channel. *IEEE Transactions on Biomedical Engineering*, 54(10), 1851–1857. doi:10.1109/TBME.2007.893498 PubMed

Ye, Y., & Ju, H. (2005). Rapid Detection of ssDNA and RNA Using Multi-Walled Carbon Nanotubes Modified Screen-Printed Carbon Electrode. *Biosensors & Bioelectronics*, 21(5), 735–741. doi:10.1016/j.bios.2005.01.004 PubMed

ADDITIONAL READING

Intanagonwiwat, C., Govindan, R., Estrin, D., Heidemann, J., & Silva, F. (2003). Directed Diffusion for Wireless Sensor Networking. *IEEE/ACM Transactions on Networking*, 11(1), 2–16. doi:10.1109/TNET.2002.808417

Naik, R., Singh, J., & Le, H. P. (2010). Intelligent Communication Module for Wireless Biosensor Networks. In P. A. Serra (Ed.), *Biosensors* (pp. 225–240). InTech., doi:10.5772/7212.

Xu, R., Zhu, H., & Yuan, J. (2011). Electric-field intrabody communication channel modeling with finite-element method. *IEEE Transactions on Biomedical Engineering*, 58(3), 705–712. doi:10.1109/TBME.2010.2093933 PubMed

Chapter 13
Exploring Internet of Things and Artificial Intelligence for Smart Healthcare Solutions

G. Yamini
Bharathidasan University, India

ABSTRACT

Artificial intelligence integrated with the internet of things network could be used in the healthcare sector to improve patient care. The data obtained from the patient with the help of certain medical healthcare devices that include fitness trackers, mobile healthcare applications, and several wireless sensor networks integrated into the body of the patients promoted digital data that could be stored in the form of digital records. AI integrated with IoT could be able to predict diseases, monitor heartbeat rate, recommend preventive maintenance, measure temperature and body mass, and promote drug administration by having a review with the patient's medical history and detecting health defects. This chapter explores IoT and artificial intelligence for smart healthcare solutions.

INTRODUCTION

In certain areas machines could be used for a certain process such as decision making, solving problems, learning and face recognition and this is possible with the help of artificial intelligence. The process of learning and making self-decisions could be done with the help of AI with the organized and unorganized data. AI is found to be ruling this digital world similar to mobile phones. Evolution in technology makes human lives easier and bringing out a new revolution in today world. Example of this advanced technology includes the Internet of Things (IoT) as shown in figure 1. Users experience is boosted up with the role of AI. The basic question regarding the IoT arises is that, "Why IoT cannot work without AI?" and "where and when IoT is required?" IoT is the technology that is composed of the combination of several sensors and embedding networks, which helps in data communication and interaction without the need for human intervention. The data gathered could be of real-time and these data could be stored in the cloud for further analysis.

DOI: 10.4018/978-1-7998-3591-2.ch013

The huge amount of data gathered with the help of IoT is possible from the varying environment. The collective data are converted into an application with the help of AI by the application of data analytics to make certain decisions. Therefore, the entire process is based on the collection and processing of data that needs an innovative way of implementing AI.

Figure 1. Popularity of Internet of Things

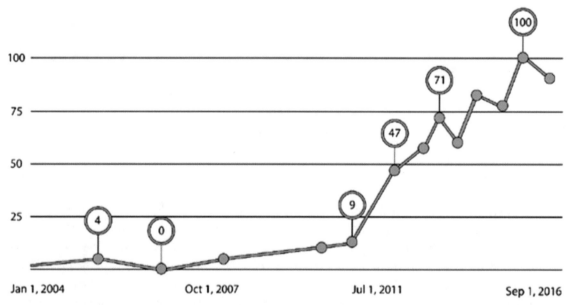

The tenability of big data had been concluded recently, which seems to be very large and incomparable. This could be unlocked with the combination of IoT with AI. According to Deloitte, the combination of AI and IoT claims to hit the market in the near future since it had already been through implementation in several markets. Several companies use the combination of AI-based IoT heads up the competitive industries with their speedy data analysis and intelligent reasoning. IoT devices are capable of transmitting millions of data that may be or may not be required for certain aspects of analysis as shown in figure 2. AI technology enhances this process by managing the IoT devices with its ultimate processing and learning abilities that are made possible with the powerful subset known as Machine learning .

IoT is said to be known as the data "supplier" and machine learning is called as data "miner". The main aim of machine learning is to predict data (Udendhran, 2017). The first process involved in machine learning is to refine the data supplied by the IoT. Numerous data arises from several sensors implemented under the IoT platform in varying environments. The correlation among the data is identified with the machine learning process and its main task is to mine the necessary data by having meaningful insights from those occurred variables and transit them for further analysis (Sharma et al. 2018). The traditional way of data analytics includes the gathering of past data, report generation and finally produce a result with data processing. Prediction is the main process that arrives with the desired outcome and searches interactions among the input variable in order to meet specific criteria (Amini, N et al 2011).

Figure 2. Applications of Internet of Things

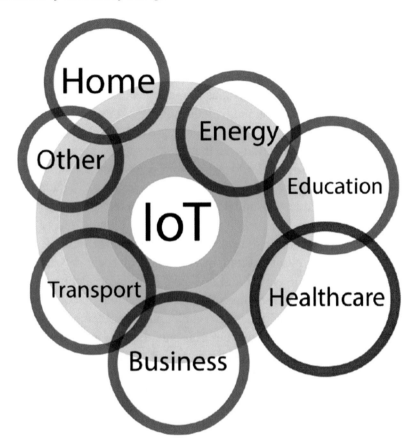

Once the data is transformed into actionable and valuable information, then big data comes into the part. This is not possible with the paper and pen records maintained manually.

AI plays a main role at this stage where the data is collected, and the meaningful extracts are obtained by data analytics. The similar patterns are revealed by AI once the data is fed and produces more informed decisions that could be done either with the help of humans or machines. Recently, business organization are trying to collaborate IoT with AI that tends to decrease the cost with the better user experience and also opens up a new pathway to these organizations(Chopra and White, 2011). In this process, the observer will promote the specific input data regarding the trigger element leading to the AEMs on the daily basis. This information will be reviewed by the doctors at a particular period of time.

Machine learning is necessary to manage certain situations occurring in IoT applications. When an error occurs in a device at an unusual situation then it is required to know whether it needs s human assistance or to respond automatically. Obviously, it needs intelligent learning with the decision-making capacity in order to make the proper decision in that particular situation. This is used as a process by Google with the RankBrain algorithm. Deep learning is used to detect the type of error and once the solution is obtained then this could function at a real-time without the need for human intervention (Chen. L et al. 2018). Smart meters for healthcare building are specially designed with sensors integrated into the IoT environment and banded in the smart grid system to upload and record the data related to electricity.

The sensors function in real-time with the bidirectional data flow and the data are collected with the relevant information that includes frequency among the various types of equipment and applications. Machine with AI learns from the obtained data and make a prediction regarding future events, which is known as the predictive analysis. This could enhance the benefit of IoT and also increases its usage among several manufacturing industries.

AI integrated with the IoT network could be used in the healthcare sector thereby improving the patient care. The data obtained from the patient with the help of certain medical healthcare devices that includes fitness trackers, mobile healthcare applications and several wireless sensor networks integrated into the body of the patients promoted digital data that could be stored in the form of digital records as shown in figure 3. AI integrated with IoT could be able to predict diseases, monitor heartbeat rate, recommend preventive maintenance, temperature, body mass and promote drug administration by having a review with the patient's medical history and detecting health defects. Accurate prediction done with the weather report helps the farmers in harvesting and farming their lands. The schedule maintained by the airplanes or trains entirely depends on the weather forecasting.

Based on the expected weather events, certain business runs by accurately maintain their resources by employing labor. Artificial Intelligence could be highly used in predicting weather conditions. AI utilizes the data that are obtained from past predictions with the actual outcomes.

By making a comparative analysis with the past predictions and obtained outcomes, the present occurring events and its capacity could be improved. This enhances the accuracy with better predictions in the future. The data is fed in the form of an algorithm that needs to be quantitative and qualitative. This makes the process to be more effective with future predictions (Udendhran R and Balamurugan M, 2020). Data is shared among the sensors as well as the higher end computer systems. Information is extracted from a device with the help of AI. Before transmitting that information, it is necessary to analyze and summarize the data. AI does the following process thereby reducing a large amount of data to the sufficient one thus promoting scalability. PSM (Practical Software and Systems Measurement) involves in the project management purposes for software measurement and is said to be a systematic information-driven approach. This was developed in the year 1994 and this had been sponsored by the US Army and US Department of Defence. This methodology has been accepted by several industries and US government organizations for the approach of managing the software-intensive system development projects. In the year 2001, the PSM process was published as the international standard ISO/IEC 15939:2007. The administrative data review, patient care observation, self-reporting, medical record reviews, clinical surveillance and the one-stage and the two-stage chart reviews are the most documented approaches. All these methods have their own advantages as well as their drawbacks based on their usage in the organizations with the available data (Holzinger A, 2016).

HEALTHCARE SUSTAINABLE MANAGEMENT

In the 21st century, recent success in healthcare sustainable building industry has gained major importance and the important goals of healthcare sustainable building design are:

- Employing non-renewable resources efficiently
- Safeguarding human health residing in building as well as lessening waste that negatively impacts the environment.

Figure 3. Internet of Things and Intelligence in smart Healthcare Solutions

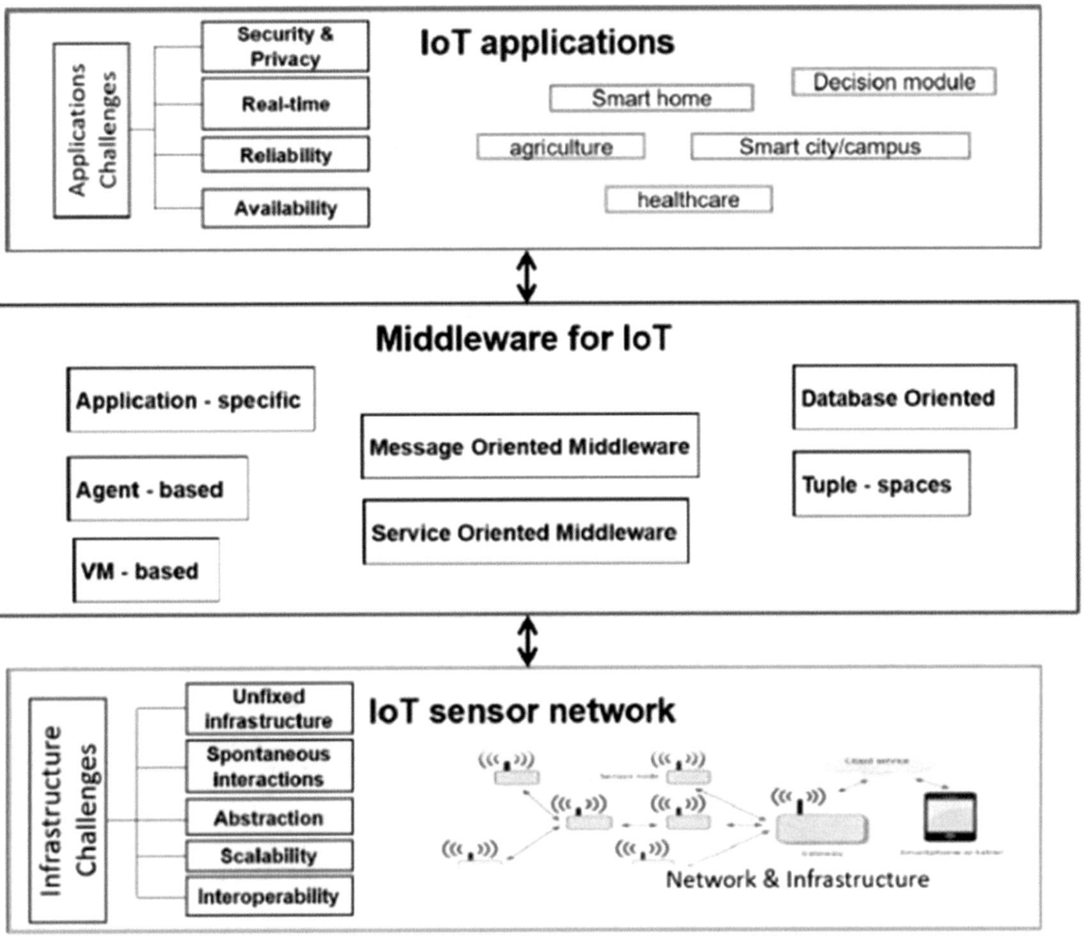

In order to provide sustainability, Green star Rating Tool which is a certification system evaluates the sustainability level in buildings. The evaluation process is based certain criteria such as energy, water, and materials. However, this widely employed tool does not provide parameters needed to implement fire safety but experts in fire protection industry have highlighted the greater fire safety risks, for instance atria allows more natural light, but can increases fire hazards because of the smoke which can spread everywhere in a faster way. In-order to make energy efficient buildings, particular type of insulation is employed for aggrandizing the structural support for thin steel walls, but this certain type of insulation can easily melt due to the high heat caused by fires which weakens the structural integrity of the building. If the structural integrity is weak, it can result in collapse of the healthcare sustainable building and cause negative impacts on the environment. Therefore, few healthcare sustainable building features can lead to fire hazards, for instance fire-extinguishing elements such as halogen gases can be dangerous to the ozone layer. Fire safety experts encourage effective collaboration between healthcare sustainable building and fire safety industry when constructing a building, instead of including regulations and requirements to enforce strict policies. It is important to collaborate early in the building design process

Figure 4. IoT Communications in multi-domains

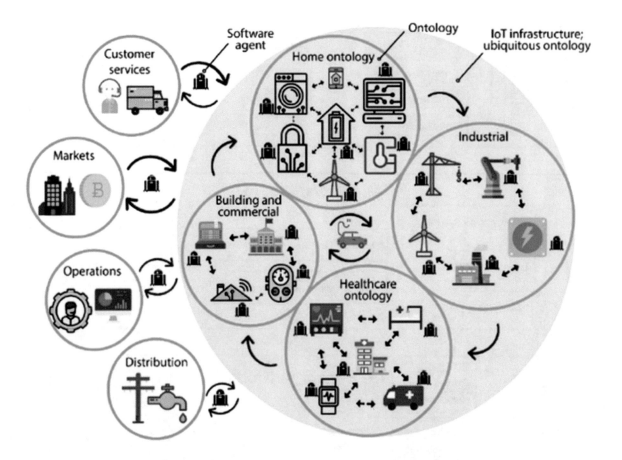

so that it will be easier to sort out disagreements since many healthcare sustainable building experts mention that t is much easier to change a building in its starting phase and create effective solutions. The following computer networks can be employed for implementing a sustainable hospital management:

- *Unshielded twisted pair:(UTP):* UTP is a widely employed cable and it is well suited for this purpose. It uses a high-speed cable and the cable contains four pairs of wires inside the jacket in which each pair is twisted such a way that the number of twists per inch varies so that it can preclude interference from adjacent pairs as well as other electrical devices. Higher the twisting, greater the supported transmission rate as well as cost per foot. The standards for the Unshielded twisted pair (UTP) can be found at EIA/TIA (Electronic Industry Association/Telecommunication Industry Association) which contains six groups of wire.
- *Network Diameter:* A hierarchical network topology is recommended for the given problem and it is important to consider the network diameter, therefore diameter is a measure of distance, however, in our case, we will measure the number of devices in which a packet has to cross to arrive at its destination. It is recommended to lower the network diameter low for predictable latency between devices. Degree of latency would be produced by each switch as well as each switch should compute destination MAC address of the frame, check its MAC address table, and forward

Figure 5. IoT Architecture

Instrumented
Measure and sense the current condition permanently

- Smart health solutions automatically capture information to proactively manage conditions and deliver preventive, therapeutic care
- Sensors that recognize physical changes such as pressure, motion, blood status, or temperature are embedded in wearable devices or other equipment

Interconnected
Devices in system communicate, share data with each other

- Smart health systems remove information barriers and seamlessly integrate data and analytics into healthcare processes to enable better decisions and comprehensive, coordinated healthcare
- Mobile and home-based devices monitor vital signs and activities in real time and communicate with personal health record services, PCs and smartphones, caregivers and healthcare professionals

Intelligent
Able to predict and to respond on unplaned events

- Smarter health systems continually analyze information from multiple devices and other sources to derive insights and recommendations for the individual's health regimes
- Analytics programs monitor device data and use rules and logic to compare against targets, track progress against goals, and send alerts when needed

the frame out the appropriate port. In the three-layer hierarchical approach, network diameter can predict the number of hops between the source as well as destination devices.

- *Bandwidth Aggregation:* Bandwidth aggregation considers the particular bandwidth needed for each part of the hierarchy. In our case, link aggregation should be created which can support multiple switch port links to be integrated for greater throughput between switches. For this purpose, EtherChannel by Cisco which is a proprietary link aggregation technology can be employed.
- *Redundancy:* Deploying redundant links can increase the cost. In cost effective healthcare management, it is unlikely to deploy redundancy at the access layer considering the cost as well as limited features in the devices, however, if needed redundancy can be created and integrated into the distribution as well as core layers of the network.

A sustainable building is the design philosophy with the physical end product to enhance the buildings performance by improving the resource efficiency in the complete lifecycle of a building (from construction, operation till demolition). The measure of these performances could be obtained by international environmental performance-rating systems for buildings. The utilization of materials is one key viewpoint that must be viewed as while developing supportable structures. There is a huge discussion about which materials will make minimal measure of harm nature over the long haul. The recognized rating systems include LEED adapted by US, Building Research Establishment Environmental Assessment Method (BREEAM) adapted in UK and Green Star adapted in Australia. The goal of all these recognized rating systems is similar; to maintain a sustainable environment in areas prone to waste, water, pollution and energy. If there is a maximum building score is then the design is said to be more sustainable. Moreover, a sustainable building is said to have a positive social impact on the local environment. More than a conventional designed building, a sustainable building contains longer lifespan, promotes social amenity, considers societal needs, maintains heritage and promotes proofing to the building at the future.

Fire has been characterized into five unique classes dependent on its source and how it tends to be contained inside a structure. These classes include:

- **Class A:** This fire includes normal ignitable materials, for example, paper, wood and material fibre. This class of fire can be extinguished by cooling, covering or wetting of the structure. Likewise, dousing specialists can be utilized to annihilate the fire.
- **Class B:** These are fires which include combustible fluids, for example, fuel, thinners, oil-based paints and oils. To kill this kind of fire, Extinguishers, for example, carbon dioxide, dry concoction and hydrogenated specialist types can be utilized.
- **Class C:** This sort of flames includes the consuming of electrical gear, where a non-leading vaporous clean specialist or covering operator is expected to put off the fire.
- **Class D:** These are fires which occur because of the ignition of metals, for example, magnesium, sodium, potassium, and aluminium. Uncommon dry powder dousing operators are required for this class of fire to be annihilated.

The hazards caused due to the fire are broadly classified from the prescriptive of the codes and standards that are initiated in a very specific manner along with the installation of the fire protection system suitable for the particular situation. These systems are initially designed and tested before they are bought under the application performance. There are several benefits by the Prescriptive codes and standards since they are very easy for the enforcement and application. Additionally, the building that are constructed with the Prescriptive codes and standards tends to manage well during the fire situation and have a good history of performance. But they do not guarantee for standard level of safety with the cost privilege. For instance, store that are bought under the classification of mercantile occupancies, with one side of the store selling the greeting cards and the other part with the liquor bottles tends to adapt the same protection guidance but the fire hazard happening at both the shops could be different.

From the above situation, it is known that the whole building should be necessarily considered for implementing the fire safety that is promoted by the "Performance-based design" tool. "Performance-based design" is an engineering approach to fire protection design that depends on the following factors:

- Based on the essential fire safety objectives and goals
- The fire scenario consideration and analysis
- Based on the performance criteria, engineering tools and methodologies, the quantitative measurements are generated with the sufficient fire safety objectives and goals.

Fire safety objectives of a building could be obtained by considering the performance-based design. Environment protection, property protection, life safety and mission continuity are all included in these objectives. Occupational Safety and Health Act, U.S. Department of Labor in the year 1970 states that "standards developed under the act are enforced to promote safety to the working men and women; by providing sufficient information, research, training and education regarding the occupational safety and health; to promote a healthy working situation and condition for the people living in the state". Through this sufficient information could be obtained by the firefighters to interact the building at the case of any emergency situations. The combined efforts made by the code officials, designers and related stakeholders promote a safe workplace for the fire-fighters.

The building codes set a minimum requirement for the fire safe structures to be designed and constructed. This works in setting up an optimum safety and economic feasibility thereby promoting a safe environment for the healthcare. Building codes are generally categorized as two types:

Specification or prescriptive codes- This explains in detail regarding the building components, assembling the structure as well as the size and the shape of the components used under construction to promote a safe environment. Certain criteria are set for the objectives to be reached; these criteria should be generally reached while the builders try to choose the components and materials for building the construction.

Building codes plays a vital role in promoting safety to the structure and also in the earthquake condition. Recently, building codes includes minimum requirements as structural stability, ventilation, built-in safety equipment, fire resistance, sanitation, lightning and means of egress (Giliker P, 2011). Additionally, building codes also involves in establishing the area and the height criteria for the building in order to prevent and protect the structure from fire based on the intended usage. These requirements could vary based on the types of occupancy from one tenancy to the other. Moreover the codes also includes 1) provisions for automatic fire suppression system 2) necessity for avoiding the flame spread with the interior design 3) enclosure of vertical openings such as pipe chases, stair shafts and elevator shafts and finally 4) promoting an exit pathway during the fire hazard.

LEVERAGING AI AND UBIQUITOUS COMPUTING FOR SMART HEALTHCARE

Because of decreasing birth rates and late deployment of advanced healthcare technologies in developing countries, integrating new technology with existing infrastructure can assist poor people to employ smart healthcare solutions, in most developing countries smart grid technology exists, therefore it is simple to integrate it with artificial intelligence for deploying smart healthcare solutions as shown in figure 5. Based on the effectiveness of IoT and Ubiquitous computing, an intelligent information system is said to be the set of software and hardware that involves the skilled people for the process of decision making and co-ordination among the organization as shown in figure 6. HIS system had been mainly designed in governing the operation carried out in the healthcare organizations such as the clinic, patient registration, activities regarding the administration, financial aspects and communication. There are several categorizations under HIS that are totally combined in managing the hospital administration and patient care. In this system, healthcare data storage with the information analysis is done systematically. These data collected by the HIS is classified into three main categories: 1) clinical information that are done during the patients visit to the hospital 2) administrative information and 3) external information. These include the data regarding the health facilities, household surveys, and administrative data, national health accounts (NHA) and health researches and civil registrations. This could have access to the health information system by describing the situations and trends followed in the healthcare unit. Further, these situations could be thoroughly analysed and could be helpful in planning facility, co-ordination and in decision making.

One of the recent concepts among the decision making involves the use of information systems since this information system increases the facility in the exchange of data among the users. The information system is used by the people and the organization to organize process, collect and distribute data. Through

Figure 6. Interaction among the healthcare actors and with the healthcare information system (HIS)

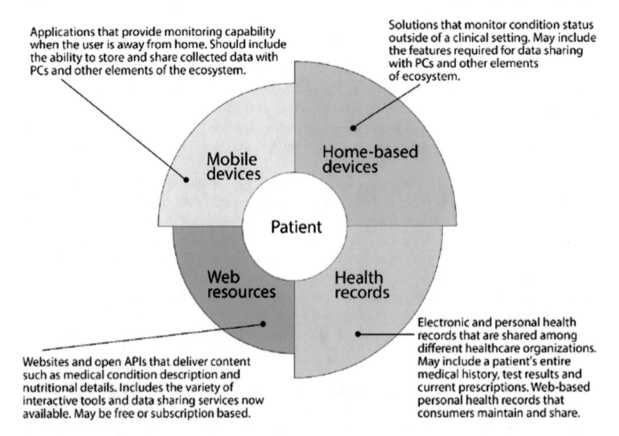

this process, quality information could be received that helps in taking the rational decisions in order to meet the needs of the customers. Rational decision making is said to be a crucial process unlike several other sectors, which offers certain provisions to the community as shown in figure 7.

Moreover, fast increase in the medical data leads to the categorization of healthcare from the discovery of drugs, evolution and increase in spreading of diseases, medical histories, exposure of patients to the environment, etc. This fast growing data is necessary to have an effective management to increase the health service delivery quality and the increase in the medical data had led to several challenges that had been faced by the hospitals in managing the information systems, which is sufficiently necessary to utilize the medical data for the rational decision-making process as shown in figure 8.

A community-based smart rehabilitation system is employed for sharpening as well as rejuvenating the functional abilities which enhances the physiological care. These systems should interact conveniently apportioning of medical resources based on the requirements of the patient employing IoT based rehabilitation system with an ontology-based automating technique as shown in figure 9.

The pulse oximeter or oxygen saturation device is employed to monitor as well as analyze the patient's oxygen saturation which follows a non-invasive pattern based on metrics as shown in figure 9. These devices are expensive but after the advent of artificial intelligence and internet of things, most healthcare devices and applications (Wolterink et al. 2016).

Figure 7. Elements in a Smart healthcare Solutions

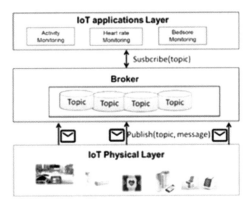

Figure 8. IoT Components

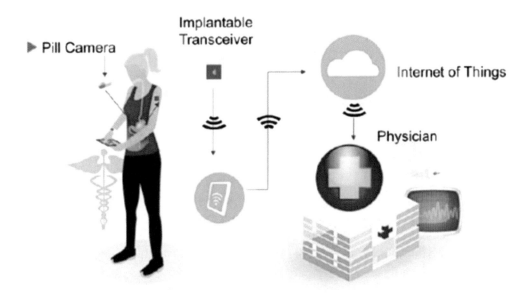

The recent success of these techniques led to major advancements in medical image processing applications as well as medical sensors for healthcare applications, for instance advance blood oxygen devices and heart rate monitoring devices are integrated with sensors and automated for effective healthcare management.

CONCLUSION

So far, it is said that the combination of healthcare imaging with the information system falls under the category of healthcare information system. Moreover, at the present scenario, certain domain such as

Figure 9. Internet of things in monitoring patients using sensors

Questions	Metrics
How many events are captured by the process?	• Total count of events registered by the system
How many patients are captured by the process?	• Total count of patients registered by the system
How many patients present events?	• Total count of patients with events
How many patients present AEs?	• Total count of patients with events categorized as AE
Out of the total of events, how many are AEs?	• Percentage of AEs over the total count of events. • Ratio of AEs over the total count of events • Total count of AE

Ubiquitous computing, Big Data, Cloud Computing, M-Commerce etc, plays a major role in this field. They form the integrated part of the healthcare organization systems. Various studies had been adapted regarding the healthcare information systems. Several researchers have followed the information system and they had developed this information system in many healthcare institutions and showed their benefits to the wide range of stake holders.

REFERENCES

Amini, N., Sarrafzadeh, M., Vahdatpour, A., & Xu, W. (2011). Accelerometer-based on-body sensor localization for health and medical monitoring applications. *Pervasive and Mobile Computing*, *7*(6), 746–760. doi:10.1016/j.pmcj.2011.09.002 PMID:22347840

Boland, G. W., Duszak, R. Jr, McGinty, G., & Allen, B. Jr. (2014). Delivery of appropriateness, quality, safety, efficiency and patient satisfaction. *Journal of the American College of Radiology*, *11*(1), 7–11. doi:10.1016/j.jacr.2013.07.016 PMID:24387963

Chen, L., Jones, A. L., Mair, G., Patel, R., Gontsarova, A., Ganesalingam, J., & ... . (2018). Rapid automated quantification of cerebral leukoaraiosis on CT images: A multicenter validation study. *Radiology*, *288*(2), 573–581. doi:10.1148/radiol.2018171567 PMID:29762091

Chopra, S., & White, L. F. (2011). *A legal theory for autonomous artificial agents*. The University of Michigan Press. doi:10.3998/mpub.356801

Giliker, P. (2011). Vicarious liability or liability for the acts of others in tort: A comparative perspective. *J Eur Tort Law.*, *2*(1), 31–56. doi:10.1515/jetl.2011.31

Holzinger, A. (2016). Project iML – proof of concept interactive machine learning. *Brain Inform*, *3*(2), 119–131.

Sharma, P., Schwemmer, C., Persson, A., Schoepf, U. J., Kepka, C., Hyun Yang, D., & Nieman, K. (2018). Diagnostic accuracy of a machine-learning approach to coronary computed tomographic angiography-based fractional flow reserve: Result from the MACHINE consortium. *Circulation: Cardiovascular Imaging*, *11*, e007217.

Udendhran, R. (2017). A Hybrid Approach to Enhance Data Security in Cloud Storage. *ICC '17 Proceedings of the Second International Conference on Internet of things and Cloud Computing at Cambridge University, United Kingdom.* Doi:10.1145/3018896.3025138

Udendhran, R., & Balamurgan, M. (2020). An Effective Hybridized Classifier Integrated with Homomorphic Encryption to Enhance Big Data Security. In *EAI International Conference on Big Data Innovation for Sustainable Cognitive Computing*. EAI/Springer Innovations in Communication and Computing. 10.1007/978-3-030-19562-5_35

ADDITIONAL READING

Ben-David, A. (1995). Monotonicity maintenance in information-theoretic machine learning algorithms. *Machine Learning*, *19*(1), 29–43. doi:10.1007/BF00994659

Bernardi, E., Carlucci, S., Cornaro, C., & Bohne, R. (2017). An analysis of the most adopted rating systems for assessing the environmental impact of buildings. *Sustainability*, *9*(7), 1226. doi:10.3390u9071226

Ghaffarianhoseini, A., Berardi, U., AlWaer, H., Chang, S., Halawa, E., Ghaffarianhoseini, A., & Clements-Croome, D. (2016). What is an intelligent building? Analysis of recent interpretations from an international perspective. *Architectural Science Review*, *59*(5), 338–357. doi:10.1080/00038628.2015.1079164

KEY TERMS AND DEFINITIONS

Data Supplier and Data Miner: Internet of things is said to be producer of data while machine learning is said to be data miner since it processes and analyzes the data produced from internet of things.

Intelligent Information System: An intelligent information system is said to be the set of software and hardware that involves the skilled people for the process of decision making and co-ordination among the organization.

Rational Decision Making: Rational decision making is said to be a crucial process unlike several other sectors, which offers certain provisions to the community.

Sensor-Cloud Infrastructure: Sensor-cloud infrastructure could be a technology that integrates cloud techniques into the WSNs. It provides users a virtual platform for utilizing the physical sensors in a clear and convenient approach.

Ubiquitous Computing (UbiComp): Ubiquitous computing (UbiComp) is characterized by the use of small, networked and portable computer products in the form of smart phones, personal digital assistants and embedded computers built into many devices, resulting in a world in which each person owns and uses many computers.

Chapter 14
The Pivotal Role of Edge Computing With Machine Learning and Its Impact on Healthcare

Muthukumari S. M.
Bharathidasan University, India

George Dharma Prakash E. Raj
Bharathidasan University, India

ABSTRACT

The global market for IoT medical devices is expected to hit a peak of 500 billion by the year 2025, which could signal a significant paradigm shift in healthcare technology. This is possible due to the on-premises data centers or the cloud. Cloud computing and the internet of things (IoT) are the two technologies that have an explicit impact on our day-to-day living. These two technologies combined together are referred to as CloudIoT, which deals with several sectors including healthcare, agriculture, surveillance systems, etc. Therefore, the emergence of edge computing was required, which could reduce the network latency by pushing the computation to the "edge of the network." Several concerns such as power consumption, real-time responses, and bandwidth consumption cost could also be addressed by edge computing. In the present situation, patient health data could be regularly monitored by certain wearable devices known as the smart telehealth systems that send an enormous amount of data through the wireless sensor network (WSN).

INTRODUCTION

The large amount of generated healthcare data could be stored in the cloud environment maintaining their privacy that could be recommended to be viewed only by the concerned users. The physiological data could be obtained by the combined machine learning and signal processing modules in the

DOI: 10.4018/978-1-7998-3591-2.ch014

traditional telecare systems. Machine learning algorithms are designed to run on the powerful servers. The connectivity problem faced by the IoT devices could be answered by the edge computing thus maintaining its efficiency even though the connectivity is poor. Several benefits are impacted with the edge computing technology (Ali Hassan Sodhro et al. 2018). First, the workload could be reduced for the medical experts by prioritizing the important task at the top level thereby managing the collected data. The second will be enabling the healthcare facility all over the remote area, where these facilities falls behind. For instance, a truck outfitted with the edge computing devices could visit certain remote villages and promote healthcare facilities by connecting the patient to the telehealth services (*A. Tanaka et al.* 2016). The third advantage is to stimulate the advancement of medical technology. Opportunities will be more with the enormous collection of data. Edge computing involves in managing and labeling the data in an efficient and uniform way that makes the data to be shared securely making the process easier (Á. Alesanco and J. García, 2010). Mining of data could also be made simple for the researchers, which was said to the difficult task in the past era. The greatest impact of edge computing will be the treatment of chronic diseases by monitoring the collected data (A. Monteiro et al. 2016). Certain disorders such as diabetics and congestive heart problems could be monitored continously with the combination of 5G cellular networks along with the IoT by in-home monitoring of patients. This chapter presents the state-of-the-art edge computing and machine learning techniques, which can enhance the biomedical applications (Bellman R. 1957).

BACKGROUND

The development in the Internet of Things (IoT) and mobile usage has found a profound impact in the enhancement of cloud storage. Hyperscale data centers had been created due to the accelerated data center traffic that has been created by the increased usage of cloud computing environment (B. M. C. Silva et al. 2015). The data storage capacity should increase significantly due to the growth of IoT based application, which plays a major role in Smart city, Healthcare and Industries. According to the prediction done by the Cisco Global Cloud Index, the workload created by the cloud environment will be increased by 94% and the traditional data center will be just 6% by the year 2021(B. Mei et al. 2015). Certain impacts limit the blooming cloud infrastructure and reduce its flexibility that is as follows:

1. **Latency/Determinism:** The time delay created among the interaction of IoT devices and the cloud is known as latency. Certain industries like Electronics Health Records (EHR) and Telemedicine has a major concern with the latency requirements and this impact must be concerned essentially.
2. **Data/Bandwidth:** According to Statistics, the installation of IoT devices could increase by 31 billion worldwide by the year 2020. In generating the medical records, at least 15-20 devices should be connected at each patient's bed, which increases the data rate. This causes a limitation in the network bandwidth, overpowering the cloud and also increases data traffic.
3. **Privacy/Security:** According to IBM data breach study, the security and the privacy of each patient's data in the healthcare line could be affected due to the increased data breach cost, which will be about $480 per patient and this cost could be three times the data cost of other industries.

Edge computing is highly advantageous as the data could be processed and filtered, thereby sending only the useful data to the cloud (C. Orwat et al. 2008). This facility brings the cloud computing to the

EDGE of the network. The drawbacks mentioned above in the cloud infrastructure will be eradicated by the edge computing. According to a study conducted by Grand View Research, Inc., the global edge computing market size is estimated to reach USD 3.24 billion by the year 2025. CloudIoT makes cognitive healthcare framework possible along 5G (R. Dawson and P.W Lavori, 2004). This facility finds a major impact in the medical application, reaching human lives anywhere at the world. The development of Smart cities promotes cutting-edge technologies in the medical line and in fact emotion recognition module embedded into a framework is also made possible (Dr. S. Mohan Kumar and Darpan Majumdar, 2018).

EDGE COMPUTING IN SMART HEALTHCARE SYSTEMS

All IoT devices and wireless devices like Wearable Sensors, Hospital Monitors are connected to the edge nodes with low latency and some other devices directly connected to the cloud which leads high latency (E. Topol, 2013). Emergency Alert and medical Storage Files are connected to the Cloud and all these are comes under Edge of Network as shown in the above figure 1.

Figure 1. Edge Computing in Smart Healthcare Systems

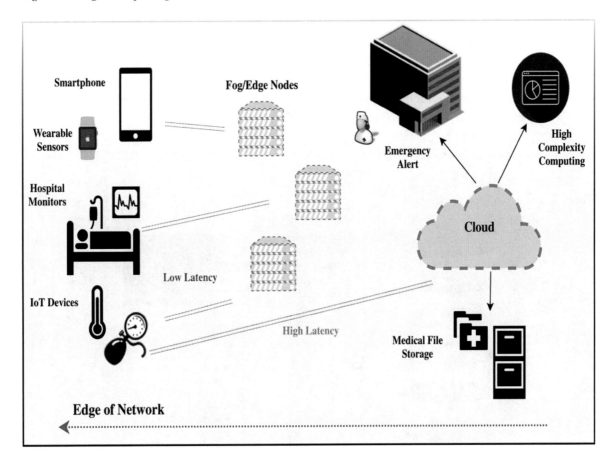

INTEGRATING EDGE COMPUTING AND MACHINE LEARNING FOR HEALTH CARE APPLICATION

In hospitals, a cloud infrastructure of Azure IoT Edge gateway (Linux server) could be registered with the Azure IoT Hub. Pipeline could be configured in the form of JSON file in the IoT Hub. For instance, the communication among the devices could be done with the data ingest container and the ML model shows the output (E. MacIntosh et al. 2016). The right container will be chosen from the container register that is handled by the edge devices, which is deployed by the JSON edge configuration file. Apple Company also introduces their version of AI integrated smart phone in iPhoneX with the new A11 Bionic chip (E. A. Oladimeji et al. 2011). Probably, many smart phones had been implemented with the AI technology, especially for the speech recognition, which enables digital assistants.

Moreover, this technology had been empowered by the cloud environment. In future, neural engine with several machine learning algorithms will be implemented in the smart phones and this heads the industries to the higher level (Frank Alexander Kraamer et al. 2017). Asserts leading to the misuse of cloud should be considered at the premature state. However, AI leads the way and has a real impact on everyday's life. Increase in the process tends to adapt the latest technology and this could also lead to cost savings with the role of edge technology with the decreased latency (G. J. Mandellos et al. 2009).

In healthcare application, a good IoT framework should intelligently prioritize and use the network resources with the trusted and secure channel. This could be possible by effectively preprocessing the input data retrieved from the sensor nodes. This could be done by connecting the leaf devices with the cloud servers at the back-end. These cloud servers has the capability to access heavy computational resources from which the incoming data could be prioritized and sends the information to the end point device.

The signatures could be extracted from the incoming data with the data mining and machine learning concepts applied at the back-end. Thus with the captured data, the medical interpretation could be made. The health condition data of the patient will be displayed at the front-end. In this concept, edge computing plays a role in avoiding the high latencies of decisions in cloud.

The computations will be done at IoT or at the "edge" of the network rather than making every decision at the cloud environment. Several sensor nodes such as ECG, EMG, EEG, Temperature and SPO2

Figure 2. IoT model for health care

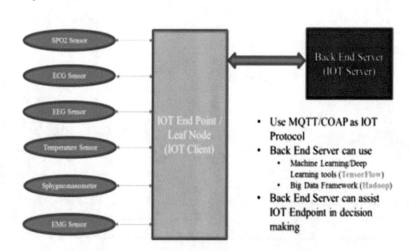

sensors could be integrated with the Leaf Node or IoT End Point that in turn communicates with the back-end server, which is shown in the figure 1. IoT protocols such as MQTT, CoAP are used in the communication of IoT End Point with the back-end server. The backend server helps in the performance of intensive tasks computation.

The main focus will be with the edge computing that could send notifications to the IoT End Point, which is managed by the MQTT client. On the other hand, the cloud environment supports the MQTT server, thereby creating a transport protocol that leads to the IoT End point and the data server communication. As mentioned earlier, for the real-time data analysis, cloud computing will not the sufficient environment due to the high network latencies (Anagnostopoulos et al. 2015).

This extends the idea to the edge computing architecture, where the edge point device guides the computation of the endpoint device through the data collected from the cloud servers. The bill gets fixed by the MQTT protocol. Instead of IoT protocol, CoAP could also be used in this process. The main role of this infrastructure is to combine the machine learning concepts along with the IoT and cloud and thereby concentrating on the edge computing techniques in the healthcare domain.

DEMYSTIFYING IMAGE PROCESSING IN HEALTHCARE

Image processing concept plays a major role in the healthcare industries. Strategies based on the Adaptive intervention had found its way in detecting breast cancer with the mammography screening test and also with the treatment of AIDS (J. Sametinger et al. 2015). But there is a scarcity of randomized trials that have employed dynamic treatment protocols, due to the historical lack of theory for the design and analysis of such a trial. The issues of sample size calculations with the randomization have been dealt recently with the better insights. The Bayesian framework discussed in deals with the multicenter design considerations and adaptive randomization of sequentially randomized trials (S.E Kahou et al. 2016). Sequentially randomized trials within the person mental illness and its theoretical innovations are dealt in and regarding cancer. With certain trials, these protocols are not considered to be sequential in nature but are rather considered to be a separate trial. Sequential decision problem are handled with the conventional backwards induction method, which is also known as Dynamic Programming (K. Wac et al. 2009). All the covariates of longitudinal distribution should be necessarily modeled in the dynamic regimes context. Treatment could be recommended incorrectly with the misspecification of the distribution due to the limited knowledge of this model. Methods do not have any such type of limitations. The time-varying covariates are adjusted with the optimal treatment discontinuation time through causal approaching. Analyzing sequentially randomized trials has been done in with the likelihood-based approach for the optimal regimes prostate cancer regions (K. Xu et al. 2016). Here the probabilities are taken from the information collected from the patient.

RECENT ADVANCEMENT IN MACHINE LEARNING

Figure 3. Applications of Machine Learning in Healthcare

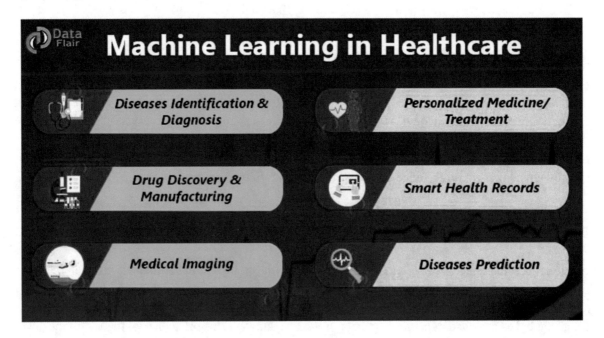

DISEASE IDENTIFICATION AND DIAGNOSIS

The major role played by the machine learning is to detect the disease through the image processing techniques, which could be very hard through the normal diagnosis. This disease could be anything like genetic disorders or cancers that is very difficult to be diagnosed at a very initial stage (K. Zhang et al. 2016).

Integration of cognitive computing with genome-based tumor sequencing done at IBM Watson Genomics is the very best example through which fastest diagnosis could be done. Therapeutic treatments are done with the AI based bio pharma followed by Berg in oncology (P.W Lavori, 2000). Moreover P1vital's PReDicT (Predicting Response to Depression Treatment) aims in promoting normal clinical conditions for the disease diagnosis and treatment.

DRUG DISCOVERY AND MANUFACTURING

Drug discovery process at the early-stage is one of the primary applications of machine learning. Precision medicine and Next-generation sequencing are certain R&D technologies included with this technique that could determine another alternative approach for the treatment of multiple diseases (L. M. Vaquero and L. Rodero-Merino, 2014). Unsupervised learning is a technique of machine learning that could identify patterns in the given data without providing any predicted output. Microsoft developed project Hanover that uses certain machine learning techniques with the AI technologies, which takes multiple initiatives in treating cancer as well as prescribing drugs for AML (Acute Myeloid Leukemia) (L. Catarinucci, 2016).

DIAGNOSING MEDICAL IMAGES

Computer vision is said to be a successful technology that works with the combination of Deep learning and machine learning. Image analysis had been done in Microsoft with the initiative technique called the Inner Eye mainly developed for the diagnosis of diseases with the images (S.A. Murphy, 2004). Due to the huge growth of medical data, machine learning plays its role in diagnosing the disease from the medical images with the role of AI.

SMART HEALTH RECORDS

Due to the update of technology, the maintenance of up-to-date medical reports could be highly possible with the special algorithms implemented for the data entry process. But still medical data entry takes time thereby increasing complexity (R. Bellman, 1957). This could be highly reduced with the machine learning techniques thereby producing high efficiency, saving cost, reducing computational time and effort.

The techniques such as ML-based OCR recognition and vector machines steps into the document classification methods and finds its place in machine learning-based handwriting recognition technology in MATLAB and Google's Cloud Vision API (G. Muhammad et al. 2014). MIT advances their development process to the next generation by stepping into the machine learning based tools that helps in the suggesting the clinical treatment with the diagnosis.

CROWD SOURCED DATA COLLECTION

In the healthcare field, crowd sourcing is found to be a rage due to the numerous amounts of medical data uploaded by the people and that helps the researchers and caretakers with the plenty of information. This type of real-time health data produces high consequences in such a way medicine will be apparent down the line (M. S. Hossain and G. Muhammad, 2015). Machine learning based face recognition had been developed in Apple's Research kit that helps with the treatment of Asperger's and Parkinson's disease.

IBM had made a partnership with the Medtronic for obtaining a real-time diabetes and insulin data with the crowd sourced information. Due to the development of IoT and mobile technologies, several researches had been carried out with the obtained crowd sourced data and also enhanced is done with the tough-to-diagnose cases data.

OUTBREAK PREDICTION

The combination of machine learning along with the AI based techniques had been implemented today for the epidemics prediction all around the world. Numerous data had been collected from the satellites, website information, social media etc, and researchers and scientist also have huge access to them all. From malaria till the impact of certain chronic disease identification and prediction could be done by the artificial neural networks through the obtained information.

This is extremely necessary for third-world countries that lacks with the medical and educational systems. ProMED-mail is the best example of this type, which is an reporting platform based n the internet that monitors the emerging diseases and the evolving ones and also promotes data at a real-time as shown in the figure 4.

Figure 4. Fog Computing in Healthcare IoT

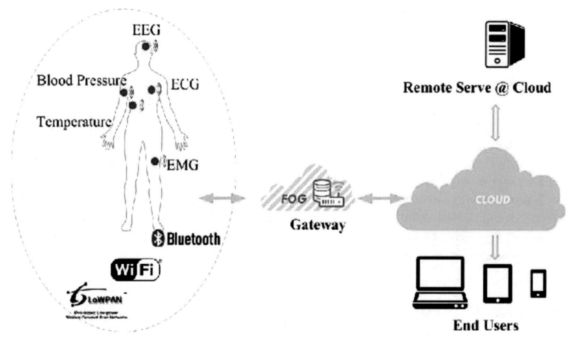

SMART e-HEALTH GATEWAYS AT THE EDGE OF HEALTHCARE MONITORING SYSTEMS

In Smart e-Health Gateways healthcare monitoring systems the Sensor Network monitor the patient body temperature, Blood pressure, Heart rate and monitor the sleep time of the patient. These sensed data's are sent to the gateway. The Gateway is then process the data and do data filtering and mining, give location notification of the patient and these processed data's are sent to Internet for data storage, data analytics, decision making and medical caregivers interface as shown in the figure

COMPONENTS OF IoT BASED HEALTHCARE MONITORING SYSYTEM

Here the Medical Devices and Hospital Rooms are monitored by Smart e-Health Gateways. And the Smart e-Health Gateways are fixed at each rooms by connecting sensors and formed Mesh-based Wireless Sensor Network which is used to sense the data and fetch information from the patients is shown in figure 6.

Figure 5. Smart e-Health Gateways at the Edge

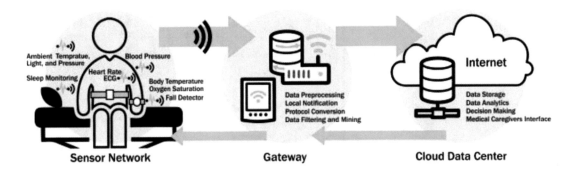

Table 1. Merits and Demerits of Edge Computing in Healthcare

Merits	Demerits
Provides telemedicine and Remote patient monitoring.	Need more security measurements to keep the patient health record safely.
Machine-to-machine communication or machine –to-human interaction.	Remote health monitoring only in smart cities.
It delivers applications and services to remote area by the use of data center locations.	Should increase the bandwidth capacity to the data center for fastest delivery of the medical services.
Smart sensors monitors the ill patients accurately.	Sometimes the range of the sensor for sensing the patient become collapsed due to long distance.
Virtualization technology brings the patients closer to the doctors.	Virtualization monitoring needs to bring more attention to the patient.
Telemedicine programs brought special care for the patients.	All suburban areas needs the telemedicine program.
Simple and Flexible to use	Limited number of services
It is powerful to use	Sometimes fails practically
Highly efficient and secure healthcare monitoring	Third party auditor requires for do operation
Very easy to use and secure	There are some challenges to secure healthcare monitoring
Data can be maintained privately	Not to resolve all issues
Easy Monitoring system	Sometimes it fails

Figure 6. IoT Based Healthcare Components

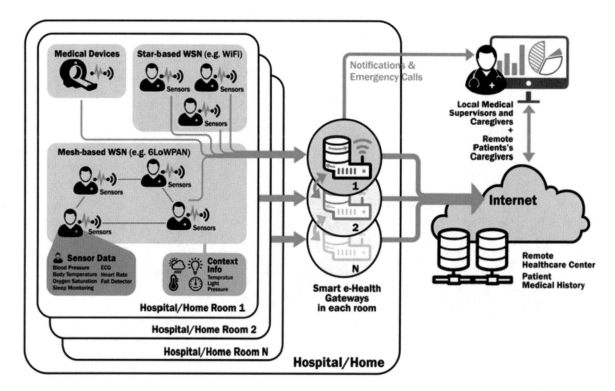

EDGE COMPUTING BASED APPLICATION FOR HEALTHCARE IMAGE PROCESSING

The integration of several smart systems along with the smart healthcare falls into the category of smart cities. This successful endeavor is possible only with the combination of cloud computing along with the edge computing process. The resources and the process are assigned with the proposed healthcare system that comes out with the perfect workflow and pool of resources. The completion of the assigned task will be notified by its own edge nodes for every resource. Then the cloud environment plays a role in reassigning the resources with the scheduling algorithm. Rather than the regular client-server computing, cloud computing is found to be more effective, reliable and it is also said to be faster. Cloud also supports the fault tolerance policies and hence the workflow will also be ensured.

The cloud-fog based workflow system had been presented in figure 7. The software related to the workflow and the databases are stored in the cloud and on the other hand assigning the task with the completion of task notification will be send from the edge node to the cloud infrastructure. Tablet and cell phones acts on the edge node, which are said to be the resources of the smart devices. When the task assigned is accomplished by the resource then the respective notification will be send by the edge node to the cloud that the particular resource is available at the pool for it to be re-assigned.

The communication mode among the medical team and the cloud in the base station is shown in the above figure. This communication is said to be mobile and wireless. In this level, the edge and fog are found to be alike. The fog and the edge computing push the computing and processing capabilities closer to the originated place of the data. In the above framework, the data is found to be closer to the smart

Figure 7. Cloud-fog based workflow system

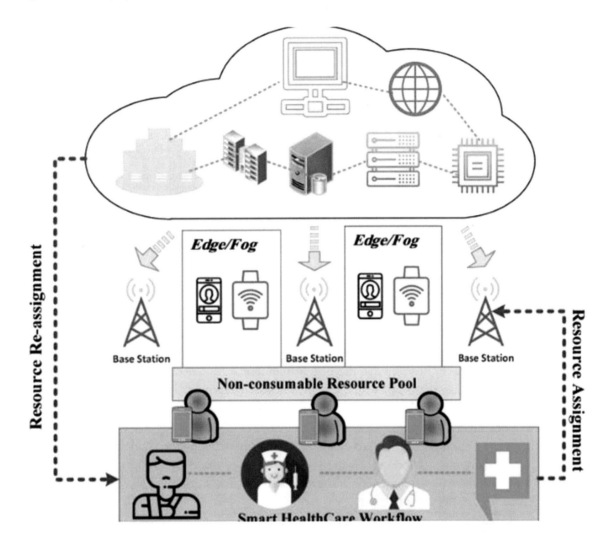

healthcare workflow. The resources are categorized as consumable and non-consumable according to the above framework. The resources is said to be consumable if it expires by time; else it is said to be non-consumable. Non-human resources such as equipments, machines etc, are said to be consumable resources. On the other hand, humans are the best example of non-consumable resources.

The process carried out in both the cloud and the edge mode is presented in figure 8. In the cloud mode, scheduling occurs and all the assignments in done at the edge mode.

CONCLUSION

Fog severs as a new architecture for control, computing, storage and networking that brings the end users very closer to the technologies. Moreover, decentralization is also achieved at the edge of the net-

Figure 8. Combination of cloud and edge computing for task scheduling

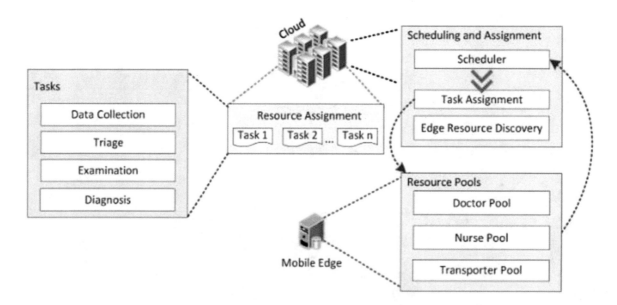

work. Due to the increasing use of wearables (sensor devices), enormous amount of data is generated. The data related to healthcare could be in the form of numerical format or images. In case of images, the affected regions are identified and with the help of machine learning algorithms, classification and preprocessing is done based on the types of diseases. With the help of the clinical procedures, the data gets assisted by the cloud and the fog services. This large pool of data serves as a huge benefit for the IoT healthcare. Egde computing techniques could be involved in reducing the latencies produced by the cloud computing environment. Moreover, with the combination of these technologies, healthcare could be highly benefitted, thereby reaching people at every corner of the world.

REFERENCES

Alemdar, H., & Ersoy, C. (2010). Wireless sensor networks for healthcare: A survey. *Computer Networks*, *54*(15), 2688–2710. doi:10.1016/j.comnet.2010.05.003

Alesanco, Á., & García, J. (2010). Clinical assessment of wireless ECG transmission in real-time cardiac telemonitoring. *IEEE Transactions on Information Technology in Biomedicine*, *14*(5), 1144–1152. doi:10.1109/TITB.2010.2047650 PMID:20378476

Anagnostopoulos, C. N., & Giannoukos, T. I. (2015). Features and classifiers for emotion recognition from speech: A survey from 2000 to 2011. *Artificial Intelligence Review*, *43*(2), 155–177. doi:10.100710462-012-9368-5

Bellman, R. (1957). *Dynamic Programming*. Princeton University Press.

Bertini, M., Marcantoni, L., Toselli, T., & Ferrari, R. (2016). Remote monitoring of implantable devices: Should we continue to ignore it? *International Journal of Cardiology*, *202*, 368–377. doi:10.1016/j.ijcard.2015.09.033 PMID:26432486

Cao, Y., Chen, S., Hou, P., & Brown, D. (2015). FAST: A fog computing assisted distributed analytics system to monitor fall for stroke mitigation. *Proc. IEEE Int. Conf. Netw., Archit. Storage (NAS)*, 2–11.

Catarinucci, L., de Donno, D., Mainetti, L., Palano, L., Patrono, L., Stefanizzi, M. L., & Tarricone, L. (2016). An IoT-aware architecture for smart healthcare systems. *IEEE Internet Things J.*, *2*(6), 515–526. doi:10.1109/JIOT.2015.2417684

I. Corporation. (2016). *Bigger data for better healthcare*. Tech. Rep.

Craciunescu, R., Mihovska, A., Mihaylov, M., Kyriazakos, S., Prasad, R., & Halunga, S. (2015). Implementation of Fog computing for reliable E-health applications. *Proc. 49th Asilomar Conf. Signals, Syst. Comput.*, 459–463. 10.1109/ACSSC.2015.7421170

Dawson, R., & Lavori, P. W. (2004). Placebo-free designs for evaluating new mental health treatments: The use of adaptive treatment strategies. *Statistics in Medicine*, *23*(21), 3249–3262. doi:10.1002im.1920 PMID:15490427

Digital Pills Make Their Way to Market. (2012). Available: http://blogs.nature.com/news/2012/07/digital-pills-make-their-wayto-market.html

(2018). Dr. S. Mohan Kumar and Darpan Majumdar (2018). Healthcare Solution Based on machine Learning Applications in IoT and Edge Computing. *International Journal of Pure and Applied Mathematics. Volume*, *16*(119), 1473–1484.

Eide, R. (2016). *Low energy wireless ECG: An exploration of wireless electrocardiography and the utilization of low energy sensors for clinical ambulatory patient monitoring* (M.S. thesis). Dept. Comput. Inf. Sci., Norwegian Univ. Sci. Technol., Trondheim, Norway.

Farandos, N. M., Yetisen, A. K., Monteiro, M. J., Lowe, C. R., & Yun, S. H. (2015). Contact lens sensors in ocular diagnostics. *Advanced Healthcare Materials*, *4*(6), 792–810. doi:10.1002/adhm.201400504 PMID:25400274

Gia, T. N., Jiang, M., Rahmani, A.-M., Westerlund, T., Liljeberg, P., & Tenhunen, H. (2015). Fog computing in healthcare Internet of Things: A case study on ECG feature extraction. *Proc. IEEE Int. Conf. Comput. Inf. Technol., Ubiquitous Comput. Commun., Dependable, Auto. Secur. Comput., Pervasive Intell. Comput. (CIT/IUCC/DASC/PICOM)*, 356–363. 10.1109/CIT/IUCC/DASC/PICOM.2015.51

Hossain, M. S., & Muhammad, G. (2015). Cloud-assisted speech and face recognition framework for health monitoring. *Mobile Networks and Applications*, *20*(3), 391–399. doi:10.100711036-015-0586-3

Hossain, M. S., & Muhammad, G. (2016). Cloud-assisted Industrial Internet of Things (IIoT) - enabled framework for health monitoring. *Computer Networks*, *101*, 192–202. doi:10.1016/j.comnet.2016.01.009

Hossain, M. S., Muhammad, G., Al-Qurishi, M., Masud, M., Almogren, A., Abdul, W., & Alamri, A. (2017). Cloud-Oriented Emotion Feedback-based Exergames Framework. *Multimedia Tools and Applications*. Advance online publication. doi:10.100711042-017-4621-1

Hossain, M. S., Muhammad, G., Alhamid, M. F., Song, B., & Al-Mutib, K. (2016). Audio-Visual Emotion Recognition Using Big Data Towards 5G. *Mobile Networks and Applications*, *221*(5), 753–763. doi:10.100711036-016-0685-9

Hosseini, M.-P., Hajisami, A., & Pompili, D. (2016). Real-time epileptic seizure detection from eeg signals via random subspace ensemble learning. *Proc. IEEE Int. Conf. Auto. Comput. (ICAC)*, 209–218. 10.1109/ICAC.2016.57

Jenke, R., Peer, A., & Buss, M. (2014). Feature Extraction and Selection for Emotion Recognition from EEG. *IEEE Transactions on Affective Computing*, *5*(3), 327–339. doi:10.1109/TAFFC.2014.2339834

Kahou, S. E., Bouthillier, X., Lamblin, P., Gulcehre, C., Michalski, V., Konda, K., Jean, S., Froumenty, P., Dauphin, Y., Boulanger-Lewandowski, N., Chandias Ferrari, R., Mirza, M., Warde-Farley, D., Courville, A., Vincent, P., Memisevic, R., Pal, C., & Bengio, Y. (2016, June). EmoNets: Multimodal deep learning approaches for emotion recognition in video. *Journal on Multimodal User Interfaces*, *10*(2), 99–111. doi:10.100712193-015-0195-2

Khushaba, R. N., Kodagoda, S., Takruri, M., & Dissanayake, G. (2012). Toward improved control of prosthetic fingers using surface electromyogram (EMG) signals. *Expert Systems with Applications*, *39*(12), 10731–10738. doi:10.1016/j.eswa.2012.02.192

Kim, Y., Lee, H., & Provost, E. M. (2013). Deep learning for robust feature generation in audiovisual emotion recognition. *2013 IEEE International Conference on Acoustics, Speech and Signal Processing*, 3687-3691. 10.1109/ICASSP.2013.6638346

Kraamer, B., Tamkittikhun, & Palma. (2017). Fog Computing in Healthcare – A Review and Discussion. *IEEE Access : Practical Innovations, Open Solutions*; Advance online publication. doi:10.1109/Access.2017.2704100

Lavori, P. W., & Dawson, R. (2000). A design for testing clinical strategies: Biased adaptive within-subject randomization. *Journal of the Royal Statistical Society. Series A (General)*, *163*(1), 29–38. doi:10.1111/1467-985X.00154

López, G., Custodio, V., & Moreno, J. I. (2016). LOBIN: E-textile and wirelesssensor-network-based platform for healthcare monitoring in future hospital environments. *IEEE Transactions on Information Technology in Biomedicine*, *14*(6), 1446–1458. doi:10.1109/TITB.2010.2058812 PMID:20643610

MacIntosh, E., Rajakulendran, N., Khayat, Z., & Wise, A. (2016). *Transforming Health: Shifting From Reactive to Proactive and Predictive Care*. Available: https://www.marsdd.com/newsand-insights/transforming-health-shifting-from-reactive-to-proactive-andpredictive-care/

Mandellos, G. J., Koutelakis, G. V., Panagiotakopoulos, T. C., Koukias, M. N., & Lymberopoulos, D. K. (2009). Requirements and solutions for advanced telemedicine applications. In *Biomedical Engineering*. InTech.

Masip-Bruin, X., Marín-Tordera, E., Alonso, A., & Garcia, J. (2016). Fog-tocloud computing (F2C): The key technology enabler for dependable ehealth services deployment. *Proc. Medit. Ad Hoc Netw. Workshop (Med-Hoc-Net)*, 1–5.

Mei, B., Cheng, W., & Cheng, X. (2015). Fog computing based ultraviolet radiation measurement via smartphones. *Proc. 3rd IEEE Workshop Hot Topics Web Syst. Technol. (HotWeb)*, 79–84.

Monteiro, A., Dubey, H., Mahler, L., Yang, Q., & Mankodiya, K. (2016). Fit: A fog computing device for speech tele-treatments. *Proc. IEEE Int. Conf. Smart Computing*, 1–3. 10.1109/SMART-COMP.2016.7501692

Muhammad, G., Alsulaiman, M., Amin, S. U., Ghoneim, A., & Alhamid, M. (2017). A Facial-Expression Monitoring System for Improved Healthcare in Smart Cities. *IEEE Access: Practical Innovations, Open Solutions*, 5(1), 10871–10881. doi:10.1109/ACCESS.2017.2712788

Murphy, S. A. (2003). Optimal dynamic treatment regimes (with discussion). *Journal of the Royal Statistical Society. Series B. Methodological*, 65(2), 331–366. doi:10.1111/1467-9868.00389

Murphy, S. A. (2004). An experimental design for the development of adaptive treatment strategies. *Statistics in Medicine*, 24(10), 1455–1481. doi:10.1002im.2022 PMID:15586395

Nazir, Ali, Ullah, & Garcia-Magarino. (2019). Internet of Things for Healthcare Using Wireless Communications or Mobile Computing. *Wireless Communications and Mobile Computing*. Doi:10.1155/2019/5931315

Obermeyer, Z., & Emanuel, E. J. (2016). Predicting the future—Big data, machine learning, and clinical medicine. *The New England Journal of Medicine*, 375(13), 1216–1219. doi:10.1056/NEJMp1606181 PMID:27682033

Oladimeji, E. A., Chung, L., Jung, H. T., & Kim, J. (2011). Managing security and privacy in ubiquitous ehealth information interchange. *Proc. 5th Int. Conf. Ubiquitous Inf. Manage. Commun. (ICUIMC)*, 26:1–26:10. Available: https://doi.acm.org/10.1145/1968613.1968645

Orwat, C., Graefe, A., & Faulwasser, T. (2008). Towards pervasive computing in health care—A literature review. *BMC Medical Informatics and Decision Making*, 8(1), 26. doi:10.1186/1472-6947-8-26 PMID:18565221

Oueida, Kotb, Aloqaily, & Jararweh. (2018). An Edge Computing Based Smart Healthcare Framework For Resource Management. *Sensors, 18*(12), 4307.

Paksuniemi, M., Sorvoja, H., Alasaarela, E., & Myllyla, R. (2005). Wireless sensor and data transmission needs and technologies for patient monitoring in the operating room and intensive care unit. *Proc. 27th Annu. Int. Conf. Eng. Med. Biol. Soc. (IEEE-EMBS)*, 5182–5185. 10.1109/IEMBS.2005.1615645

Perera, C., Zaslavsky, A., Christen, P., & Georgakopoulos, D. (2014). Context aware computing for the internet of things: A survey. *IEEE Commun. Surveys Tuts.*, 16(1), 414–454. doi:10.1109/SURV.2013.042313.00197

'Philips' Intelligent Pill Targets Drug Development and Treatment for Digestive Tract Diseases. (2008). Available: https://phys.org/news/2008-11-philips-intelligent-pill-drug-treatment.html

Robins, J. M. (1994). Correcting for non-compliance in randomized trials using structural nested mean models. *Communications in Statistics*, 23(8), 2379–2412. doi:10.1080/03610929408831393

Sametinger, J., Rozenblit, J., Lysecky, R., & Ott, P. (2015). Security challenges for medical devices. *Communications of the ACM*, *58*(4), 74–82. doi:10.1145/2667218

Schneider, L. S., Tariot, P. N., Lyketsos, C. G., Dagerman, K. S., Davis, K. L., Davis, S., Hsiao, J. K., Jeste, D. V., Katz, I. R., Olin, J. T., Pollock, B. G., Rabins, P. V., Rosenheck, R. A., Small, G. W., Lebowitz, B., & Lieberman, J. A. (2001). National Institute of Mental Health Clinical Antipsychotic Trials of Intervention Effectiveness (CATIE): Alzheimer disease trial methodology. *American Journal of Geriatric Psychology*, *9*(4), 346–360. doi:10.1097/00019442-200111000-00004 PMID:11739062

Silva, B. M. C., Rodrigues, J. J. P. C., de la Torre Díez, I., López-Coronado, M., & Saleem, K. (2015). Mobile-health: A review of current state in 2015. *Journal of Biomedical Informatics*, *56*, 265–272. doi:10.1016/j.jbi.2015.06.003 PMID:26071682

Sodhro, Luo, & Arunkumar. (2018). Mobile Edge Computing Based QoS Optimization in Medical Healthcare Applications. *International Journal of Information Management*. DOI: .2018.08.004 doi:10.1016/j.ijinfomgt

Steele, R., & Lo, A. (2013). Telehealth and ubiquitous computing for bandwidthconstrained rural and remote areas. *Personal and Ubiquitous Computing*, *17*(3), 533–543. doi:10.100700779-012-0506-5

Tanaka, A., Utsunomiya, F., & Douseki, T. (2016). Wearable self-powered diaper-shaped urinary-incontinence sensor suppressing response-time variation with 0.3 V start-up converter. *IEEE Sensors Journal*, *16*(10), 3472–3479. doi:10.1109/JSEN.2015.2483900

Thall, P. F., Millikan, R. E., & Sung, H.-G. (2000). Evaluating multiple treatment courses in clinical trials. *Statistics in Medicine*, *19*(8), 1011–1028. doi:10.1002/(SICI)1097-0258(20000430)19:8<1011::AID-SIM414>3.0.CO;2-M PMID:10790677

Thall, P. F., & Wathen, J. K. (2005). Covariate-adjusted adaptive randomization in a sarcoma trial with multistate treatments. *Statistics in Medicine*, *24*(13), 1947–1964. doi:10.1002im.2077 PMID:15806621

The Cloud Standards Customer Council (CSCC). (2016). *Impact of cloud computing on healthcare*. Reference architecture, Version 1.0. Available: http://cloud-council.org

Topol, E. (2018). *The Creative Destruction of Medicine*. Basic Books.

Vaquero, L. M., & Rodero-Merino, L. (2014). Finding your way in the fog. *ACM SIGCOMM Comput. Commun. Rev.*, *44*(5), 27–32. doi:10.1145/2677046.2677052

Wac, K., Bargh, M. S., Beijnum, B. J. F. V., Bults, R. G. A., Pawar, P., & Peddemors, A. (2009). Power- and delay-awareness of health telemonitoring services: The mobihealth system case study. *IEEE Journal on Selected Areas in Communications*, *27*(4), 525–536. doi:10.1109/JSAC.2009.090514

Xu, K., Li, Y., & Ren, F. (2016). An energy-efficient compressive sensing framework incorporating online dictionary learning for long-term wireless health monitoring. *Proc. IEEE Int. Conf. Acoust., Speech Signal Process*, 804–808. 10.1109/ICASSP.2016.7471786

Yin, Y., Zeng, Y., Chen, X., & Fan, Y. (2016). The Internet of Things in healthcare: An overview. *J. Ind. Inf. Integr.*, *1*, 3–13. doi:10.1016/j.jii.2016.03.004

Yogesh, H., Ngadiran, A., Yaacob, B., & Polat, K. (2017). A new hybrid PSO assisted biogeography-based optimization for emotion and stress recognition from speech signal. *Expert Systems with Applications, 69*, 149–158. doi:10.1016/j.eswa.2016.10.035

Zhang, K., Liang, X., Baura, M., Lu, R., & Shen, X. (2014). PHDA: A priority based health data aggregation with privacy preservation for cloud assisted WBANs. *Inf. Sci., 284*, 130–141. doi:10.1016/j.ins.2014.06.011

Zhang, R., Bernhart, S., & Amft, O. (2016). Diet eyeglasses: Recognising food chewing using emg and smart eyeglasses. *Proc. IEEE 13th Int. Conf. Wearable Implant. Body Sensor Netw.*, 7–12. 10.1109/BSN.2016.7516224

ADDITIONAL READING

Chen, H., & Liu, H. (2016). A remote electrocardiogram monitoring system with good swiftness and high reliablility. *Computers & Electrical Engineering, 53*, 191–202. doi:10.1016/j.compeleceng.2016.02.004

Chen, M., Li, W., Hao, Y., Qian, Y., & Humar, I. (2018). Edge cognitive computing based smart healthcare system. *Journal of Future Generation Computer Systems, Elsevier, 86*, 403–411. doi:10.1016/j.future.2018.03.054

Huang, Y. M., Hsieh, M. Y., Chao, H. C., Hung, S. H., & Park, J. H. (2009). Pervasive, secure access to a hierarchical sensor-based healthcare monitoring architecture in wireless heterogeneous networks. *IEEE Journal on Selected Areas in Communications, 27*(4), 400–411. doi:10.1109/JSAC.2009.090505

Nejati, H., Pomponiu, V., Do, T.-T., Zhou, Y., Iravani, S., & Cheung, N.-M. (2016). Smartphone and mobile image processing for assisted living: Healthmonitoring apps powered by advanced mobile imaging algorithms. *IEEE Signal Processing Magazine, 33*(4), 30–48. doi:10.1109/MSP.2016.2549996

Øyri, K., Balasingham, I., Samset, E., Høgetveit, J. O., & Fosse, E. (2006). Wireless continuous arterial blood pressure monitoring during surgery: A pilot study. *Anesthesia and Analgesia, 102*(2), 478–483. doi:10.1213/01.ane.0000195232.11264.46 PMID:16428546

Pace, P., Aloi, G., & Caliciuri, G. (2018). *An edge-based architecture to support efficient applications for healthcare industry 4.0.* IEEE Transactions.

Sodhro, A. H., & Baik, S. W. (2019). Mobile edge computing based QoS optimization in medical healthcare applications. *International Journal of Information Management, Elsevier, 45*, 308–318. doi:10.1016/j.ijinfomgt.2018.08.004

Wang, H., Gong, J., & Zhuang, Y. (2017). Healthedge: Task scheduling for edge computing with health emergency and human behavior consideration in smart homes. *IEEE International Conference on Big Data.*

KEY TERMS AND DEFINITIONS

Cloud IoT: It integrates cloud computing and Internet of Things to alleviate the quality of service in healthcare organizations and improve the medical facilities in clinics. And do interaction among medical staff and general practitioners.

Edge Computing: Due to long distance causes number of risks factors. It may cause bandwidth congestion and network latency. To concern these things, foremost all healthcare organizations moves forward to edge computing, which analyze the data and send it to the nearby system situated at cloud.

Healthcare Applications: A good IoT framework should intelligently prioritize and use the network resources with the trusted and secure channel. This could be possible by effectively preprocessing the input data retrieved from the sensor nodes. This could be done by connecting the leaf devices with the cloud servers at the backend.

Image Processing: Image processing concept plays a major role in the healthcare industries. Strategies based on the Adaptive intervention had found its way in detecting breast cancer with the mammography screening test and also with the treatment of AIDS.

IoT Edge Gateway: It can generate and exchange data within single framework. It offered Remote patient monitoring system which reduces the time duration.

Machine Learning: The major role played by the machine learning is to detect the disease through the image processing techniques, which could be very hard through the normal diagnosis. This disease could be anything like genetic disorders or cancers that is very difficult to be diagnosed at a very initial stage.

Telehealth Service: For instance, a truck outfitted with the edge computing devices could visit certain remote villages and promote healthcare facilities by connecting the patient to the telehealth services.

Chapter 15
Neuro–Fuzzy–Based Smart Irrigation System and Multimodal Image Analysis in Static–Clustered Wireless Sensor Network for Marigold Crops

Karthick Raghunath K. M.

ⓘD https://orcid.org/0000-0002-6948-5240

Malla Reddy Institute of Engineering and Technology, India

Anantha Raman G. R.

Malla Reddy Institute of Engineering and Technology, India

ABSTRACT

As a decorative flower, marigolds have become one of the most attractive flowers, especially on the social and religious arena. Thus, this chapter reveals the potential positive resultants in the production of marigold through neuro-fuzzy-based smart irrigation technique in the static-clustered wireless sensor network. The entire system is sectionalized into clustering phase and operational phase. The clustering phase comprises three modules whereas the operational phase also includes three primary modules. The neuro-fuzzy term refers to a system that characterizes the structure of a fuzzy controller where the fuzzy sets and rules are adjusted using neural networks iteratively tuning techniques with input and output system data. The neuro-fuzzy system includes two distinct way of behavior. The vital concern of the system is to prevent unnecessary or unwarranted irrigation. Finally, on the utilization of multimodal image analysis and neuro-fuzzy methods, it is observed that the system reduces the overall utilization of water (~32-34%).

DOI: 10.4018/978-1-7998-3591-2.ch015

INTRODUCTION

In India, marigold is one of the most commonly grown flowers and used extensively during religious and social functions in various forms. Based on the cultivation report (Sujitha & Shanmugasundaram, 2017), the southern state namely Tamil Nadu which ranks number one among the marigold production, that covers 6733 hectare with an estimated production range of ~100995 tonnes annually, and the productivity ranges around 15 tonnes/hectare. Most of the time, the marigold flowers are intentionally grown as decorative crops that could be marketed as loose flowers or ornamental garlands for specialty usage. It is also one of the prominent instinctive origins of xanthophylls as an artificial additive to brighten poultry skin and egg yolks (Sujitha & Shanmugasundaram, 2017). Apart from these usages, the most part of the flowers also being used efficaciously for commercial dyestuff fabrics (Sujitha & Shanmugasundaram, 2017). High necessities of marigold flower either as a extracted products or as cut flower is always peak in most of the developed and developing countries (Italy, Taiwan, South Korea, Japan, UK, Mexico, Spain, United States). Hence, exportation of marigold flower plays a crucial role in the uplift of farmer's economy. Although the entire cultivation of flower is systemically exercised in an open field environment, still the irrigation process lacks the regulated and efficient utilization of water which is the firm demand for the raise in yield.

Smart irrigation is rightfully became a crucial practice of modern days agriculture. For majority of the cropping strategies especially in arid and semiarid regions require an efficient and systematical irrigation system to yield the maximum production, and to elevate the socio-economic condition. Beside linear-movable irrigation systems, Self-propelled centre pivot systems are commonly applied to irrigate the agri-field quite systematically; however, crucial variances in soil properties and water availableness exist all over the fields (Crow & Wood 2003). Automated irrigation systems are efficient to irrigate plants to an appropriate level for a normal growth of plants. Conventional method results in the reduction of productivity and energy.

The modern techniques of Neuro-fuzzy have found application in almost all the fields. The "Neuro-Fuzzy" term entails a kind of system characterized for a similar structure of a fuzzy controller where the fuzzy sets and rules are adjusted using neural networks tuning techniques in an iterative way with input and output system data. The system includes two distinct way of behavior. In the first phase called learning phase, it behaves like neural network that learns its internal parameter. In the second phase called the execution phase, it behaves like a fuzzy logic system.

Neural-fuzzy networks are often referred as connectionist models that are trained as neural networks, but their system is structurized as a fuzzy rule interpreter to obtain an unambiguous, crisp resultant. A Neuro-fuzzy inference system comprises a set of linguistic rules and an inference operation that are substantiated or aggregated with a connectionist framework for better adaption.

The neuro-fuzzy controller which is developed can determine the quantity of water required by plants in well defined depth using sensor nodes such as soil moisture sensor, humidity sensor and temperature sensor. Automatic irrigation scheduling system can be used to supply the water automatically based on the measurements specified through sensors. Sensors fixed in the field are used to ensure the water level in the soil. A solution to the problem of excessive supply of water can prevent damage of crops.

The combined analysis of multi-images of the same specimen adopted in various imaging modalities is typically used to acquire more selective information about the flower growth progress than is potentially analyzed from just a single modality. Thus, after the successive execution of operational phase, to analyze the growth progress of marigold flower; Non- Subsampled Contourlet Transform (NSCT) is employed for fusing multi-modal images (L. da Cunha et al., 2006).

The proposed neuro-fuzzy based smart irrigation technique in precision agriculture of marigold cultivation is addressed in Section 2 of this chapter. The complete evaluation of smart irrigation system based upon the multimodal image analyzing is also discussed in Section 3. Justifiable and reasonable conclusion and the future research work of this proposed technique is discoursed in the Section 4 of this chapter.

PROPOSED SMART IRRIGATION SYSTEM

To study the effect of smart irrigation system, the demonstration and research analysis was carried out in marigold crop field at Hosur region in Tamil Nadu state of India, which is located at the longitude of 77.8597148 and latitude of 12.7499988. The proposed smart irrigation system will collect temperature, soil moisture and humidity data from the marigold farming environment through deployed WSN in each hourly basis with respect to the current time stamp. Soil moisture evaporation rate depends on the above parameters. These parametric values are sensed by corresponding wireless sensors node and further the real-time data are processed through neuro-fuzzy system. Based on the final resultant of the process the irrigation procedures are carried out. The entire block diagram of the process is described in Section 3.1.

Process Flow of the System

The process flow of the proposed system is depicted in the figure 1, apart from that the internal process of the neuro-fuzzy schema is shown in figure 2. Initially, all the sensed data from the layer 1(temperature value, current soil moisture value, and present humidity value) were collected and provided as input parametric value to the Layer 2 part of neuro-fuzzy system for Fuzzification process. Then several fuzzified values are treated with well framed rule-base in layer 3 which produces a specified output fuzzified values to each of the input fuzzified value set in layer 4. Finally, to obtain a crisp output; a defuzzification methodology is applied on those output values in layer 5.

Clustering Phase

System Model

The following parameters are considered in the demonstrative analysis. (1) Temperature sensor, humidity sensor, and soil moisture sensor are considered for monitoring and irrigation process. In this experimental demonstration one hectare of marigold crop field is chosen and nodes are dispersed randomly in a 100 × 100 square unit of that region following a uniform distribution. (2) Base station is fixed in the centre of the experimental field and all the nodes after their deployment send join messages to the base station containing their local information such as location coordinates, energy status, and neighboring node details. (4) The base station (BS) is a node with no energy constraint and enhanced computation capabilities.

Figure 1. Block diagram of WSN based Smart Irrigation Schema

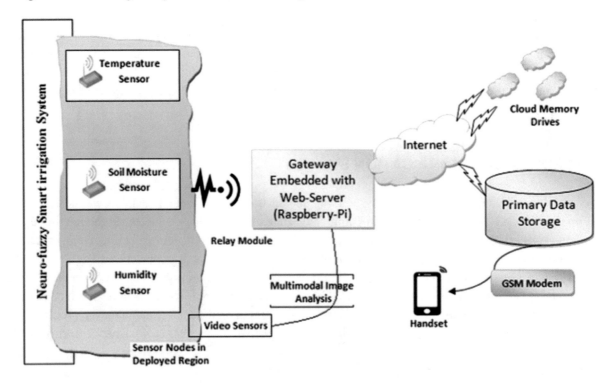

Figure 2. Block diagram of Neuro-fuzzy inference Module

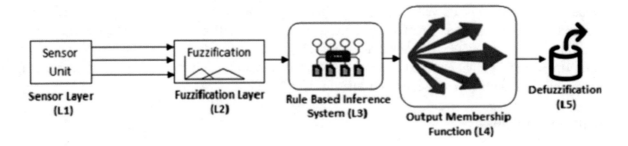

Cluster Formation and Cluster-Head (CH) Selection Module

Without the involvement of separate algorithm for clustering, BS forms the static clusters based on the horizontal and vertical lines that are originated from fixed coordinates which is refereed as Cluster Segregation Point (CSP). Thus, approximately four cluster zones namely UL, UR, LL, and LR zones are formed which is depicted in figure 3.

Since all the nodes in the network are of homogenous type, each and every parametric values and processing capabilities are also found to be same. Thus, initially, a randomized node which is very close to the CSP is chosen as Cluster Head (CH) in each zone. After this, all the Chosen CHs broadcast its selection information as an advertisement messages to all other non-CH nodes within their zonal range

Figure 3. Cluster Formation

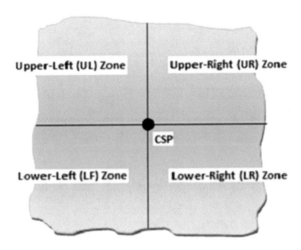

of the network. The non-CH nodes which receives the advertisement message, replies to the corresponding CH along with their essential information like location coordinates, and current energy status. Now each CH node estimates the distance of all the cluster heads from its current coordinates.

Data Transmission Scheduler Module

The proposed scheme employs conventional TDMA (Falconer et al., 1995) technique for synchronized data transmission. The selected CHs in each zone allocate data transmission slots to all nodes within the cluster zone. Thus, except slot_0, the odd numbered slots to all of its non-CH nodes for data transmission whereas all the even numbered slots are assigned to itself. During the broadcast of slot allocation, all CHs embed its local time to all non-CHs for synchronization purpose. Hence, through this procedure, the proposed scheme prevents collision and enables traffic-free communication. At the session of slot_0 all the non-CHs nodes adjust its local clock to the timestamp of CHs. Figure 4 shows the sample slot allocation strategy.

Operational Phase

Sensing Module

The initial part of operation phase involves sensing module that further comprises soil moisture, temperature and sensor modules. Here, BE-000030 Xcluma Soil moisture sensor module is utilized. Soil moisture module is most sensible device, usually employed to monitor the moisture cognitive content of the soil. In other words, it assesses the volumetric cognitive content of water present in the soil and denotes the moisture measure in terms of voltage. It also has the tendency to provide both analog and digital resultants. Most of the analog output measures are in term of percentage mapped between the range of 0 and 100. Figure 5.a, depicts the sample device of BE-000030 Xcluma Soil moisture sensor module.

Figure 4. Data Transmission Scheduling

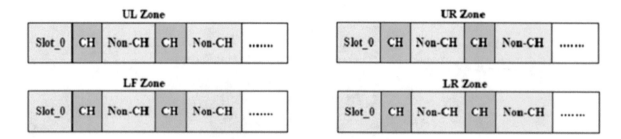

Figure 5. (a) Soil Moisture Sensor Module; (b) Temperature and Humidity Sensor Module

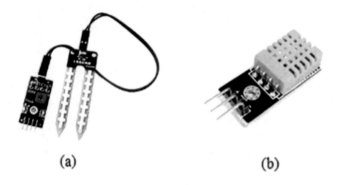

(a)   (b)

Next, DHT11 Humidity and Temperature Sensor are employed for further investigation that yields calibrated digital outcomes. It can be interfaced with any microcontroller like, Raspberry Pi, Arduino, etc. and renders instantaneous outcome. Moreover, it is a low cost temperature and humidity sensor which enables long term stability and high reliability. Figure 5.b, depicts the sample device of DHT11 Temperature and Humidity Sensor Module.

Low-Cost Controller Module

To adapt low cost experimental setup, we preferred Raspberry Pi 3 Model A+ as low-cost controller module which is smaller in size that can be utilized for multi-utility purpose. This device weighs around 29g which simplifies the programming platform (especially Python) for developing applications and very easy to compute. It comprises 64-bit quad core processor executing at 1.4 GHz, 5 GHz WLAN, 4.2V Bluetooth, and 2.4 GHz dual-band. Figure 6, represents the sample of Raspberry Pi 3 Model A+ low-cost controller module.

Neuro-Fuzzy System

The major part of the operation phase includes the computation part of Neuro-fuzzy system which is always referred as learning machine that further determines the parameters of a fuzzy system (i.e., fuzzy sets and rules) by tapping estimated methodologies from neural networks. Developing neuro-fuzzy infer-

Figure 6. Raspberry Pi 3 Model A+

ence systems involves both the cognition and the evolving inference mechanisms, and samples regularly vary in time. For the better understanding purpose of this system, table 1 explores relevant and essential studies of the system to the maximum extent.

Comparative Study of Relevant Neuro-Fuzzy Inference Module

Neuro-Fuzzy Inference Module

The entire computational block diagram of Neuro-Fuzzy system is depicted in figure 7 where Layer 1 consists of sensor nodes which senses the values and is known to be the sensor unit. Layer 2 comprises all the input membership functions that perform fuzzification. Layer 3 is rule based inference system. Layer 4 exhibits all of the output membership functions and layer 5 referred as output layer or defuzzification unit that provide crisp resultant.

Here, Layer 3 executes an AND operation which is computed as the (min) function on the inflow activation measures of the membership function nodes. Moreover, the implemented membership functions are of triangular type. An OR operation is manipulated in Layer 4 that are calculated as the (max) function. This (max) functions are computed to the weighted activation values of the rule nodes connected in that layer.

$$\mathbf{Oy}^{4L} = \mathbf{max}\{\mathbf{Ox}^{3L} \bullet \mathbf{w}_{x,y}\} \tag{1}$$

where, Oy^{4L} is the activation of the yth node of layer 4.

Membership Function

A Membership Function (MF) is defined as a curve that characterizes the fuzzy set by attributing the corresponding degree of membership or membership value for each element. It deals with the mapping of input values to a membership value between the unit interval 0 and 1. The initial step of fuzzification is to define the fuzzy set which contains linguistic variables, linguistic terms with the degree of membership ranging from 0 to 1. Most commonly, the elements are described between entirely belonging to the fuzzy set and entirely not belonging to the fuzzy set because each set does not constitute a crisp, unambiguous boundary. Herein, to regulate the smart irrigation, membership function like temperature, soil moisture,

Table 1. (Ravi & Prasad, 2010)

Methodology	Parameters			
	Learning Algorithm	No. of Layers	Purpose	Area of Utilization
Adaptive Neuro Fuzzy Inference System (Jang R, 1992)	Back-propagation	One	· Ensures that each linguistic term is represented by only one FS. · One of the best tradeoff between neural and fuzzy systems, providing smoothness, due to the FC interpolation, adaptability, due to the NN Back-propagation. · Strong computational complexity restrictions.	· Agricultural economic variables forecasting. · Automatic control and signal processing.
Generalized Approximate Reasoning based Intelligent Control (Bherenji & Khedkar, 1992)	Gradient descent and reinforcement learning	One	Provides a well balanced method for combining the well qualitative knowledge of human experts in terms of symbolic and learning strengths of ANN.	Cart-pole balancing system.
Fuzzy Adaptive learning Control Network (Lin CT & Lee CSG, 1991)	Unsupervised and gradient descent	Two	· Provides a framework for structure and parameter adaptation for designing neuro-fuzzy systems. · Able to partition the input and output spaces and then find proper fuzzy rules and optimal membership functions dynamically.	· Pattern Classification. · Control Engineering.
Neuro-Fuzzy Control (Nauck & Kruse, 1997)	Reinforcement and Back-propagation learning	Two	NEFCON can be used to learn an initial rule base, if no prior knowledge about the system is available or even to optimize a manually defined rule base.	· Control for Flexible Robotic Arm · Tracking of Aircraft and Ground Vehicles.
Self Constructing Neural Fuzzy Inference Network (JC Feng & LC Teng, 1998)	Supervised and Back-propagation	One	· Sequential learning scheme is of on-line operation. · To provide an optimal way for determining the consequent part of fuzzy if-then rules during the structure learning phase.	Control, communication, and signal processing.

and humidity are considers as linguistic input variable to neural network. Low (L), Optimal (OP) and high (H) are applied as linguistic terms, and mapped with different ranges of parametric values. Table 2 represents the membership function mapped with different sets of linguistics terms.

Figure 7. Computational Block diagram of Neuro-Fuzzy System

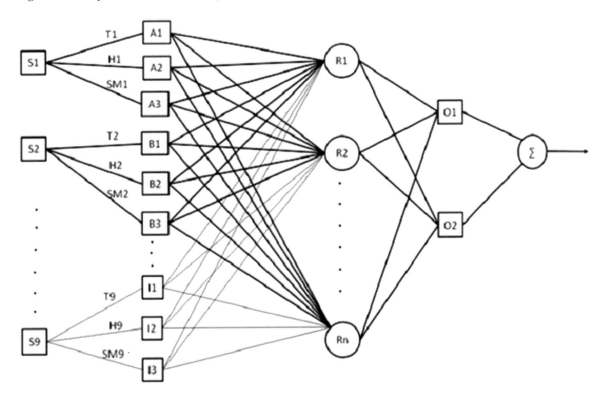

The fuzzy set is characterized by different forms of membership functions like trapezoidal, linear, triangular singleton Gaussian or piecewise, etc (Nguyen HT, 1995). The research work utilizes trapezoidal MF which efficiently fuzzifies the uncertain, limited input parameters. Figure 8a, b, and c represent the trapezoidal MF for the linguistic variables temperature, soil moisture and humidity respectively. Furthermore, the fuzzy sets of three linguistic variables are formulated in the Equation (2) to (10).

$$T_L(X) = \begin{cases} 1, & X < 12 \\ \dfrac{(15-X)}{0.5}, & 12 \le X \le 15 \\ 0, & X \ge 15 \end{cases} \tag{2}$$

Table 2. Membership Function Table

Membership Function	Linguistic Terms		
	Low(L)	Optimal (OP)	High(H)
Temperature (in ºCelsius)	1 – 12	15 - 28	>30
Soil Moisture (in pH)	0 – 4.0	5.0 – 7.0	>7.5
Humidity (%)	0 – 50	65 – 80	>85

Figure 8. Membership Function of (a) Temperature, (b) Soil Moisture, and (c) Humidity

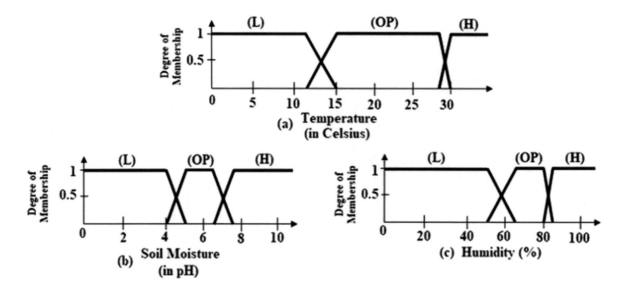

$$T_{OP}(X) = \begin{cases} 0, & X \le 12 \\ \dfrac{X-12}{0.5}, & 12 \le X \le 15 \\ 1, & 15 \le X \le 28 \\ \dfrac{30-X}{0.5}, & 28 \le X \le 30 \end{cases} \qquad (3)$$

$$T_H(X) = \begin{cases} 0, & X \le 28 \\ \dfrac{X-28}{0.5}, & 28 \le X \le 30 \\ 1, & X \ge 30 \end{cases} \qquad (4)$$

$$SH_L(X) = \begin{cases} 1, & X < 4 \\ \dfrac{(5-X)}{0.5}, & 4 \le X \le 5 \\ 0, & X \ge 5 \end{cases} \qquad (5)$$

$$SH_{OP}(X) = \begin{cases} 0, & X \leq 4 \\ \dfrac{X-4}{0.5}, & 4 \leq X \leq 5 \\ 1, & 5 \leq X \leq 7 \\ \dfrac{7.5-X}{0.5}, & 7 \leq X \leq 7.5 \end{cases} \qquad (.6)$$

$$SH_{H}(X) = \begin{cases} 0, & X \leq 7 \\ \dfrac{X-7}{0.5}, & 7 \leq X \leq 7.5 \\ 1, & X \geq 7.5 \end{cases} \qquad (7)$$

$$H_{L}(X) = \begin{cases} 1, & X < 50 \\ \dfrac{(65-X)}{0.5}, & 50 \leq X \leq 65 \\ 0, & X \geq 65 \end{cases} \qquad (8)$$

$$H_{OP}(X) = \begin{cases} 0, & X \leq 50 \\ \dfrac{X-50}{0.5}, & 50 \leq X \leq 65 \\ 1, & 65 \leq X \leq 80 \\ \dfrac{85-X}{0.5}, & 80 \leq X \leq 85 \end{cases} \qquad (9)$$

$$H_{H}(X) = \begin{cases} 0, & X \leq 80 \\ \dfrac{X-80}{0.5}, & 80 \leq X \leq 85 \\ 1, & X \geq 85 \end{cases} \qquad (10)$$

The systematical smart irrigation procedure is exclusively associated with distinguished "rules". The experimental research work employs Mamdani rule inference method in neuro-fuzzy inference module. Here, the rule base exhibits a simple "IF 'a' is X Then 'b' is Y" approach (Karthick Raghunath & Thirukumaran, 2019). Here, the combination of three input variable intends to frame 27 rules that are depicted in the figure 9. Eventually, the each and every rules are classified for training purpose and resultants are

Figure 9. Rule Mapping

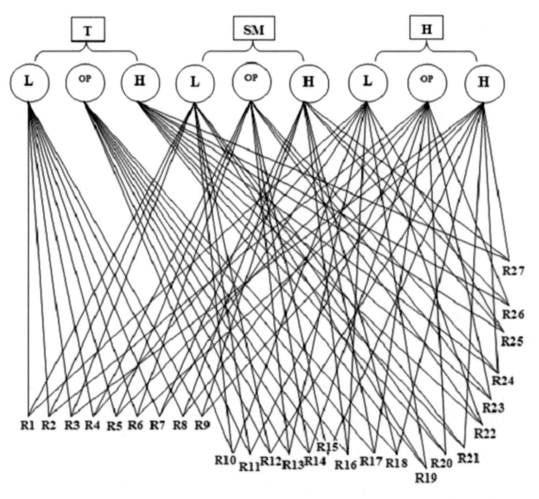

determined for decision-making during irrigation procedure. The type of irrigation facilities that are to be opted as per the results of the rules are represented in the table 3.

Multimodal Image Analyzing

Multimodal image analysis is referred as the combined investigation of multiple images of the same instance gained from various imaging modalities is generally employed to acquire more data or knowledge about the selected instance than that are possible from a individual modality.

To identify the infected plants (due to flower bud rot or powdery mildew or inflorescence blight or damping off or leaf spots and blight) and to overcome from the effects of predictable marigold plant disease, employing multimodal image analysis system is found to be essential. Majority of the marigold crop diseases happens due to fungal agametes. To be preventive from such a kind of scenario is to adapt an efficient multimodal image analysis system for optimal irrigation which avoids or prevent crop from fungal agametes. Sometime identifying and removing the infected crops may also aid in limiting the dispersion of disease effects. For the precise identification of such crop again demands an essential

Table 3. Rule classification and resultant

Rule Classification	Resultant
R1, R2, R3, R4, R5, R6, R10, R11, R12, R13, R14, R15	Low Flow (micro spray, drip lines, or drip emitters,)
R7, R8, R9, R16, R17, R20, R23	Moderate Flow (rotor, fixed spray, soaker hose, or bubbler
R18, R9, R21, R22, R24, R25, R26, R27	High Flow (Subsurface)

multimodal image analyzing system. Thus, Imote2 is utilized to regularly capture the images of the crops during its entire farming tenure. Further, the captured images are imposed to a multimodal image fusion methodology namely Non-Subsampled Contourlet Transform (NSCT) (L. da Cunha et al., 2006) which provide much detailed information regarding the current status of the crop for precise and better analysis.

The general NSCT-based image fusion access comprises the following sequence of steps:

Figure 10. Schematic pictorial representation of NSCT

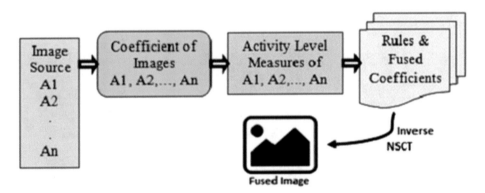

Initially, for each scale and each direction it is essential to acquire and band-pass directional sub-band coefficients and low-pass sub-band coefficients. So, NSCT is performed on captured image source to obtain those coefficients. Later, several required fusion rules are applied to choose NSCT coefficients of the processing image. Eventually, inversed NSCT is applied to the chosen coefficients and as a result the fused image is obtained. Figure 10 showcases the process flow of NSCT.

PERFORMANCE ANALYSIS OF PROPOSED SYSTEM

Evaluation of NSCT Approach

Figure 11 show that the NSCT formally recovers maximum edge contents of the captured images. The betterment of the technique can be studied from figure 11.a, where the marigold leaves are mostly prone to leaf spot which has to be monitored for proper countermeasures. Thus, the collections of different images are further denoised to enhance the quality of the image. These enhancements provide detailed information which is essential for feature extraction and for future analysis. Similarly, from figure 12, the collection of images of marigold flowers which are affected by the flower spot are denoised and then enhanced the image in reconstruction (especially around the petals). For the reconstruction process, the noise standard deviations and noise variance of the collected image are estimated. Eventually, the maximum magnitude and mean of respective coefficient in all sub-band directions are computed and classified those levels into strong or weak edges. Thus the local adaptive shrinkage in each sub-bands are performed including threshold as (L. da Cunha et al., 2006),

Figure 11. (a) Original image of marigold leaf affected by 'leaf spot', (b) Denoised with NSCT, and (c) Enhanced image by NSCT

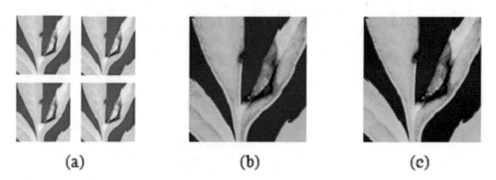

(a)　　　　　(b)　　　　　(c)

Figure 12. (a) Original image of marigold flower affected by 'Flower Spot', (b) Denoised with NSCT, and (c) Enhanced image by NSCT

(a)　　　　　(b)　　　　　(c)

$$T_{(i,j)} = \sigma^2_{U_{(i,j)}} / \sigma_{(i,j,u)} \tag{3.1}$$

where, $\sigma_{(i,j,u)}$ represents the uth coefficient variance at i^{th} directional sub-band of jth scale. $\sigma^2_{U_{(i,j)}}$ denotes the noise variance at direction i and scale j.

Evaluation of Smart Irrigation System

Marigold crop was farmed in the observational plot during 22 November 2018 to 30 February 2018 at the 12.7499988 Latitude and 77.8597148 Longitude of Hosur, Tamil Nadu. The observations are fully inspected based on the implementation proposed smart irrigation system and the resultants are compared with the absence factors of proposed system. Figure 13 represents the experimental plot along with the physical properties of the plot.

Figure 13. Experimental Plot of Marigold farming

After some of the biometric examinations for around 100 days, this research marks a unique identification of systematical irrigational system that shows the betterment of applying the proposed system. It's been recorded that around 185.5 mm of water is supplied to the crop throughout the farming period that is around 10% addition water consumption by the crops is reduced. The observations carried out or measured with respect to the plant spread, plant height, number of branches and minimum and maximum number of flower diameter, and flowers per plant.

The obtained results are compared with results of smart irrigation in absentia scenarios. The resultants denoted in the figure 14, 15, 16, 17 shows that the imposed system not only enhanced the production but also the productivity of the crop. Moreover, the cultivators gain the knowledge regarding degradation or disease factors of the plant through effective multimodal image fusion techniques.

It was observed that average flower yield per plant is 0.6 kg for the experimental plot adopted smart irrigation system, and 0.4 kg for normal flower cultivation plot (without smart irrigation system). Eventually, the production of quality flower per hectare was found to be around 21.5 tons which is approximately 20% higher than with normal flower cultivation procedures.

Figure 14. Comparative measurement and analysis of Plant height

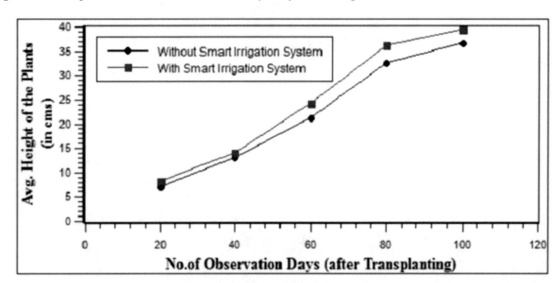

Figure 15. Comparative measurements and analysis of Number of Branches

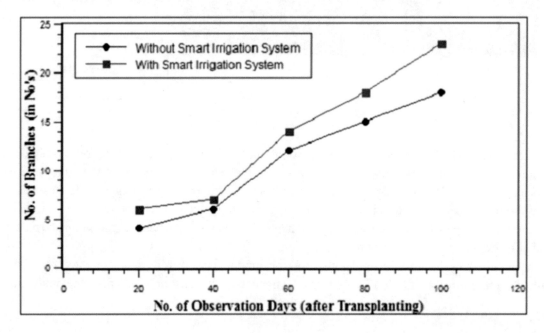

CONCLUSION

Figure 16. Comparative analysis of Plant dispersion

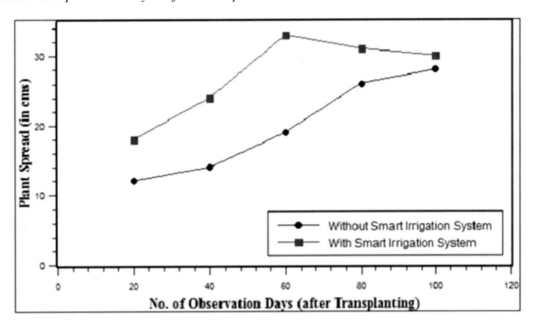

Figure 17. Comparative analysis of plant parameters like flowers per plant and flower diameter

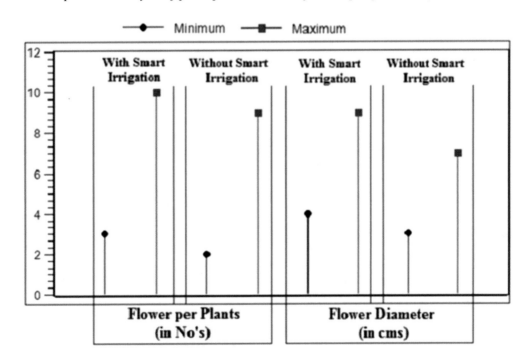

For growing marigold, the neuro-fuzzy smart irrigation system outperforms the normal farming system in terms of water use efficiency, production, and productivity. Utilization of multimodal image analysis technique for monitoring and identification of diseases in the early stage prevents the degradation of productivity and provide prior information to the cultivators. It was also exposed that all the biometric parameters are experimented as per the demands and evaluated with the resultant of the normal farming procedures. The Comparative resultants were found to be higher with smart irrigation system. It was also observed that the average flower yield per plant is 0.6 kg for farming supported by proposed system whereas it is 0.4 kg in case of normal farming. Likewise the productivity of flower/hectare was 21.5 tons for farming aided by smart irrigation system while it is 17.54 tones for normal farming. Yield was 20% higher, and utilization of water was 34% less compared to the normal farming procedures.

REFERENCES

Akyildiz, I. F., Su, W., Sankarasubramaniam, Y., & Cayirci, E. (2002). Wireless sensor networks: A survey, computer networks. *Computer Networks*, *38*(4), 393–422. doi:10.1016/S1389-1286(01)00302-4

Bherenji, H. R., & Khedkar, P. (1992). Learning and Tuning Fuzzy Logic Controllers through Reinforcements. *IEEE Transactions on Neural Networks*, *3*(5), 724–740. doi:10.1109/72.159061 PMID:18276471

Crow, W. T., & Wood, E. F. (2003). The assimilation of remotely sensed soil brightness temperature imagery into a land surface model using ensemble Kalman filtering: A case study based on ESTAR measurements during SGP97. *Advances in Water Resources*, *26*(2), 137–149. doi:10.1016/S0309-1708(02)00088-X

da Cunha, L., Zhou, J., & Do, M. N. (2006). The Non subsampled contourlet transform: Theory, design, and applications. *IEEE Transactions on Image Processing*, *15*(10), 3089–3101. doi:10.1109/TIP.2006.877507 PMID:17022272

Falconer, D., Adachi, F., & Gudmundson, B. (1995). Time division multiple access methods for wireless personal Communications. *IEEE Communications Magazine*, *33*(1), 50–57. doi:10.1109/35.339881

Jang, R. (1992). *Neuro-Fuzzy Modeling: Architectures, Analyses and Applications* (PhD Thesis). University of California, Berkeley, CA.

Juang, C. F., & Lin, C. T. (1998). An Online Self Constructing Neural Fuzzy Inference Network and its Applications. *IEEE Transactions on Fuzzy Systems*, *6*(1), 12–32. doi:10.1109/91.660805

Karthick Raghunath, K. M., & Thirukumaran, S. (2019). "Fuzzy-based fault-tolerant and instant synchronization routing technique in wireless sensor network for rapid transit system", Automatika: Journal for Control, Measurement, Electronics. *Computer Communications*, *60*(5), 547–554.

Lin, C. T., & Lee, C. S. G. (1991). Neural Network based Fuzzy Logic Control and Decision System. *IEEE Transactions on Computers*, *40*(12), 1320–1336. doi:10.1109/12.106218

Nauck, D., & Kruse, R. (1997). Neuro-Fuzzy Systems for Function Approximation. *4th International Workshop Fuzzy-Neuro Systems*.

Nguyen, H. T. (1995). *Theoretical aspects of fuzzy control* (1st ed.). John Wiley.

Ravi, S. R., & Prasad, T. V. (2010). Exploration of Hybrid Neuro Fuzzy Systems. *Conference on Advances in Knowledge Management*. DOI: 10.13140/RG.2.1.3570.0327

Sujitha, E., & Shanmugasundaram, K. (2017). Irrigation Management of Greenhouse Marigold Using Tensiometer: Effects on Yield and Water Use Efficiency. *International Journal of Plant and Soil Science*, *19*(3), 1–9. doi:10.9734/IJPSS/2017/36437

Chapter 16
Use of Eggshell as a Partial Replacement for Sand in Concrete Used in Biomedical Applications

Sebastin S.

National Engineering College, Thoothukudi, India

Murali Ram Kumar S. M.

National Engineering College, Thoothukudi, India

ABSTRACT

The recent researches show that cement mortar containing eggshell as a partial replacement of sand contained radiation absorption property. In these cement mortar samples, 5%, 10%, 15%, and 20% by mass of sand was replaced by crushed eggshells, and there was increase in radiation absorption. The result also showed that using eggshell as a partial replacement for sand leads to decrease in compressive strength of the cement mortar. Waste of any kind in the environment when its concentration is in excess can become a critical factor for humans, animals, and vegetation. The utilization of the waste is a priority today in order to achieve sustainable development. So, we have planned to increase that compressive strength by means of using seashell as a partial replacement for cement. The main objective of the project is to maintain radiation absorption by means of using eggshell along with seashell as a partial replacement for cement to increase the compressive strength.

INTRODUCTION

Waste of any kind in the environment when its concentration is in excess can become a critical factor for humans, animals, and vegetation as stated by (Dr.Amarnath Yerramala, 2014). The utilization of the waste is a priority today in order to achieve sustainable development. In this developing world, many countries are going to face the severe environmental problems due to the rapid population growth. Nowa-

DOI: 10.4018/978-1-7998-3591-2.ch016

days, Indians are generating waste products at a rather alarming rate. It is much faster than the natural degradation process and they are using up resources at a speed exceeding the rate these materials are being replaced (Amu, O.O., A.B. Fajobi and B.O. Oke, 2005). As a way to disposal the waste, we need to recycle it to become useful product for the Indians. Despite the massive amount and complexity of waste produced, the standards of waste management in India are still cannot able to control waste generation rates and its composition, disposal of municipal wastes with toxic and hazardous waste, indiscriminate disposal or dumping of waste and inefficient utilization of disposal site results in severe health hazards to both humans and animals. To reduced this problems the waste product such as oil palm shell, fly ash and bottom ash are being used in construction industries as an additional material or replacement of the material in the concrete to decrease the cost while reducing the amount of waste.

The construction industries are searching for the alternative product that can increase their profit and in the same time can be an environmentally friendly material (ASTM Annual Book of Standards, 2004). The conversion of waste obtained from agricultural processes into biocompatible materials (biomaterials) used in medical surgery is a strategy that will add more value in waste utilization. This strategy has successfully turned the rather untransformed wastes into high value products. Eggshell is an agricultural waste largely considered as useless and is discarded mostly because it contributes to pollution. This waste has potential for producing hydroxyapatite, a major component found in bone and teeth. Hydroxyapatite is an excellent material used in bone repair and tissue regeneration. The use of eggshell to generate hydroxyapatite will reduce the pollution effect of the waste and the subsequent conversion of the waste into a highly valuable product. In this paper, we reviewed the utilization of this agricultural waste (eggshell) in producing hydroxyapatite (ASTM C 642–82, 1995). The process of transforming eggshell into hydroxyapatite and nanohydroxyapatite is an environmentally friendly process. Eggshell based hydroxyapatite and nanohydroxyapatite stand as good chance of reducing the cost of treatment in bone repair or replacement with little impact on the environment (ASTM C 642–82, 1995).

Chicken's egg is one of the waste materials that contribute lot to the environmental problem. Eggshells are available in large quantity due to their large consumptions majority of them are by products from other processes or natural materials (Bonavetti, V., Donza, H., Menédez, G., Cabrera, O and Irassar, E.F, 2003) . The major benefit of egg shell is its ability to replace certain amount of fine aggregate and still able to display fine aggregate property, thus reducing the weight of concrete, where seashell can be used as a replacement for cement to a certain amount because of its cementitious properties. The use of such by products in concrete construction not only prevents these products from being land-filled but also enhances the properties of concrete in the fresh and hardened states. Crushed Eggshell obtained by crushing the shells using the crusher has been established to be good partial replacement for fine.

Background of Study

Agricultural waste is any waste being generated from different farming processes in accumulative concentration. Adequate utilization of agricultural waste reduces environmental problems caused by irresponsible disposal of the waste. The management of agricultural wastes is indispensable and a crucial strategy in global waste management. Waste of any kind in the environment when its concentration is in excess can become a critical factor for humans, animals, and vegetation. The nature, quantity, and type of agricultural waste generated vary from country to country. The search for an effective way to properly manage agricultural waste will help protect the environment and the health quality. For sustainable development, wastes should be recycled, reused, and channelled towards the production of value added products. This is to protect the environment on one side and on the other side to obtain value added products while establishing a zero waste standard. The utilization of the waste is a priority today in order to achieve sustainable development (Markle Foundation, 2003) .

One way that adds great value to agricultural waste is its utilization as a biomaterial used in medical surgery and therapeutics. The production of biocompatible material or biomaterial from agro waste has added a different dimension to the utilization of agricultural waste for value added product. This is possible because some of this waste contains active compounds that have value in medical applications. This is a novel practice that is expected to have value in medical sciences. Most researches on agricultural waste focused mainly on its energy potentials or its use as effective chemical feedstock and as renewable raw materials because of its abundance, its cheap availability (Markle Foundation, 2003) . This conversion into valuable products or energy sources is carried out by microorganisms or their components. Many agricultural wastes were reported to be effective feedstock in making useful products. The waste is readily available and cheap. Agricultural waste has been proven to serve as a good replacement option which can be used as biomaterials in therapies that replaces bone for the growth of osteoblasts. Avian eggshell is an agricultural waste that received attention and has the potential of being used in medical and dental therapy. The use of other wastes as biomaterials has also been reported.

The development of biomaterials for bone tissue replacements has increased and attracted a lot of interest due to the rise in the number of patients that requires bone replacements, especially in those suffering from bone cancer, trauma, and ageing. The biomaterials must be biocompatible with sufficient mechanical strength to support the weight of human body before being used as bone implants.

This paper will give an insight into utilization of avian eggshell, an agricultural waste, to produce hydroxyapatite. Eggshells remain largely unutilised and untransformed as they are discarded wrongly. These shells are made up of calcium carbonate that can be used to produce hydroxyapatite, the major inorganic part in bones which is used in bone and dental therapy (Metzger and Fortin, 2003) .

From the previous research, the eggshell had been found as a good material to be an additive material or replacement material in the concrete (Mitchell, 1999). The eggshell can be found easily and it also found that have good strength characteristic when mixed with the concrete (Wise, 2003)). Eggshells are agriculture waste materials generated from chick hatcheries, bakeries, fast-food restaurant among other which can litter the environmental and consequently constituting environmental problem or pollution which would require proper handling. The uses of eggs are very high in the food making process and the waste of eggshell is usually disposed in landfills without any pre-treatment since it is normally useless. The US food industry generates 150,000 tons of shell waste a year. The disposal methods for waste eggshells are 26.6% as fertilizer, 21.1% as animal feed ingredients, 26.3% discarded in municipal dumps, and 15.8% used in other ways. Many landfills are unwilling to take the waste because the shells and the attached membrane attract vermin. Together the calcium carbonate eggshell and protein rich membrane are useless.

Recent inventions have allowed for the egg cracking industry to separate the eggshell from the eggshell membrane. The eggshell is mostly made up of calcium carbonate and the membrane is valuable protein. When separated both products have an array of uses. Scientifically, seashells known that it is mainly composed of compound of calcium which is very similar to cement. Literature has shown that the eggshell primarily contains lime, calcium and protein where it can be used as an alternative raw material in the production of wall tile, concrete, cement paste and other. Therefore, the use of eggshell in the concrete can decrease the cost of raw material and can give a profit to the construction industries while saving the environment. Therefore, eggshell can be used to reduce the cost of construction and produced as a new material in developing the construction industries.

MATERIALS

Ordinary Portland Cement

The raw materials used for manufacture of Portland cement are calcareous materials, such as limestone or chalk, and argillaceous materials such as shale or clay. The process of manufacture of cement consists of grinding the raw materials, mixing them intimately in certain proportions depending upon their purity and composition and burning them in a kiln at a temperature of about 1300 to1500°C, at which temperature, the materials sinkers and partially fuses to form nodular shaped clinker. Ordinary Portland cement (OPC) 53 grade cement was used for all mortar mixtures. The chemical composition of the crushed eggshell is shown in Table 1.

Table 1. Chemical Composition of Cement

COMPONENTS	%
Lime (CaO)	60 – 63%
Silica (SiO$_2$)	17 – 25%
Alumina (Al$_2$O$_3$)	03 – 08%

CHEMICAL COMPOSITION

Sand

Sand is a naturally occurring granular material. It is defined by size, being finer than gravel and coarser than silt. Sand can also refer to a textural class of soil or soil type; i.e. a soil containing more than 85% sand-sized particles by mass. The composition of sand varies, depending on the local rock sources and conditions, but the most common constituent of sand in inland continental settings and non-tropical coastal settings is silica (silicon dioxide, or SiO2), usually in the form of quartz (Hawkins, P., Tennis, P. and Detwiler, R 2003) .

ISO 14688 grades sands as fine, medium and coarse with ranges 0.063 mm to 0.2 mm to 0.63 mm to 2.0 mm. In the United States, sand is commonly divided into five sub-categories based on size: very fine sand (1¤16 – 1¤8 mm diameter), fine sand (1¤8 mm – 1¤4 mm), medium sand (1¤4 mm – 1¤2 mm), coarse sand (1¤2 mm – 1 mm), and very coarse sand (1 mm – 2 mm). These sizes are based on the Krumbein phi scale. Ennore standard sand passing through 4.75mm sieve was used for all mortar mixtures.

Eggshell

Fresh eggshell consists of a typically three-layered structure; the foamy cuticle layer on the outer surface resembles a ceramic; the middle layer is spongy; the inner layer consists of lamellar layers. It represents almost 11% of the egg total weight. Calcium carbonate (calcite) is the main component in eggshells and is the major inorganic substance found in an egg and it makes up about 94% of chemical composition of eggshell.

Aggregates passing through 4.75 mm sieve and predominately retained on 75 μm sieve are classified as fine aggregate. Finely grinded eggshell is used as fine aggregate is used as a partial replacement for fine sand. The main ingredient in eggshells is calcium carbonate (the same brittle white stuff that chalk, limestone, cave stalactites, sea shells, coral, and pearls are made of). The shell itself is about 95% CaCO3 (which is also the main ingredient in sea shells). The remaining 5% includes Magnesium, Aluminium, Phosphorous, Sodium, Potassium, Zinc, Iron, Copper, Ironic acid and Silica acid. Eggshell has a cellulosic structure and contains amino acids; thus, it is expected to be a good bio-sorbent and it was reported that large amounts of eggshells are produced in some countries, as waste products and disposed in landfills annually.

The following are some of the advantages of egg shell

1. Considerable reduction in alkali-silica and sulphate expansions.
2. Meets the most stringent environmental regulations nationwide.
3. Ideal for painting in occupied spaces.
4. Excellent durability and washable finish.
5. Resist mould and mildew on the paint film.
6. Saves money; less material required.
7. Meets strict performance and aesthetic requirements.

The eggshells are collected and cleaned properly. The cleaned eggshells are finely grounded and sieved. The crushed eggshell passing through 1.18mm and retains on 600μ is used as a partial replacement for sand.

The chemical composition of the crushed eggshell is shown in Table 2.

Table 2. Chemical Composition of Eggshell

COMPONENTS	%
Calcium carbonate	95
Magnesium	0.8
Calcium Phosphate	0.8
Organic Materials	3.4

SAMPLE PREPARATION

Preparation of Cement Mortar

In the preparation of cement mortar cubes, cement conforming with IS 3535-1986, standard Ennore sand conforming to IS: 650- 196, Eggshells and Seashells were used. The practical work is done based on the "BS EN 1015: Methods of test for mortar for masonry", the formula therein has been used to calculate manually. Initially the mix is made by taking the regular quantity of cement and sand which is of CM mix ratio 1:3 for a 7.1cm x 7.1cm x 7.1cm cube. The quantity of cement and sand taken are 200 g and 600 g respectively with water cement ratio as 0.5 which is 100 g weight of cement. The cement is of OPC 53 grade and the sand is of the size passes through 4.75mm sieve and retains on 2.36mm is used.

Preparation of Ordinary Cement Mortar Sample

The practical work is done based on the "BS EN 1015: Methods of test for mortar for masonry", the formula therein has been used to calculate manually. In manual mixing, OPC of 53 grade and sand is mixed. A depression is then made in the middle of mix and required amount of water is added. Then the mixture is mixed well. The resultant mix should be consumed within 30 minutes from the instant of adding water to the mix.

Preparation of Cement Mortar Sample Containing Eggshell

In the preparation of cement mortar sample containing seashell, cement conforming to BS EN 1015, standard sand and eggshell were used. The sample names, the mixing ratios and the materials used are given in table 3. While preparing the sample mortars, 5%, 10%, 15% and 20% [2] by mass of sand was replaced by crushed eggshells. The materials are weighted accurately and mixed properly, and then compacted by using the vibrating machine for 2 minutes. The 3 no's of samples were allowed to set for a day, after which the samples were taken and kept in the curing tank for the period of 7 days and 28 days. After the 7 days and 28 days of curing, samples were taken from the curing tank and dried. Next the cube sample is tested for its compressive strength by using the CTM of 200 ton and their corresponding values were noted.

Table 3. Mix proportion of cement mortar containing eggshell

SAMPLE NAME	MORTAR COMPONENTS		
	CEMENT (g)	SAND (g)	EGGSHELL (g)
M	200	600	-
ME_1	200	570	30 (10%)
ME_2	200	540	60 (10%)
ME_3	200	510	90 (15%)
ME_4	200	480	120 (20%)

MIX PROPORTION

The mix proportion of cement mortar sample containing eggshell is indicated in Table 3,

Compressive Strength of Cement Mortar

Compressive strength of cement mortar samples were determined (A. J. Olarewaju, M. O. Balogun and S. O. Akinlolu 2011). The compressive strength of the cement mortar sample was tested by means of using the Compression Testing Machine (CTM) of 200 ton.

The compressive strength of cement mortar samples is shown in the following Figure 1,

DISCUSSION

28-day compressive strength of mortar samples are given in Figure.1.

Figure 1. Compressive strength of cement mortar sample containing Eggshell

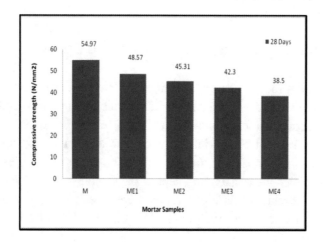

Compressive Strength of Cement Mortar Containing Eggshell

From the Figure.1, it shows that increase in percentage of eggshell results in decrease in compressive strength of cement mortar and decrease in self weight of cement mortar. The 28-day compressive strength of cement mortar sample containing Eggshell is shown in Table 4.

Table 4. Compressive strength of cement mortar containing Eggshell

SAMPLES NAME	AVERAGE ULTIMATE LOAD (kN)	COMPRESSIVE STRENGTH (N /mm²)
M	277.10	54.97
ME₁	244.84	48.57
ME₂	228.41	45.31
ME₃	213.23	42.3
ME₄	194.08	38.5

COMPARISON OF WEIGHT OF ORDINARY CEMENT MORTAR WITH THE CEMENT MORTAR CONTAINING EGGSHELL

Comparison of weight of ordinary cement mortar with the cement mortar containing eggshell shown in the Figure 2, we can conclude that the increase in percentage of eggshell results in decrease in self weight of cement mortar. Hence the eggshell can be used as partial replacement of sand in the place where the light weight cement mortar with permissible compressive strength is required.

Figure 2. Comparison of weight of ordinary cement mortar with the cement mortar containing eggshell

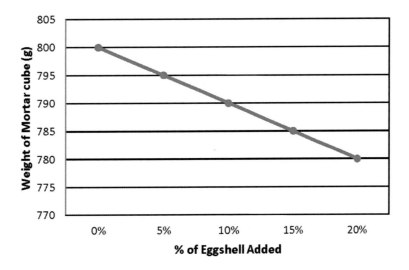

CONCLUSION

The addition of eggshell as a partial replacement of sand results in decrease in compressive strength of cement mortar. But the advantage of using eggshell was that, the increase in percentage of eggshell results in the decrease in self weight of cement mortar. Also, the chemical composition of eggshell have a property to resist the radioactivity to the certain extent, hence it can be used for radiotherapy rooms, nuclear reactors and in the buildings where the radioactive materials are used. In future, the cement mortar containing eggshell can be used for paver blocks, radioactivity resisting bricks and other biomedical applications.

REFERENCES

Amu, O. O., Fajobi, A. B., & Oke, B. O. (2005). Effect of eggshell powder on the stabilizing potential of lime on an expansive clay soil. *Research Journal of Agriculture and Biological Sciences*, *1*, 80–84.

ASTM Annual Book of Standards. (2004). *Cement; Lime* (Vol. 4). Gypsum.

ASTM C 642–82 (1995) Test method for specific gravity, absorption and voids in hardened concrete. Annual book of ASTM standards, vol. 04.02.

Bonavetti, V., Donza, H., Menédez, G., Cabrera, O., & Irassar, E. F. (2003). Limestone Filler Cement in Low w/c Concrete: A Rational Use of Energy. *Cement and Concrete Research*, *33*(6), 865–871. doi:10.1016/S0008-8846(02)01087-6

Hawkins, P., Tennis, P., & Detwiler, R. (2003). The use of limestonein Portland cement: A state-of-the-art review, EB227. *Portland Cement Association*.

Markle Foundation. (2003). *Connecting for Health—A Public–Private Collaborative: Key Themes and Guiding Principles*. Markle Foundation.

Metzger, J., & Fortin, J. (2003). *California Health Care Foundation. Computerized Physician Order Entry in Community Hospitals*. http://www.fcg.com/healthcare/ps-cpoe-research-and-publications.asp

Mitchell, T. (1999). Machine learning and data mining. *Communications of the ACM*, *42*(11), 31–36. doi:10.1145/319382.319388

Olarewaju, A. J., Balogun, M. O., & Akinlolu, S. O. (2011). Suitability of Eggshell Stabilized Lateritic Soil as Subgrade Material for RoadConstruction. *EJGE*, *16*, 899–908.

Yerramala. (2014). Properties of concrete with eggshell powder as cement replacement. *Indian Concrete Journal*.

Compilation of References

(2018). Dr. S. Mohan Kumar and Darpan Majumdar (2018). Healthcare Solution Based on machine Learning Applications in IoT and Edge Computing. *International Journal of Pure and Applied Mathematics. Volume, 16*(119), 1473–1484.

'Philips' Intelligent Pill Targets Drug Development and Treatment for Digestive Tract Diseases. (2008). Available: https://phys.org/news/2008-11-philips-intelligent-pill-drug-treatment.html

Abiodun, O. I., Jantan, A., Omolara, A. E., Dada, K. V., Mohamed, N. A., & Arshad, H. (2018). State-of-the-art in artificial neural network applications: A survey. *Heliyon, 4*(11), e00938. doi:10.1016/j.heliyon.2018.e00938 PMID:30519653

Agatonovic-Kustrin, S., & Beresford, R. (2000). Basic concepts of Artificial Neural Network (ANN) modeling and its application in pharmaceutical research. *Journal of Pharmaceutical and Biomedical Analysis, 22*(5), 717–727. doi:10.1016/S0731-7085(99)00272-1 PMID:10815714

Ahmed, M., Yamany, S., Mohamed, N., Farag, A. A., & Moriarty, T. (2002). A modified fuzzy c-means algorithm for bias field estimation and segmentation of MRI data. *IEEE Transactions on Medical Imaging, 21*(3), 1939. doi:10.1109/42.996338 PMID:11989844

Akselrod-Ballin, A., Karlinsky, L., Alpert, S., Hasoul, S., Ben-Ari, R., & Barkan, E. (2016). A region based convolutional network for tumor detection and classification in breast mammography. In *Deep learning and data labeling for medical applications* (pp. 197–205). Springer. doi:10.1007/978-3-319-46976-8_21

Akyildiz, I. F., Su, W., Sankarasubramaniam, Y., & Cayirci, E. (2002). Wireless sensor networks: A survey, computer networks. *Computer Networks, 38*(4), 393–422. doi:10.1016/S1389-1286(01)00302-4

Akyildiz, I. F., Su, W., Senkorasubramanioam, Y., & Cayirci, E. (2002). Wireless Sensor Networks a Survey. *Computer Networks, 28*(4), 383–422.

Alemdar, H., & Ersoy, C. (2010). Wireless sensor networks for healthcare: A survey. *Computer Networks, 54*(15), 2688–2710. doi:10.1016/j.comnet.2010.05.003

Alesanco, Á., & García, J. (2010). Clinical assessment of wireless ECG transmission in real-time cardiac telemonitoring. *IEEE Transactions on Information Technology in Biomedicine, 14*(5), 1144–1152. doi:10.1109/TITB.2010.2047650 PMID:20378476

Alex Krizhevsky, I. S. (2012). ImageNet Classification with Deep Convolutional Neural Networks. *Advances in Neural Information Processing Systems 25(2).*

Al-Fuqaha, Guizani, Mohammadi, Aledhari, & Ayyash. (2015). Internet of Things: A Survey on Enabling Technologies, Protocols, and Applications. IEEE Communication Surveys & Tutorials, 17(4).

Ali & Smith. (2016). On Learning Algorithm Selection for Classification. *Journal on Applied Soft Computing, 6*(2), 119–138.

Alpaydin, E. (2010). *Introduction to machine learning* (2nd ed.). MIT Press.

Alvarez, L., Guichard, F., Lions, P.L., & Morel, J.M. (1992). Axiomes et ´equations fundamentals du traitement d'images. *Paris 315. C. R. Acad. Sci.*, 135–138.

Alvarez, L., Lions, P. L., & Morel, J. M. (1992). Image selective smoothing and edge detection by nonlinear diffusion. *SIAM Journal on Numerical Analysis, 29*(3), 845–866. doi:10.1137/0729052

Amato, F., López, A., Peña-Méndez, E. M., Hamp, P. A., & Havel, J. (2013). Artificial Neural Networks in medical diagnosis, *Journal of Applied Biomedicine. Appl. Biomed, 11*(2), 47–58. doi:10.2478/v10136-012-0031-x

Amu, O. O., Fajobi, A. B., & Oke, B. O. (2005). Effect of eggshell powder on the stabilizing potential of lime on an expansive clay soil. *Research Journal of Agriculture and Biological Sciences, 1*, 80–84.

Anagnostopoulos, C. N., & Giannoukos, T. I. (2015). Features and classifiers for emotion recognition from speech: A survey from 2000 to 2011. *Artificial Intelligence Review, 43*(2), 155–177. doi:10.100710462-012-9368-5

Ando, S. (2000). Consistent gradient operators. *IEEE Transactions on Pattern Analysis and Machine Intelligence, 22*(3), 252–265. doi:10.1109/34.841757

Angenent, S., Haker, S., & Tannenbaum, A. (2003). Minimizing flows for the Monge-Kantorovich problem. *SIAM Journal on Mathematical Analysis, 35*(1), 61–97. doi:10.1137/S0036141002410927

Antiochia, R., Lavagnini, I., & Magno, F. (2004). Amperometric Mediated Carbon Nanotube Paste Biosensor for Fructose Determinatio. *Analytical Letters, 37*(8), 1657–1669. doi:10.1081/AL-120037594

Asada, H. H., Shaltis, P., Reisner, A., Rhee, S., & Hutchinson, R. C. (2003). Mobile monitoring with wearable photoplethysmographic biosensors. IEEE Engineering in Medicine and Biology Magazine, 2003(22), 28–40. doi:10.1109/MEMB.2003.1213624 PubMed

Ashburner, J. (2007). A fast diffeomorphic image registration algorithm. *NeuroImage, 38*(1), 95–113. doi:10.1016/j.neuroimage.2007.07.007 PMID:17761438

ASTM Annual Book of Standards. (2004). *Cement; Lime* (Vol. 4). Gypsum.

ASTM C 642–82 (1995) Test method for specific gravity, absorption and voids in hardened concrete. Annual book of ASTM standards, vol. 04.02.

Aubreville, M., Krappmann, M., Bertram, C., Klopfleisch, R., & Maier, A. (2017). A guided spatial transformer network for histology cell differenti- ation. In T. E. Association (Ed.), Eurographics workshop on visual computing for biology and medicine (pp. 21–25). Academic Press.

Aubreville, M., Knipfer, C., Oetter, N., Jaremenko, C., Rodner, E., Denzler, J., Bohr, C., Neumann, H., Stelzle, F., & Maier, A. (2017). Automatic classification of cancerous tissue in laser endomicroscopy images of the oral cavity using deep learning. *Scientific Reports, 7*(1), 41598–017. doi:10.103841598-017-12320-8 PMID:28931888

Aubreville, M., Stöve, M., Oetter, N., de Jesus Goncalves, M., Knipfer, C., Neumann, H., & ... . (2018). Deep learning-based detection of motion artifacts in probe-based confocal laser endomicroscopy images. *International Journal of Computer Assisted Radiology and Surgery.* Advance online publication. doi:10.100711548-018-1836-1 PMID:30078151

Beck, T., & Levine, R. (2012). Industry growth and capital allocation: Does having a market- or bank-based system matter? *Journal of Financial Economics, 64*(2), 147–180. doi:10.1016/S0304-405X(02)00074-0

Bellman, R. (1957). *Dynamic Programming*. Princeton University Press.

Bengio, Y. (1994). Learning long-term dependencies with gradient descent is difficult. *IEEE Transactions on Neural Networks*, 157-66. doi:10.1109/72.279181

Bengio, P. L. (2007). *Greedy layerwise training of deep networks*. NIPS.

Bergveld, P. (1996). The Future of Biosensors. *Sensors and Actuators. A, Physical, 56*, 65–73.

Bertini, M., Marcantoni, L., Toselli, T., & Ferrari, R. (2016). Remote monitoring of implantable devices: Should we continue to ignore it? *International Journal of Cardiology, 202*, 368–377. doi:10.1016/j.ijcard.2015.09.033 PMID:26432486

Bherenji, H. R., & Khedkar, P. (1992). Learning and Tuning Fuzzy Logic Controllers through Reinforcements. *IEEE Transactions on Neural Networks, 3*(5), 724–740. doi:10.1109/72.159061 PMID:18276471

Bilitewski, U., & Turner, A. P. F. (Eds.). (2000). *Biosensors for Environmental Monitoring*. Harwood Academic.

Bishop, C. M. (1995). *Neural Networks for Pattern Recognition*. Oxford University Press.

Bishop, C. M. (2006). *Pattern recognition and machine learning*. Springer.

Bloch, F. (1946). Nuclear induction. *Physical Review, 70*, 460474.

Bloembergen, N., Purcell, E. M., & Pound, R. V. (1948). Relaxation effects in nuclear magnetic resonance absorption. *Physical Review, 73*(7), 679–712. doi:10.1103/PhysRev.73.679

BodyBugg. (n.d.). http://www.bodybugg.com/

Boland, G. W., Duszak, R. Jr, McGinty, G., & Allen, B. Jr. (2014). Delivery of appropriateness, quality, safety, efficiency and patient satisfaction. *Journal of the American College of Radiology, 11*(1), 7–11. doi:10.1016/j.jacr.2013.07.016 PMID:24387963

Bonavetti, V., Donza, H., Menédez, G., Cabrera, O., & Irassar, E. F. (2003). Limestone Filler Cement in Low w/c Concrete: A Rational Use of Energy. *Cement and Concrete Research, 33*(6), 865–871. doi:10.1016/S0008-8846(02)01087-6

Breininger, K., & Würfl, T. (2018). *Tutorial: how to build a deep learning framework*. https://github.com/kbreininger/tutorial-dlframework

Brett, C. M. A., & Brett, A. M. O. (1994). *Electrochemistry Principles, Methods, And Applications*. Oxford University Press.

Briot, J., Hadjeres, G., & Pachet, F. (2017). *Deep learning techniques for music generation – a survey*. CoRR abs/1709.01620

Brownlee, J. (2019). Applications of Deep Learning for Computer Vision. *Machine Learning Mastery*.

Brownlee, J. (2019). How to Visualize Filters and Feature Maps in Convolutional Neural Networks. *Deep Learning for Computer Vision*. Retrieved from https://machinelearningmastery.com/how-to-visualize-filters-and-feature-maps-in-convolutional-neural-networks/

Cai, H., Xu, Y., He, P., & Fang, Y. (2003). Indicator Free DNA Hybridization Detection by Impedance Measurement Based on the DNA-Doped Conducting Polymer Film Formed on the Carbon Nanotube. Modified Electrode Electroanalysis, 9(15), 1864–1870. doi:10.1002/elan.200302755

Cai, H., Cao, X., Jiang, Y., He, P., & Fang, Y. (2003). Carbon Nanotube- Enhanced Electrochemical DNA Biosensor for DNA Hybridization Detection. *Analytical and Bioanalytical Chemistry, 375*(2), 287–293. doi:10.1007/s00216-002-1652-9 PubMed

Camgoz, S. H. (2016). Using Convolutional 3D Neural Networks for User-independent continuous gesture recognition. *International Conference on Pattern Recognition (ICPR)*, 49-54. 10.1109/ICPR.2016.7899606

Canet, D. (1996). *Nuclear Magnetic Resonance: Concepts and Methods*. John Wiley & Sons.

Cao, Y., Chen, S., Hou, P., & Brown, D. (2015). FAST: A fog computing assisted distributed analytics system to monitor fall for stroke mitigation. *Proc. IEEE Int. Conf. Netw., Archit. Storage (NAS)*, 2–11.

Catarinucci, L., de Donno, D., Mainetti, L., Palano, L., Patrono, L., Stefanizzi, M. L., & Tarricone, L. (2016). An IoT-aware architecture for smart healthcare systems. *IEEE Internet Things J.*, 2(6), 515–526. doi:10.1109/JIOT.2015.2417684

Chang, K., Hightower, J., & Kveton, B. (2009). Inferring identity using accelerometers in television remote controls. PerCom '09.

Chen, K., & Seuret, M. (2017). *Convolutional Neural Networks for page segmentation of historical document images*. Academic Press.

Chen, S., Zhong, X., Hu, S., Dorn, S., Kachelriess, M., & Lell, M. (2018). Automatic multi-organ segmentation in dual energy CT using 3D fully convolutional network. In B. van Ginneken & M. Welling (Eds.), MIDL. Academic Press.

Chen, L., Jones, A. L., Mair, G., Patel, R., Gontsarova, A., & Ganesalingam, J. (2018). Rapid automated quantification of cerebral leukoaraiosis on CT images: A multicenter validation study. *Radiology*, 288(2), 573–581. doi:10.1148/radiol.2018171567 PMID:29762091

Chen, L., Papandreou, G., Kokkinos, I., Murphy, K., & Yuille, A. L. (2018b). DeepLab: Semantic Image 15 Segmentation with Deep Convolutional Nets, Atrous Convolution, and Fully Connected CRFs. IEEE 16. *IEEE Transactions on Pattern Analysis and Machine Intelligence*, 40(4), 834–848. doi:10.1109/TPAMI.2017.2699184 PubMed

Chopra, S., & White, L. F. (2011). *A legal theory for autonomous artificial agents*. The University of Michigan Press. doi:10.3998/mpub.356801

Clark, K., Vendt, B., Smith, K., Freymann, J., Kirby, J., Koppel, P., Moore, S., Phillips, S., Maffitt, D., Pringle, M., Tarbox, L., & Prior, F. (2013). The Cancer Imaging Archive (TCIA): Maintaining and operating a public information repository. *Journal of Digital Imaging*, 26(6), 1045–1057. doi:10.100710278-013-9622-7 PMID:23884657

Clevert, D.-A., & Unterthiner, T. H. (2016). *Fast and Accurate Deep Network Learning by Exponential Linear Units (ELUs)*. ICLR.

Coenen, A., Kim, Y. H., Kruk, M., Tesche, C., De Geer, J., Kurata, A., Lubbers, M. L., Daemen, J., Itu, L., & Rapaka Giliker, P. (2011). Vicarious liability or liability for the acts of others in tort: A comparative perspective. *J Eur Tort Law.*, 2(1), 31–56.

Collobert, Kavukcuoglu, & Farabet. (2011). *Torch7: A Matlab-like Environment for Machine Learning*. Academic Press.

Craciunescu, R., Mihovska, A., Mihaylov, M., Kyriazakos, S., Prasad, R., & Halunga, S. (2015). Implementation of Fog computing for reliable E-health applications. *Proc. 49th Asilomar Conf. Signals, Syst. Comput.*, 459–463. 10.1109/ACSSC.2015.7421170

Crow, W. T., & Wood, E. F. (2003). The assimilation of remotely sensed soil brightness temperature imagery into a land surface model using ensemble Kalman filtering: A case study based on ESTAR measurements during SGP97. *Advances in Water Resources*, 26(2), 137–149. doi:10.1016/S0309-1708(02)00088-X

Cuirel, L B., Romero, S.M., Cuireses, A., & Alvarez, R. L. (2019). Deep learning and Big data in health care: A double review for critical learners. *Appl.Sci, 9*(11), 2331.

D'Souza, S. F. (1999). Immobilized enzymes in bioprocess. *Current Science, 77,* 69–79.

D'Souza, S. F. (2001). Microbial biosensors. *Biosensors & Bioelectronics, 16*(6), 337–353. doi:10.1016/S0956-5663(01)00125-7 PubMed

da Cunha, L., Zhou, J., & Do, M. N. (2006). The Non subsampled contourlet transform: Theory, design, and applications. *IEEE Transactions on Image Processing, 15*(10), 3089–3101. doi:10.1109/TIP.2006.877507 PMID:17022272

Dahl, G. E., Yu, D., Deng, L., & Acero, A. (2012). Context-dependent pre-trained deep neural networks for large-vocabulary speech recognition. *IEEE Trans Actions Audio Speech Lang Process, 20*(1), 30–42. doi:10.1109/TASL.2011.2134090

Dakopoulos, D., & Bourbakis, N. G. (2010). Wearable obstacle avoidance electronic travel aids for blind: A survey. *IEEE Transactions on Systems, Man, and Cybernetics, 40*(1), 25–35. doi:10.1109/TSMCC.2009.2021255

Damodaram, R., & Valarmathi, M. L. (2012). *Experimental study on meta heuristic optimization algorithms.* Academic Press.

Davenport, T. H., & Lucker, J. (2015). Running on data. *Deloitte Review,* (16), 5–15.

Dawson, R., & Lavori, P. W. (2004). Placebo-free designs for evaluating new mental health treatments: The use of adaptive treatment strategies. *Statistics in Medicine, 23*(21), 3249–3262. doi:10.1002im.1920 PMID:15490427

De Leon, D. (2012). Unconstrained optimization with several variables, in math 232- mathematical models with technology, spring 2012 lecture notes. Department of Mathematics, California State University, Fresno.

Deb, K. (2014). Optimization for engineering design. Algorithms and examples. PHI Learning Private Limited.

Deb, K., Pratap, A., Agarwal, S., & Meyarivan, T. (2002). A fast and elitist multiobjective genetic algorithm: Nsga-ii. *IEEE Transactions on Evolutionary Computation, 6*(2), 182–197. doi:10.1109/4235.996017

Deb, K., Pratap, A., & Meyarivan, T. (2001). Constrained test problems for multi-objective evolutionary optimization. In *Evolutionary Multi-Criterion Optimization* (pp. 284–298). Springer. doi:10.1007/3-540-44719-9_20

Diamant, I., Bar, Y., Geva, O., Wolf, L., Zimmerman, G., Lieberman, S., & ... . (2017). Chest radiograph pathology categorization via transfer learning. In *Deep learning for medical image analysis* (pp. 299–320). Elsevier. doi:10.1016/B978-0-12-810408-8.00018-3

Digital Pills Make Their Way to Market. (2012). Available: http://blogs.nature.com/news/2012/07/digital-pills-make-their-wayto-market.html

Dolz, J., Gopinath, K., Yuan, J., Lombaert, H., Desrosiers, C., & Ben Ayed, I. (2019). Hyper Dense-Net: A Hyper-Densely Connected CNN for Multi-Modal Image Segmentation. *IEEE Transactions on Medical Imaging, 38*(5), 1116–1126. doi:10.1109/TMI.2018.2878669 PubMed

Dou, Q. C. H., Chen, H., Yu, L., Qin, J., & Heng, P.-A. (2017). Multilevel Contextual 3-D CNNs for False Positive Reduction in Pulmonary Nodule Detection. *IEEE Transactions on Biomedical Engineering, 64*(7), 1558–1567. doi:10.1109/TBME.2016.2613502 PMID:28113302

Eberhart, R. C., & Shi, Y. (2001). Particle swarm optimization: developments, applications and resources. In *Evolutionary Computation (CEC'01), Proceedings of the 2001 Congress on,* (vol. 1, pp. 81–86). 10.1109/CEC.2001.934374

Efford, N. (2000). *Segmentation. Digital Image Processing: A Practical Introduction Using Java*™. Pearson Education.

Eide, R. (2016). *Low energy wireless ECG: An exploration of wireless electrocardiography and the utilization of low energy sensors for clinical ambulatory patient monitoring* (M.S. thesis). Dept. Comput. Inf. Sci., Norwegian Univ. Sci. Technol., Trondheim, Norway.

Ekram, H., & Bhargava, V. K. (2007). *Cognitive Wireless Communications Networks*. Springer Science & Business Media.

European Commission. (2016). *Protection of personal data*. https://ec.europa.eu/justice/data-protection/

Falconer, D., Adachi, F., & Gudmundson, B. (1995). Time division multiple access methods for wireless personal Communications. *IEEE Communications Magazine*, *33*(1), 50–57. doi:10.1109/35.339881

Farandos, N. M., Yetisen, A. K., Monteiro, M. J., Lowe, C. R., & Yun, S. H. (2015). Contact lens sensors in ocular diagnostics. *Advanced Healthcare Materials*, *4*(6), 792–810. doi:10.1002/adhm.201400504 PMID:25400274

Ferguson, J. (2019). Neural Networks in Healthcare. *Journal of Healthcare*. Retrieved from https://royaljay.com/healthcare/neural-networks-in-healthcare/

Financial Services Institute of Europe. (2016). *A discussion paper on Regulating Foreign Direct Investment in Europe*. Author.

Firouzi, F., Rahmani, A. M., Mankodiya, K., Badaroglu, M., Merrett, G. V., Wong, P., & Farahani, B. (2018). Internet-of-Things and big data for smarter healthcare: From device to architecture, applications and analytics. *Future Generation Computer Systems*, *78*(2), 583–586. doi:10.1016/j.future.2017.09.016

FitBit. (n.d.). http://www.tbit.com/

Frakt, A. B., & Pizer, S. D. (2016). The promise and perils of big data in health care. *The American Journal of Managed Care*, *22*(2), 98. PMID:26885669

Franklin, A., & Douglas, G. (2010). *Comparing □nancial systems*. MIT Press.

Friedman, U. (2012). Big Data: A Short History. *Foreign Policy*.

Fukushima, K. (1979). Neural network model for a mechanism of pattern recognition unaffected by shift in position. *Neocognitron*, 658–665.

Ganesh Babu, R., Karthika, P., & AravindaRajan, V. (2020). Secure IoT Systems Using Raspberry Pi Machine Learning Artificial Intelligence. *Lecture Notes on Data Engineering and Communications Technologies, 44*, 797-805.

Ganesh Babu, R., & Amudha, V. (2016). Cluster Technique Based Channel Sensing in Cognitive Radio Networks. *International Journal of Control Theory and Applications*, *9*(5), 207–213.

Ganesh Babu, R., & Amudha, V. (2018c). A Survey on Artificial Intelligence Techniques in Cognitive Radio Networks. *Proceedings of 1st International Conference on Emerging Technologies in Data Mining and Information Security, (IEMIS)in association with Springer Advances in Intelligent Systems and Computing Series*, 99-110.

Garcia-Garcia, A., Orts-Escolano, S., Oprea, S. O., Villena-Martinez, V., & Garcia-Rodriguez, J. (2017). *A Review on Deep Learning Techniques Applied to Semantic Segmentation*, arXiv: 1704.06857

Gharehchopogh, F. S., Maryam, M., & Freshte, D. M. (2013). Using Artificial Neural Network in Diagnosis of Thyroid Disease: A Case Study. *International Journal on Computational Sciences and Applications, 3*.

Gia, T. N., Jiang, M., Rahmani, A.-M., Westerlund, T., Liljeberg, P., & Tenhunen, H. (2015). Fog computing in healthcare Internet of Things: A case study on ECG feature extraction. *Proc. IEEE Int. Conf. Comput. Inf. Technol., Ubiquitous Comput. Commun., Dependable, Auto. Secur. Comput., Pervasive Intell. Comput. (CIT/IUCC/DASC/PICOM)*, 356–363. 10.1109/CIT/IUCC/DASC/PICOM.2015.51

Gietzelt, M., Wolf, K. H., & Haux, R. (2011). A nomenclature for the analysis of continuous sensor and other data in the context of health-enabling technologies. *Studies in Health Technology and Informatics*, *169*, 460–464. PubMed

Glorot, X. B., & Bengio, Y. (2010). *Understanding the difficulty of training deep feed forward neural*. http://proceedings.mlr.press/v9/glorot10a/glorot10a.pdf

Gubbi, J., Buyya, R., Marusic, S., & Palaniswami, M. (2013). Internet of Things (IoT): A vision, architectural elements, and future directions. *Future Generation Computer Systems*, *29*(7), 1645–1660. doi:10.1016/j.future.2013.01.010

Hammernik, K., Klatzer, T., Kobler, E., Recht, M. P., Sodickson, D. K., Pock, T., & Knoll, F. (2018). Learning a variational network for reconstruction of accelerated mri data. *Magnetic Resonance in Medicine*, *79*(6), 3055–3071. doi:10.1002/mrm.26977 PMID:29115689

Hank, P., Muller, S., Vermesan, O., & Van Den Keybus, J. (2013). Automotive ethernet: in-vehicle networking and smart mobility. In *Proceedings of the Conference on Design, Automation and Test in Europe*. EDA Consortium. 10.7873/DATE.2013.349

Hashemi, R. H., & Bradley, W. G. (1997). *MRI: The Basics*. Baltimore, MD: Williams & Williams. www.cis.rit.edu/htbooks/mri/inside.htm

Hawkins, P., Tennis, P., & Detwiler, R. (2003). The use of limestone in Portland cement: A state-of-the-art review, EB227. *Portland Cement Association*.

He, X., & Zhang, S. R. (2015). *Deep residual learning for image recognition*. ArXiv preprint arXiv:1512.03385

He, D., & Zeadally, S. (2015). *An Analysis of RFID Authentication Schemes for Internet of Things in Healthcare Environment Using Elliptic Curve Cryptography*. doi:10.1109/JIOT.2014.2360121

Heinzelman, W., Chandrakasan, A., & Balakrishnan, H. (2000). Energy-Efficient Communication Protocol for Wireless Microsensor Networks. Proceedings of the Hawaii Conference on System Sciences, 1-10. doi:10.1109/HICSS.2000.926982

Hesamian, M. H., Jia, W., He, X., & Kennedy, P. (2019). Deep Learning Techniques for Medical Image Segmentation: Achievements and Challenges. *Journal of Digital Imaging*, *32*(4), 582–596. doi:10.100710278-019-00227-x PMID:31144149

Hinkelmann, K. (2016). *Neural Networks*. University of Applied Sciences Northwestern Switzerland.

Hinton, G. E., Osindero, S., & Teh, Y.-W. (2006). A fast learning algorithm for deep belief nets. *Neural Computation*, *18*(7), 1527–1554. doi:10.1162/neco.2006.18.7.1527 PMID:16764513

Hinton, G. E., & Salakhutdinov, R. R. (2006). Reducing the dimensionality of data with neural networks. *Science*, *313*(5786), 504–507. doi:10.1126cience.1127647 PMID:16873662

Hoang, D. B., & Chen, L. (2010). Mobile Cloud for Assistive Healthcare (MoCAsH). Services Computing Conference (APSCC), 2010 IEEE Asia-Pacific, 325-332.

Holzinger, A. (2016). Project iML – proof of concept interactive machine learning. *Brain Inform*, *3*(2), 119–131.

Holzinger, A. (2016). Project iML – proof of concept interactive machine learning. *Brain Informatics*, *3*(2), 119–131. doi:10.100740708-016-0042-6 PMID:27747607

Hossain, M. S., & Muhammad, G. (2015). Cloud-assisted speech and face recognition framework for health monitoring. *Mobile Networks and Applications*, *20*(3), 391–399. doi:10.100711036-015-0586-3

Hossain, M. S., & Muhammad, G. (2016). Cloud-assisted Industrial Internet of Things (IIoT) - enabled framework for health monitoring. *Computer Networks*, *101*, 192–202. doi:10.1016/j.comnet.2016.01.009

Hossain, M. S., Muhammad, G., Alhamid, M. F., Song, B., & Al-Mutib, K. (2016). Audio-Visual Emotion Recognition Using Big Data Towards 5G. *Mobile Networks and Applications*, *221*(5), 753–763. doi:10.100711036-016-0685-9

Hossain, M. S., Muhammad, G., Al-Qurishi, M., Masud, M., Almogren, A., Abdul, W., & Alamri, A. (2017). Cloud-Oriented Emotion Feedback-based Exergames Framework. *Multimedia Tools and Applications*. Advance online publication. doi:10.100711042-017-4621-1

Hosseini, M.-P., Hajisami, A., & Pompili, D. (2016). Real-time epileptic seizure detection from eeg signals via random subspace ensemble learning. *Proc. IEEE Int. Conf. Auto. Comput. (ICAC)*, 209–218. 10.1109/ICAC.2016.57

Huang, G., Liu, Z., van der Maaten, L., & Weinberger, K. Q. (2017). *Densely Connected Convolutional Networks*. https://arxiv.org/abs/1608.06993v5

Huang, G. S. (2016). Deep networks with stochastic depth. *European Conference on Computer Vision*, 646–661.

I. Corporation. (2016). *Bigger data for better healthcare*. Tech. Rep.

IBM. (2015). *Watson for a Smarter Planet: Healthcare*. Available at: https://www-03.ibm.com/innovation/us/watson/

Ihaque, I. R., & Neubert, J. (2020). Deep leaning approaches to biomedical image segmentation. *Informatics Medicine Unlocked, 18*.

Ioffe, S., & Szegedy, C. (2015). *Batch normalization: Accelerating deep network training by reducing internal covariate shift*. http://proceedings.mlr.press/v37/ioffe15.pdf

Jang, R. (1992). *Neuro-Fuzzy Modeling: Architectures, Analyses and Applications* (PhD Thesis). University of California, Berkeley, CA.

Jenke, R., Peer, A., & Buss, M. (2014). Feature Extraction and Selection for Emotion Recognition from EEG. *IEEE Transactions on Affective Computing*, *5*(3), 327–339. doi:10.1109/TAFFC.2014.2339834

Jevcak, A., Setzer, R., & Suardi, M. (2010). Determinants of capital flows to the new EU member states before and during the financial crisis. European Commission. *Economic Papers*.

Jia, Y., Shelhamer, E., Donahue, J., Karayev, S., Long, J., Girshick, R., Guadarrama, S., & Darrell, T. (2014). Caffe: Convolutional architecture for fast feature embedding. In *Proceedings of the 22nd ACM international conference on Multimedia*. ACM. 10.1145/2647868.2654889

Johansen, S. (2011). Statistical analysis of cointegrating vectors. *Journal of Economic Dynamics & Control*, *12*(2-3), 231–254. doi:10.1016/0165-1889(88)90041-3

Johnson, K. W., Soto, J. T., & Benjamin, S. (2018). Artificial Intelligence in Cardiology. *Journal of the American College of Cardiology*, *71*(23), 2668–2679. doi:10.1016/j.jacc.2018.03.521 PMID:29880128

Juang, C. F., & Lin, C. T. (1998). An Online Self Constructing Neural Fuzzy Inference Network and its Applications. *IEEE Transactions on Fuzzy Systems*, *6*(1), 12–32. doi:10.1109/91.660805

Kahou, S. E., Bouthillier, X., Lamblin, P., Gulcehre, C., Michalski, V., Konda, K., Jean, S., Froumenty, P., Dauphin, Y., Boulanger-Lewandowski, N., Chandias Ferrari, R., Mirza, M., Warde-Farley, D., Courville, A., Vincent, P., Memisevic, R., Pal, C., & Bengio, Y. (2016, June). EmoNets: Multimodal deep learning approaches for emotion recognition in video. *Journal on Multimodal User Interfaces*, *10*(2), 99–111. doi:10.100712193-015-0195-2

Karl, H., & Willig, A. (2003). *A Short Survey of Wireless Sensor Networks*. Technical University Berlin, Telecommunication Networks Group.

Karthick Raghunath, K. M., & Thirukumaran, S. (2019). "Fuzzy-based fault-tolerant and instant synchronization routing technique in wireless sensor network for rapid transit system", Automatika: Journal for Control, Measurement, Electronics. *Computer Communications*, *60*(5), 547–554.

Karthika & Vidhyasaraswathi. (2018). *Digital Video Copy Detection Using Steganography Frame Based Fusion Techniques*. Academic Press.

Karthika & Vidhyasaraswathi. (2020). *Raspberry Pi: A Tool for Strategic Machine Learning Security Allocation in IoT*. Academic Press.

Karthika, P., Ganesh Babu, R., & Nedumaran, A. (2019). Machine Learning Security Allocation in IoT. *IEEE International Conference on Intelligent Computing and Control Systems*.

Karthika, P., & Vidhyasaraswathi, P. (2017). *Content based video copy detection using frame based fusion technique*. Academic Press.

Karthika, P., & Vidhyasaraswathi, P. (2019). *Image Security Performance Analysis for SVM and ANN Classification Techniques*. Academic Press.

Ker, J., Wang, L., Rao, J., & Lim, T. (2017). Deep Learning Applications in Medical Image Analysis. *IEEE Access: Practical Innovations, Open Solutions*, *6*, 9375–9389. doi:10.1109/ACCESS.2017.2788044

Khushaba, R. N., Kodagoda, S., Takruri, M., & Dissanayake, G. (2012). Toward improved control of prosthetic fingers using surface electromyogram (EMG) signals. *Expert Systems with Applications*, *39*(12), 10731–10738. doi:10.1016/j.eswa.2012.02.192

Kim, J., Hong, J., Park, H., Kim, J., Hong, J., & Park, H. (2018). Prospects of deep learning for medical imaging. *Precis Future Med.*, *2*(2), 37–52. doi:10.23838/pfm.2018.00030

Kim, Y., Lee, H., & Provost, E. M. (2013). Deep learning for robust feature generation in audiovisual emotion recognition. *2013 IEEE International Conference on Acoustics, Speech and Signal Processing*, 3687-3691. 10.1109/ICASSP.2013.6638346

King, B. F. Jr. (2018). Artificial intelligence and radiology: What will the future hold? *Journal of the American College of Radiology*, *15*(3), 1–3. doi:10.1016/j.jacr.2017.11.017 PMID:29371088

King, S. B. III. (2016). Big Trials or Big Data. *JACC: Cardiovascular Interventions*, *9*(8), 869–870. doi:10.1016/j.jcin.2016.03.003 PMID:27101918

Kohli, M., Summers, R., & Geis, R. (2017). Medical image data and datasets in the era of machine learning—Whitepaper from the 2016 C-MIMI meeting dataset session. *Journal of Digital Imaging*, *30*(4), 392–399. doi:10.100710278-017-9976-3 PMID:28516233

Komodakis, S. Z. (2016). *Wide residual networks*. BMVC.

Kraamer, B., Tamkittikhun, & Palma. (2017). Fog Computing in Healthcare – A Review and Discussion. *IEEE Access : Practical Innovations, Open Solutions*; Advance online publication. doi:10.1109/Access.2017.2704100

Krauss, Do, & Huck. (2017). Deep neural networks,gradient boosted trees, random forests: Statistical arbitrage on the S&P 500. *European Journal of Operational Research.*

Krittanawong, C., Zhang, H., Wang, Z., Aydar, M., & Kitai, T. (2017). Artificial intelligence in precision cardiovascular medicine. *Journal of the American College of Cardiology, 69*(21), 2657–2664. doi:10.1016/j.jacc.2017.03.571 PMID:28545640

Krizhevsky, A., Sutskever, I., & Hinton, G. E. (2012). ImageNET classification with deep convolutional neural networks. Advances in Neural Information Processing Systems, 1097–105.

Krizhevsky, A., Sutskever, I., & Hinton, G. E. (2012). ImageNet classification with deep convolutional neural networks. *Proceedings of the 25th International Conference on Neural Information Processing Systems*, 1097-1105.

Krstevska, A., & Petrovska, M. (2012). The economic impacts of the foreign direct investments: Panel estimation by sectors on the case of Macedonian economy. Journal of Central Banking Theory and Practice, 55-73.

Kumar, A., Fulham, M., Feng, D., & Kim, J. (2018). Co-Learning Feature Fusion Maps from PET-CT Images of Lung Cancer. IEEE Transactions on Medical Imaging. Advance online publication. PubMed doi:10.1109/TMI.2019.2923601

Kumar, L. D., Grace, S. S., Krishnan, A., Manikandan, V., Chinraj, R., & Sumalatha, M. (2012). Data ðltering in wireless sensor networks using neural networks for storage in cloud. In *Recent Trends In Information Technology (ICRTIT), 2012 International Conference on.* IEEE.

Kuo, A. M.-H. (2011). Opportunities and challenges of cloud computing to improve health care services. *Journal of Medical Internet Research, 13*(3), e67. doi:10.2196/jmir.1867 PMID:21937354

Lakhani, P., Gray, D. L., Pett, C. R., Nagy, P., & Shih, G. (2018). Hello world deep learning in medical imaging. *Journal of Digital Imaging, 31*(3), 283–289. doi:10.100710278-018-0079-6 PMID:29725961

Lambrianou, A., Demin, S., & Hall, E. A. (2008). Protein Engineering and Electrochemical Biosensors. *Advances in Biochemical Engineering/Biotechnology, 109*, 65–96. doi:10.1007/10_2007_080 PubMed

Lavori, P. W., & Dawson, R. (2000). A design for testing clinical strategies: Biased adaptive within-subject randomization. *Journal of the Royal Statistical Society. Series A (General), 163*(1), 29–38. doi:10.1111/1467-985X.00154

Lavrac, N. (1999). Selected techniques for data mining in medicine. *Journal of Artificial Intelligence in Medicine, 16*(1), 3–23. doi:10.1016/S0933-3657(98)00062-1 PMID:10225344

LeCun, Y., Bengio, Y., & Hinton, G. (2015). Deep learning. *Nature, 521*(7553), 436–444. doi:10.1038/nature14539 PMID:26017442

Lee, J. G., Jun, S., Cho, Y.-W., Lee, H., Kim, G. B., Seo, J. B., & Kim, N. (2017). Deep learning in Medical imaging: General overview. *Korean Journal of Radiology, 18*(4), 570–584. doi:10.3348/kjr.2017.18.4.570 PMID:28670152

Liang, Z.P., & Lautenberg, P.C. (1999). *Principles of Magnetic Resonance Imaging: a signal processing perspective.* IEEE Press.

Liang, Z., Zhang, G., Huang, J. X., & Hu, Q. V. 2014. Deep learning for healthcare decision making with emrs. *Proc. Int. Conf. Bioinformat. Biomed.*, 556–559. 10.1109/BIBM.2014.6999219

Li, G., Liao, J. M., Hu, G. Q., Ma, N. Z., & Wu, P. J. (2005). Study of Carbon Nanotube Modified Biosensor Formonitoring Total Cholesterol in Blood. *Biosensors & Bioelectronics, 20*(10), 2140–2144. doi:10.1016/j.bios.2004.09.005 PubMed

Lin, C. T., & Lee, C. S. G. (1991). Neural Network based Fuzzy Logic Control and Decision System. *IEEE Transactions on Computers*, *40*(12), 1320–1336. doi:10.1109/12.106218

Link, J. A. B., Fabritius, G., Alizai, M. H., & Wehrle, K. B. (2010). Seeing the world through the eyes of rats. The 2nd IEEE International Workshop on Information Quality and Quality of Service for Pervasive Computing.

Litjens, G., Kooi, T., Bejnordi, B. E., Adiyoso Setio, A. A., Ciompi, F., Ghafoorian, M., van der Laak, J. A., van Ginneken, B., & Sánchez, C. I. (2017). A Survey on Deep Learning in Medical Image Analysis. *Medical Image Analysis*, *42*, 60–88. doi:10.1016/j.media.2017.07.005 PMID:28778026

Liu, J., Liu, J., Xu, L., & Jin, W. (2010). Electronic travel aids for the blind based on sensory substitution. Proceedings of the 2010 5th International Conference on Computer Science and Education (ICCSE). doi:10.1109/ICCSE.2010.5593738

Logan, K. (2016). What happens when big data blunders? *Communications of the ACM*, *59*(6), 15–16. doi:10.1145/2911975

Looney, Stevenson, Nicolaides, Plasencia, & Molloholli, Natsis, & Collins. (2018). Fully automated, real-time 3D ultrasound segmentation to estimate first trimester placental volume using deep learning. *JCI Insight*, *3*(10). Advance online publication. doi:10.1172/jci.insight.120178

López, G., Custodio, V., & Moreno, J. I. (2016). LOBIN: E-textile and wirelesssensor-network-based platform for healthcare monitoring in future hospital environments. *IEEE Transactions on Information Technology in Biomedicine*, *14*(6), 1446–1458. doi:10.1109/TITB.2010.2058812 PMID:20643610

Lundefvold, A. S., & Lundervold, A. (2019). An overview of deep learning in medical imaging focusing on MRI. *Zeitschrift fur Medizinische Physik*, *29*(2), 102–127. doi:10.1016/j.zemedi.2018.11.002 PMID:30553609

MacIntosh, E., Rajakulendran, N., Khayat, Z., & Wise, A. (2016). *Transforming Health: Shifting From Reactive to Proactive and Predictive Care*. Available: https://www.marsdd.com/newsand-insights/transforming-health-shifting-from-reactive-to-proactive-andpredictive-care/

Maes, F., Collignon, A., Vandermeulen, D., Marchal, G., & Suetens, P. (1997). Multimodality image registration by maximization of mutual information. *IEEE Transactions on Medical Imaging*, *16*(2), 187–198. doi:10.1109/42.563664 PMID:9101328

Maha Lakshmi, T., Swathi, R., & Srinivas, A. (2016). MATLAB Implementation of an Efficient Technique for Detection of Brain Tumor by using Watershed Segmentation and Morphological Operation. *GRD Journals- Global Research and Development Journal for Engineering, 1*(4).

Maier, A., Syben, C., Lasser, T., & Rless, C. (2019). A gentle introduction to deep learning in medical image processing. *Zeitschrift fur Medizinische Physik*, *29*(2), 86–101. doi:10.1016/j.zemedi.2018.12.003 PMID:30686613

Malhotra, B. D., Singhal, R., Chaubey, A., Sharma, S. K., & Kumar, A. (2005). Recent Trends in Biosensors. *Current Applied Physics*, *5*(2), 92–97. doi:10.1016/j.cap.2004.06.021

Mamoshina, P., Vieira, A., Putin, E., & Zhavoronkov, A. (2016). Applications of Deep Learning in Biomedicine. *Molecular Pharmaceutics*, *13*(5), 1445–1454. doi:10.1021/acs.molpharmaceut.5b00982 PMID:27007977

Mandellos, G. J., Koutelakis, G. V., Panagiotakopoulos, T. C., Koukias, M. N., & Lymberopoulos, D. K. (2009). Requirements and solutions for advanced telemedicine applications. In *Biomedical Engineering*. InTech.

Manjeshwar, A., & Agarwal, D. P. (2001). Teen: a Routing Protocol for Enhanced Efficiency in Wireless Sensor Networks. Proc. 15th Int. Parallel and Distributed Processing Symp., 2009–2015. doi:10.1109/IPDPS.2001.925197

Maolood, I. Y., Abdulridha Al-Salhi, Y. E., & Lu, S. (2018). Thresholding for medical image segmentation for cancer using fuzzy entropy with level set algorithm. *De Gruyter. Open Medicine: a Peer-Reviewed, Independent, Open-Access Journal*, *13*(1), 374–383. doi:10.1515/med-2018-0056 PMID:30211320

Markle Foundation. (2003). *Connecting for Health—A Public–Private Collaborative: Key Themes and Guiding Principles*. Markle Foundation.

Marschollek, M., Gietzelt, M., Schulze, M., Kohlmann, M., Song, B., & Wolf, K. H. (2012). Wearable Sensors in Healthcare and Sensor-Enhanced Health Information Systems: All Our Tomorrows? *Healthcare Informatics Research*, *18*(2), 97–104. doi:10.4258/hir.2012.18.2.97 PubMed

Masip-Bruin, X., Marín-Tordera, E., Alonso, A., & Garcia, J. (2016). Fog-tocloud computing (F2C): The key technology enabler for dependable ehealth services deployment. *Proc. Medit. Ad Hoc Netw. Workshop (Med-Hoc-Net)*, 1–5.

Masood, S., Gupta, S., Wajid, A., Gupta, S., & Ahmed, M. (2017). Prediction of human ethnicity from facial images using neural networks. *Data Engineering and Intelligent Computing. Advances in Intelligent Systems and Computing*, *542*, 217–226. doi:10.1007/978-981-10-3223-3_20

Matrubutham, U., & Sayler, G. S. (1998). *Enzyme and Microbial Biosensors: Techniques and Protocols*. Humana Press.

Maviglia, S. M., Zielstorff, R. D., Paterno, M., Teich, J. M., Bates, D. W., & Kuperman, D. J. (2003). Automating complex guidelines for chronic disease: Lessons learned. *Journal of the American Medical Informatics Association*, *10*(2), 154–165. doi:10.1197/jamia.M1181 PMID:12595405

Mazura, J. C., Juluru, K., Chen, J. J., Morgan, T. A., John, M., & Siegel, E. L. (2012). Facial recognition software success rate for the identification of 3D surface reconstructed facial images: Implications for patient privacy and security. *Journal of Digital Imaging*, *25*(3), 347–351. doi:10.100710278-011-9429-3 PMID:22065158

Mei, B., Cheng, W., & Cheng, X. (2015). Fog computing based ultraviolet radiation measurement via smartphones. *Proc. 3rd IEEE Workshop Hot Topics Web Syst. Technol. (HotWeb)*, 79–84.

Metzger, J., & Fortin, J. (2003). *California Health Care Foundation. Computerized Physician Order Entry in Community Hospitals*. http://www.fcg.com/healthcare/ps-cpoe-research-and-publications.asp

Miao, L., Djouani, K., Kurien, A., & Noel, G. (2012). Network Coding and Competitive Approach for Gradient Based Routing in Wireless Sensor Networks. *Ad Hoc Networks*, *10*(6), 990–1008. Advance online publication. doi:10.1016/j.adhoc.2012.01.001

Miluzzo, E., Zheng, X., Fodor, K., & Campbell, A. T. (2008). Radio characterization of 802.15.4 and its impact on the design of mobile sensor networks. Proc 5th European Conf on Wireless Sensor Networks (EWSN 08, 171–188. doi:10.1007/978-3-540-77690-1_11

Miluzzo, Lane, Fodor, Peterson, Lu, Musolesi, Eisenman, Zheng, & Campbell. (2008). Sensing meets mobile social networks: the design, implementation and evaluation of the cenceme application. In *Proceedings of the 6th ACM conference on Embedded network sensor systems*, (pp. 337–350). ACM. 10.1145/1460412.1460445

Misra, S., & Dias Thomasinous, P. (2010). A Simple, Least-Time and Energy-Efficient Routing protocol with One-Level Data Aggregation for Wireless Sensor Networks. *Journal of Systems and Software*, *83*(5), 852–860. doi:10.1016/j.jss.2009.12.021

Mitchell, T. (1999). Machine learning and data mining. *Communications of the ACM*, *42*(11), 31–36. doi:10.1145/319382.319388

Mitrovic, N., Petronijevic, M., Kostic, V., & Bankovic, B. (2011). Active Front End Converter in Common DC Bus Multidrive Application. XLVI Proc. of Inter. Conf. ICEST 2011, 3(2), 989-992.

Mnih, V., Kavukcuoglu, K., Silver, D., Rusu, A. A., Veness, J., Bellemare, M. G., Graves, A., Riedmiller, M., Fidjeland, A. K., Ostrovski, G., Petersen, S., Beattie, C., Sadik, A., Antonoglou, I., King, H., Kumaran, D., Wierstra, D., Legg, S., & Hassabis, D. (2015). Human-level control through deep reinforcement learning. *Nature*, *518*(7540), 529–533. doi:10.1038/nature14236 PMID:25719670

Moeskops, P., Viergever, M. A., Mendrik, A. M., de Vries, L. S., Benders, M. J., & Isgum, I. (2016). Ïsgum I (2016). Automatic segmentation of MR brain images with a convolutional neural network. *IEEE Transactions on Medical Imaging*, *35*(5), 1252–1261. doi:10.1109/TMI.2016.2548501 PMID:27046893

Moghimi, F. H., & Wickramasinghe, N. (2014). *Artificial Neural Network Excellence to Facilitate Lean Thinking Adoption in Healthcare Contexts.* Springer Science + Business Media. DOI doi:10.1007/978- 1-4614-8036-5_2

Monteiro, A., Dubey, H., Mahler, L., Yang, Q., & Mankodiya, K. (2016). Fit: A fog computing device for speech teletreatments. *Proc. IEEE Int. Conf. Smart Computing*, 1–3. 10.1109/SMARTCOMP.2016.7501692

Mordvintsev A, Olah C, Tyka M. (2015). Inceptionism: going deeper into neural networks. *Google Research Blog*.

Muhammad, G., Alsulaiman, M., Amin, S. U., Ghoneim, A., & Alhamid, M. (2017). A Facial-Expression Monitoring System for Improved Healthcare in Smart Cities. *IEEE Access: Practical Innovations, Open Solutions*, *5*(1), 10871–10881. doi:10.1109/ACCESS.2017.2712788

Murphy, S. A. (2003). Optimal dynamic treatment regimes (with discussion). *Journal of the Royal Statistical Society. Series B. Methodological*, *65*(2), 331–366. doi:10.1111/1467-9868.00389

Murphy, S. A. (2004). An experimental design for the development of adaptive treatment strategies. *Statistics in Medicine*, *24*(10), 1455–1481. doi:10.1002im.2022 PMID:15586395

Naik, R., Singh, J., & Le, H. P. (2010). Intelligent Communication Module for Wireless Biosensor Networks. In P. A. Serra (Ed.), *Biosensors* (pp. 225–240). InTech., doi:10.5772/7212.

Nauck, D., & Kruse, R. (1997). Neuro-Fuzzy Systems for Function Approximation. *4th International Workshop Fuzzy-Neuro Systems*.

Nazir, Ali, Ullah, & Garcia-Magarino. (2019). Internet of Things for Healthcare Using Wireless Communications or Mobile Computing. *Wireless Communications and Mobile Computing*. Doi:10.1155/2019/5931315

Neff, G. (2013). Why big data won't cure us. *Big Data*, *1*(3), 117–123. doi:10.1089/big.2013.0029 PMID:25161827

NessAiver, M. (1997). *All You Really Need to Know about MRI Physics.* Baltimore, MD: Simply Physics.

Nguyen, H. T. (1995). *Theoretical aspects of fuzzy control* (1st ed.). John Wiley.

Nirschl, J. J., Janowczyk, A., Peyster, E. G., Frank, R., Margulies, K. B., Feldman, M. D., & ... . (2017). Deep learning tissue segmentation in cardiac histopathology images. In *Deep learning for medical image analysis* (pp. 179–195). Elsevier. doi:10.1016/B978-0-12-810408-8.00011-0

Nishimura, D.G. (1996). *Principles of Magnetic Resonance Imaging.* Academic Press.

Obermeyer, Z., & Emanuel, E. J. (2016). Predicting the future—Big data, machine learning, and clinical medicine. *The New England Journal of Medicine*, *375*(13), 1216–1219. doi:10.1056/NEJMp1606181 PMID:27682033

OECD. (2012). Glossary of foreign direct investment terms and definitions. Retrieved from http://www.oecd.org/investment/investmentfordevelopment/2487495.pdf

OECD. (2017). Economic survey – Europe. doi:10.1787/eco_surveys-aus-2017-en

Oladimeji, E. A., Chung, L., Jung, H. T., & Kim, J. (2011). Managing security and privacy in ubiquitous ehealth information interchange. *Proc. 5th Int. Conf. Ubiquitous Inf. Manage. Commun. (ICUIMC),* 26:1–26:10. Available: https://doi.acm.org/10.1145/1968613.1968645

Olarewaju, A. J., Balogun, M. O., & Akinlolu, S. O. (2011). Suitability of Eggshell Stabilized Lateritic Soil as Subgrade Material for RoadConstruction. *EJGE, 16,* 899–908.

Orwat, C., Graefe, A., & Faulwasser, T. (2008). Towards pervasive computing in health care—A literature review. *BMC Medical Informatics and Decision Making, 8*(1), 26. doi:10.1186/1472-6947-8-26 PMID:18565221

Oueida, Kotb, Aloqaily, & Jararweh. (2018). An Edge Computing Based Smart Healthcare Framework For Resource Management. *Sensors, 18*(12), 4307.

Paksuniemi, M., Sorvoja, H., Alasaarela, E., & Myllyla, R. (2005). Wireless sensor and data transmission needs and technologies for patient monitoring in the operating room and intensive care unit. *Proc. 27th Annu. Int. Conf. Eng. Med. Biol. Soc. (IEEE-EMBS),* 5182–5185. 10.1109/IEMBS.2005.1615645

Paul, A. K., Banerje, I., Snatra, B. K., & Neogi, N. (2009). Application of AC Motors and Drives in Steel Industries. *XV National Power System Conference,* 6(2), 159-163.

Pereira, S., Pinto, A., Alves, V., & Silva, C. A. (2015). Deep convolutional neural networks for the segmentation of gliomas in multi-sequence MRI. *International Workshop on Brain lesion: Glioma, Multiple Sclerosis, Stroke and Traumatic Brain Injuries,* 131–143.

Perera, C., Zaslavsky, A., Christen, P., & Georgakopoulos, D. (2014). Context aware computing for the internet of things: A survey. *IEEE Commun. Surveys Tuts., 16*(1), 414–454. doi:10.1109/SURV.2013.042313.00197

Perkins, C., Das, S. R., & Royer, E. M. (2000). Performance Comparison of Two On-demand Routing Protocols for Ad Hoc Networks. *IEEE Infocom, 2000,* 3–12.

Petronijevic, M., Veselic, B., Mitrovic, N., Kostic, V., & Jeftenic, B. (2011, May). Comparative Study of Unsymmetrical Voltage Sag Effects on Adjustable Speed Induction Motor Drives. Electric Power Applications, IET, 5(5), 432–442. doi:10.1049/iet-epa.2010.0144

Pluim, J. P. W., Maintz, J. B. A., & Viergever, M. A. (2003). Mutualinformation-based registration of medical images: A survey. *IEEE Transactions on Medical Imaging, 22*(8), 986–1004. doi:10.1109/TMI.2003.815867 PMID:12906253

Ponomarev, G. V., Gelfand, M. S., & Kazanov, M. D. (2012). A multilevel thresholding combined with edge detection and shape-based recognition for segmentation of fetal ultrasound images. In *Proc. Chall. US biometric Meas. from fetal ultrasound images* (pp. 17–19). ISBI.

Prathap, M. (2020). Analysis for Time-Synchronized Channel Swapping in Wireless Sensor Network. Academic Press.

Prathap, Sharma, & Arthi. (2010). *Http server and client performances of multithreaded packets.* Academic Press.

Prior, F. W., Brunsden, B., Hildebolt, C., Nolan, T. S., Pringle, M., Vaishnavi, S. N., & Larson-Prior, L. J. (2009). Facial recognition from volume rendered magnetic resonance imaging data. *IEEE Transactions on Information Technology in Biomedicine, 13*(1), 5–9. doi:10.1109/TITB.2008.2003335 PMID:19129018

Prisilla, J., Iyyanki, M.K., & Aruna, M. (2018). Semantic Segmentation Using Fully Convolutional Net: A Review. *IOSR Journal of Engineering, 8*(9), 62-68.

Prisilla, J., & Iyyanki, M. K. (2019). Convolution Neural Networks: A case study on brain tumor segmentation in medical care. *Proc. International Conference on ISMAC-CVB. Springer Nature Switzerland AG*. 10.1007/978-3-030-00665-5_98

Putin, E., Mamoshina, P., Aliper, A., Korzinkin, M., Moskalev, A., Kolosov, A., Ostrovskiy, A., Cantor, C., Vijg, J., & Zhavoronkov, A. (2016). Deep biomarkers of human aging: Application of deep neural networks to biomarker development. *Aging (Albany NY), 8*(5), 1–21. PMID:27191382

Rao, G. S. V. R. K., & Sundararaman, K. (2010). A Pervasive Cloud initiative for primary healthcare services. Intelligence in Next Generation Networks (ICIN), 2010 14th International Conference on, 1-6.

Ravi, D., Wong, C., Deligianni, F., Berthelot, M., Andreu-Perez, J., Lo, B., & Yang, G.-Z. (2017). Deep Learning for Health Informatics. *IEEE Journal of Biomedical and Health Informatics, 21*(1), 4–21. doi:10.1109/JBHI.2016.2636665 PMID:28055930

Ravi, S. R., & Prasad, T. V. (2010). Exploration of Hybrid Neuro Fuzzy Systems. *Conference on Advances in Knowledge Management*. DOI: 10.13140/RG.2.1.3570.0327

Renier, L., & De Volder, A. G. (2010). Vision substitution and depth perception: Early blind subjects experience visual perspective through their ears. *Disability and Rehabilitation. Assistive Technology, 5*(3), 175–183. doi:10.3109/17483100903253936 PubMed

Rivas, G. A., & Pedano, M. L. (2006). Electrochemical DNA Biosensors. In *Craig Encyclopedia of Sensors*. American Scientific Publishers.

Robins, J. M. (1994). Correcting for non-compliance in randomized trials using structural nested mean models. *Communications in Statistics, 23*(8), 2379–2412. doi:10.1080/03610929408831393

Rondeau, T. W., & Bostain, C. W. (2009). *Artificial Intelligence in Wireless Communication*. Artech House.

Rosenblatt, F. (1958). The perceptron: A probabilistic model for information storage and organization in the brain. *Psychological Review, 65*(6), 386–408. doi:10.1037/h0042519 PMID:13602029

Roth, H. R., Lu, L., Farag, A., Shin, H.-C., Liu, J., & Turkbey, E. B. (2015). DeepOrgan: multi-level deep convolutional networks for automated pancreas segmentation. In *International conference on medical image computing, computer-assisted intervention*. Springer. 10.1007/978-3-319-24553-9_68

Rueckert, D., Sonoda, L. I., Hayes, C., Hill, D. L., Leach, M. O., & Hawkes, D. J. (1996). Nonrigid registration using freeform deformations: Application to breast MR images. *IEEE Transactions on Medical Imaging, 18*(8), 712–721. doi:10.1109/42.796284 PMID:10534053

Rueda, S., Fathima, S., Knight, C. L., Yaqub, M., Papageorghiou, A. T., Rahmatullah, B., Foi, A., Maggioni, M., Pepe, A., Tohka, J., Stebbing, R. V., McManigle, J. E., Ciurte, A., Bresson, X., Cuadra, M. B., Sun, C., Ponomarev, G. V., Gelfand, M. S., Kazanov, M. D., ... Noble, J. A. (2014). Evaluation and comparison of current fetal ultrasound image segmentation methods for biometric measurements: A grand challenge. *IEEE Transactions on Medical Imaging, 33*(4), 797–813. doi:10.1109/TMI.2013.2276943 PubMed

Rumelhart, D., Hinton, G. E., & Williams, R. J. (1986). Learning representations by backpropagating error. *Nature, 323*(6088), 533–536. doi:10.1038/323533a0

Russakovsky, O. D., Deng, J., Su, H., Krause, J., Satheesh, S., Ma, S., Huang, Z., Karpathy, A., Khosla, A., Bernstein, M., Berg, A. C., & Fei-Fei, L. (2015). ImageNet Large Scale Visual Recognition Challenge. *International Journal of Computer Vision, 115*(3), 211–252. doi:10.100711263-015-0816-y

Salakhutdinov, N. S. (2014). Dropout: A Simple Way to Prevent Neural Networks from Overfitting. *Journal of Machine Learning Research*, 1929–1958.

Sametinger, J., Rozenblit, J., Lysecky, R., & Ott, P. (2015). Security challenges for medical devices. *Communications of the ACM, 58*(4), 74–82. doi:10.1145/2667218

Sánchez, J., & Elías, M. (2007). Guidelines for designing mobility and orientation software for blind children. Proceedings of the IFIP Conference on Human-Computer Interaction, 10–14. doi:10.1007/978-3-540-74796-3_35

Sarasvathi, V., & Santhakumaran, A. (2011). Towards Artificial Neural Network Model to Diagnose Thyroid Problems, Global. *Journal of Computer Science and Technology, 11*, 53–55.

Saravanan, K., & Sasithra, S. (2014). Review on Classification Based on Artificial Neural Networks. *International Journal of Ambient Systems and Applications, 2*(4), 11–18. doi:10.5121/ijasa.2014.2402

Schempp, W. G. (1998). *Magnetic Resonance Imaging: Mathematical Foundations and Applications.* John Wiley and Sons.

Schier, R. (2018). Artificial intelligence and the practice of radiology: An alternative view. *Journal of the American College of Radiology, 15*(7), 1004–1007. doi:10.1016/j.jacr.2018.03.046 PMID:29759528

Schmidhuber, J. (2015). Deep learning in neural networks: An overview. *Neural Networks, 61*, 85–117. doi:10.1016/j.neunet.2014.09.003 PMID:25462637

Schneider, L. S., Tariot, P. N., Lyketsos, C. G., Dagerman, K. S., Davis, K. L., Davis, S., Hsiao, J. K., Jeste, D. V., Katz, I. R., Olin, J. T., Pollock, B. G., Rabins, P. V., Rosenheck, R. A., Small, G. W., Lebowitz, B., & Lieberman, J. A. (2001). National Institute of Mental Health Clinical Antipsychotic Trials of Intervention Effectiveness (CATIE): Alzheimer disease trial methodology. *American Journal of Geriatric Psychology, 9*(4), 346–360. doi:10.1097/00019442-200111000-00004 PMID:11739062

Schurgers, C., & Srivastava, M. B. (2001). Energy Efficient Routing in Wireless Sensor Networks. Proc. Communications for Network-Centric Operations: Creating the Information Force. IEEE Military Communications Conf. MILCOM, 2001(1), 357–361.

Selvadoss & Thanamani. (2018). *Investigating the Effects of Routing protocol and Mobility Model and their Interaction on MANET Performance Measures.* Academic Press.

Selvadoss, A., & Thanamani, P. M. (2016). *An Active Surveillance Approach for Performance Monitoring of Computer Networks.* Academic Press.

Sermanet, D. E. (2014). *OverFeat: Integrated Recognition, Localization and Detection using Convolutional Networks.* ICLR.

Shah, R. C. (2008). On the performance of bluetooth and ieee 802.15.4 radios in a body area network. *BODYNETS, 08*, 1–9.

Sharma, P. (2019). *Image Classification vs. Object Detection vs. Image Segmentation.* Analytics Vidhya.

Sharma, S., Schwemmer, P., Persson, C., Schoepf, A., Kepka, U. J., Hyun, C., Yang, D., & Nieman, K. (2018). Diagnostic accuracy of a machine-learning approach to coronary computed tomographic angiography-based fractional flow reserve: Result from the MACHINE consortium. *Circulation: Cardiovascular Imaging, 11*, e007217. PMID:29914866

Shelhamer, E., Long, J., & Darrell, T. (2016). *Fully Convolutional Networks for Semantic Segmentation,* arXiv: 1605.06211

Shen, D., Wu, G., & Il Shuk, H. (2017). Deep Learnoing in medical image analysis. *Annual Review of Biomedical Engineering, 19*(1), 221–248. doi:10.1146/annurev-bioeng-071516-044442 PMID:28301734

Shin, H. C., Roth, H. R., Gao, M., Lu, L., Xu, Z., Nogues, I., Yao, J., Mollura, D., & Summers, R. M. (2011). Deep convolutional neural networks for computer-aided detection: CNN architectures, dataset characteristics and transfer learning. *IEEE Transactions on Medical Imaging, 35*(5), 1285–1298. doi:10.1109/TMI.2016.2528162 PMID:26886976

Shi, X., Schillings, P., & Boyd, D. (2004). Applying Artificial Neural Networks and Virtual experimental design to quality improvement of two industrial processes. *International Journal of Production Research, 42*(1), 101–118. doi:10.1080/00207540310001602937

Shuiwang Ji Wei Xu, M. Y. (2012). 3D Convolutional Neural Networks for Human Action Recognition. *2nd International Conference of Signal Processing and Intelligent Systems (ICSPIS).*

Sichitiu, M. L. (2004). *Cross Layer Scheduling for Power Efficiency in Wireless Sensor Networks.* Academic Press.

Silva, B. M. C., Rodrigues, J. J. P. C., de la Torre Díez, I., López-Coronado, M., & Saleem, K. (2015). Mobile-health: A review of current state in 2015. *Journal of Biomedical Informatics, 56*, 265–272. doi:10.1016/j.jbi.2015.06.003 PMID:26071682

Silver, D., Huang, A., Maddison, C. J., Guez, A., Sifre, L., Van Den Driessche, G., Schrittwieser, J., Antonoglou, I., Panneershelvam, V., Lanctot, M., Dieleman, S., Grewe, D., Nham, J., Kalchbrenner, N., Sutskever, I., Lillicrap, T., Leach, M., Kavukcuoglu, K., Graepel, T., & Hassabis, D. (2016). Mastering the game of go with deep neural networks and tree search. *Nature, 529*(7587), 484–489. doi:10.1038/nature16961 PMID:26819042

Simonyan, K., & Zisserman, A. (2014). *Very Deep Convolutional Networks for Large-Scale Image Recognition.* arXiv 2014, arXiv:1409.1556

Sodhro, Luo, & Arunkumar. (2018). Mobile Edge Computing Based QoS Optimization in Medical Healthcare Applications. *International Journal of Information Management.* DOI: .2018.08.004 doi:10.1016/j.ijinfomgt

Steele, R., & Lo, A. (2013). Telehealth and ubiquitous computing for bandwidthconstrained rural and remote areas. *Personal and Ubiquitous Computing, 17*(3), 533–543. doi:10.100700779-012-0506-5

Sujitha, E., & Shanmugasundaram, K. (2017). Irrigation Management of Greenhouse Marigold Using Tensiometer: Effects on Yield and Water Use Efficiency. *International Journal of Plant and Soil Science, 19*(3), 1–9. doi:10.9734/IJPSS/2017/36437

Sukumar, S. R., Natarajan, R., & Ferrell, R. K. (2015). Quality of Big Data in health care. *International Journal of Health Care Quality Assurance, 28*(6), 621–634. doi:10.1108/IJHCQA-07-2014-0080 PMID:26156435

Suresh, A. (2017). Heart Disease Prediction System Using ANN, RBF and CBR. *International Journal of Pure and Applied Mathematics, 117*(21), 199–216.

Suresh, A., Udendhran, R., & Balamurgan, M. (2019). Hybridized neural network and decision tree based classifier for prognostic decision making in breast cancers. Springer - Journal of Soft Computing. doi:10.100700500-019-04066-4

Suresh, A., Udendhran, R., Balamurgan, M., & Varatharajan, R. (2019). A Novel Internet of Things Framework Integrated with Real Time Monitoring for Intelligent Healthcare Environment. Springer-Journal of Medical System, 43(6), 165. doi:10.1007/s10916-019-1302-9 PubMed

Suresh, A. (2017). Heart Disease Prediction System Using ANN, RBF and CBR. *International Journal of Pure and Applied Mathematics, 117*(21), 199–216.

Suresh, A., Udendhran, R., & Balamurgan, M. (2019). A Novel Internet of Things Framework Integrated with Real Time Monitoring for Intelligent Healthcare Environment. *Springer-Journal of Medical System*, *43*, 165. doi:10.100710916-019-1302-9

Suzuki, K. (2017). Survey of deep learning applications to medical image analysis. *Med Imaging Technol*, *35*, 212–226.

Szegedy, C., Liu, W., Jia, Y., Sermanet, P., Reed, S. E., Anguelov, D., Erhan, D., Vanhoucke, V., & Rabinovich, A. (2014). *Going Deeper with Convolutions*. arXiv 2014, arXiv:1409.4842

Szegedy, C., Liu, W., Jia, Y., Sermanet, P., Reed, S., Anguelov, D., Erhan, D., Vanhoucke, V., & Rabinovich, A. (2015). Going deeper with convolutions. *IEEE Conference on Computer Vision and Pattern Recognition (CVPR)*, 1 - 9. doi:doi:10.1109/CVPR.2015.7298594

Tanaka, A., Utsunomiya, F., & Douseki, T. (2016). Wearable self-powered diaper-shaped urinary-incontinence sensor suppressing response-time variation with 0.3 V start-up converter. *IEEE Sensors Journal*, *16*(10), 3472–3479. doi:10.1109/JSEN.2015.2483900

Tang, A., Tam, R., Cadrin-Chênevert, A., Guest, W., Chong, J., Barfett, J., Chepelev, L., Cairns, R., Mitchell, J. R., Cicero, M. D., Poudrette, M. G., Jaremko, J. L., Reinhold, C., Gallix, B., Gray, B., Geis, R., O'Connell, T., Babyn, P., Koff, D., ... Shabana, W. (2018). Canadian association of radiologists white paper on artificial intelligence in radiology. *Canadian Association of Radiologists Journal*, *69*(2), 120–135. doi:10.1016/j.carj.2018.02.002 PMID:29655580

Tan, W. R., Chan, C. S., Aguirre, H. E., & Tanaka, K. (2017). ArtGAN: artwork synthesis with conditional categorical GANs. *2017 IEEE International Conference on Image Processing (ICIP)*, 3760–4. 10.1109/ICIP.2017.8296985

Tapu, R., Mocanu, B., & Tapu, E. (2014) A survey on wearable devices used to assist the visual impaired user navigation in outdoor environments. Proceedings of the 2014 11th International Symposium on Electronics and Telecommunications (ISETC). doi:10.1109/ISETC.2014.7010793

Tashev. (n.d.). Recent Advances in Human-Machine Interfaces for Gaming and Entertainment. Int'l J. Information Technology and Security, 3(3), 6976.

Teodorrescu, T., Kandel, A., & Jain, L. C. (1998). *Fuzzy and Neuro-fuzzy systems in medicine*. CRC Press.

Thall, P. F., Millikan, R. E., & Sung, H.-G. (2000). Evaluating multiple treatment courses in clinical trials. *Statistics in Medicine*, *19*(8), 1011–1028. doi:10.1002/(SICI)1097-0258(20000430)19:8<1011::AID-SIM414>3.0.CO;2-M PMID:10790677

Thall, P. F., & Wathen, J. K. (2005). Covariate-adjusted adaptive randomization in a sarcoma trial with multistate treatments. *Statistics in Medicine*, *24*(13), 1947–1964. doi:10.1002im.2077 PMID:15806621

The Cloud Standards Customer Council (CSCC). (2016). *Impact of cloud computing on healthcare*. Reference architecture, Version 1.0. Available: http://cloud-council.org

Thevenot, D. R., Toth, K., Durst, R. A., & Wilson, G. S. (1999). Electrochemical Biosensors: Recommended Definitions and Classification. *Pure and Applied Chemistry*, *7*, 2333–2348.

Tiponut, V., Popescu, S., Bogdanov, I., & Caleanu, C. (2008). Obstacles Detection System for Visually-impaired Guidance. New Aspects of system. Proceedings of the 12th WSEAS International Conference on SYSTEMS, 350–356.

Tiponut, V., Ianchis, D., Bash, M., & Haraszy, Z. (2011). Work Directions and New Results in Electronic Travel Aids for Blind and Visually Impaired People. *Latest Trends Syst.*, *2*, 347–353.

Topol, E. (2018). *The Creative Destruction of Medicine*. Basic Books.

Turing, A. M. (1950). Computing machinery and intelligence. Mind. *New Series.*, *59*(236), 433–460. doi:10.1093/mind/LIX.236.433

Udendhran, R. (2017). A Hybrid Approach to Enhance Data Security in Cloud Storage. ICC '17 Proceedings of the Second International Conference on Internet of things and Cloud Computing at Cambridge University, United Kingdom. Doi:10.1145/3018896.3025138

Udendhran, R., & Balamurgan, M. (2020). An Effective Hybridized Classifier Integrated with Homomorphic Encryption to Enhance Big Data Security. In *EAI International Conference on Big Data Innovation for Sustainable Cognitive Computing*. EAI/Springer Innovations in Communication and Computing. 10.1007/978-3-030-19562-5_35

Vahdatpour, A., Amini, N., Xu, W., & Sarrafzadeh, M. (2011). Accelerometer-based on-body sensor localization for health and medical monitoring applications. *Pervasive and Mobile Computing*, *7*(6), 746–760. doi:10.1016/j.pmcj.2011.09.002 PubMed

van den Heuvel, T. L. A., de Bruijn, D., de Korte, C. L., & van Ginneken, B. (2018). Automated measurement of fetal head circumference using 2D ultrasound images. *PLoS One*, *13*(8). doi:10.1371/journal.pone.0200412

van Gerven, M., & Bohte, S. (2017). Editorial: Artificial Neural Networks as Models of Neural Information Processing. *Frontiers in Computational Neuroscience*, *11*, 114–114. doi:10.3389/fncom.2017.00114 PMID:29311884

Van Leemput, K., Maes, F., Vandermeulen, D., & Suetens, P. (1999). Automated model-based tissue classification of MR images of the brain. *IEEE Transactions on Medical Imaging*, *18*(10), 897–908. doi:10.1109/42.811270 PMID:10628949

Vaquero, L. M., & Rodero-Merino, L. (2014). Finding your way in the fog. *ACM SIGCOMM Comput. Commun. Rev.*, *44*(5), 27–32. doi:10.1145/2677046.2677052

Vayena, E., Salathé, M., Madoff, L. C., & Brownstein, J. S. (2015). Ethical challenges of big data in public health. *PLoS Computational Biology*, *11*(2), e1003904. doi:10.1371/journal.pcbi.1003904 PMID:25664461

Vest, J. R., & Gamm, L. D. (2010). Health information exchange: Persistent challenges and new strategies. *Journal of the American Medical Informatics Association*, *17*(3), 288–294. doi:10.1136/jamia.2010.003673 PMID:20442146

Viceconti, M., Henney, A., & Morley-Fletcher, E. (2016). *In silico clinical trials: how computer simulation will transform the biomedical industry*. Avicenna Consortium.

Vidhya Saraswathi, K. P. P. (2017). A Survey of Content based Video Copy Detection using. *Big Data*, *3*(5), 114–118.

Vinge, V. (1993). The coming technological singularity: How to survive in the post-human era. NASA, Lewis Research Center. *Vision (Basel)*, *21*, 11–22.

Wac, K., Bargh, M. S., Beijnum, B. J. F. V., Bults, R. G. A., Pawar, P., & Peddemors, A. (2009). Power- and delay-awareness of health telemonitoring services: The mobihealth system case study. *IEEE Journal on Selected Areas in Communications*, *27*(4), 525–536. doi:10.1109/JSAC.2009.090514

Wang, P., & Liu, Q. (2011). Biomedical Sensors and Measurement. Zheijang University. doi:10.1007/978-3-642-19525-9

Wang, W., & Krishnan, E. (2014). Big data and clinicians: A review on the state of the science. *JMIR Medical Informatics*, *2*(1), e1. doi:10.2196/medinform.2913 PMID:25600256

Ward, J. S., & Barker, A. (2013). *Undefined by data, a survey of big data definitions*. arXiv preprint arXiv:1309.5821

Wegmueller, M. S., Kuhn, J. F. A., Oberle, M., Felber, N., Kuster, N., & Fichtner, W. (2007). An attempt to model the human body as a communication channel. *IEEE Transactions on Biomedical Engineering*, *54*(10), 1851–1857. doi:10.1109/TBME.2007.893498 PubMed

Werbos, P. J. (1981). Applications of advances in nonlinear sensitivity. *Proceedings of the 10th IFIP Conference*, 762-770.

Wolterink, J. M., Leiner, T., de Vos, B. D., Coatrieux, J. L., Kelm, B. M., Kondo, S., Salgado, R. A., Shahzad, R., Shu, H., Snoeren, M., Takx, R. A., van Vliet, L. J., van Walsum, T., Willems, T. P., Yang, G., Zheng, Y., Viergever, M. A., & Išgum, I. (2016). An evaluation of automatic coronary artery calcium scoring methods with cardiac CT using the orCaScore framework. *Medical Physics*, *43*(5), 2361–2373. doi:10.1118/1.4945696 PMID:27147348

Wu, G., Kim, M., Wang, Q., Munsell, B. C., & Shen, D. (2016). Scalable High-Performance Image Registration Framework by Unsupervised Deep Feature Representations Learning. *IEEE Transactions on Biomedical Engineering*, *63*(7), 1505–1516. doi:10.1109/TBME.2015.2496253 PMID:26552069

Würfl, T., Hoffmann, M., Christlein, V., Breininger, K., Huang, Y., Unberath, M., & Maier, A. K. (2018). Deep learning computed tomography: Learning projection-domain weights from image domain in limited angle prob- lems. *IEEE Transactions on Medical Imaging*, *37*(6), 1454–1463. doi:10.1109/TMI.2018.2833499 PMID:29870373

Xie, S., Girshick, R., Dollar, P., Tu, Z., & He, K. (2016). *Aggregated Residual Transformations for Deep Neural Networks*. https://openaccess.thecvf.com/content_cvpr_2017/papers/Xie_Aggregated_Residual_Transformations_CVPR_2017_paper.pdf

Xu, K., Li, Y., & Ren, F. (2016). An energy-efficient compressive sensing framework incorporating online dictionary learning for long-term wireless health monitoring. *Proc. IEEE Int. Conf. Acoust., Speech Signal Process*, 804–808. 10.1109/ICASSP.2016.7471786

Xu, L., Tetteh, G., Lipkova, J., Zhao, Y., Li, H., Christ, P., Piraud, M., Buck, A., Shi, K., & Menze, B. H. (2018). Automated Whole-Body Bone Lesion Detection for Multiple Myeloma on 68Ga-Pentixa for PET/CT Imaging Using Deep Learning Methods. *Contrast Media & Molecular Imaging*, *23*, 919–925.

Yerramala. (2014). Properties of concrete with eggshell powder as cement replacement. *Indian Concrete Journal*.

Ye, Y., & Ju, H. (2005). Rapid Detection of ssDNA and RNA Using Multi-Walled Carbon Nanotubes Modified Screen-Printed Carbon Electrode. *Biosensors & Bioelectronics*, *21*(5), 735–741. doi:10.1016/j.bios.2005.01.004 PubMed

Yin, Peng, Li, Zhang, You, Fischer, Furth, Tasian, & Fan. (2018). Automatic kidney segmentation in ultrasound images using subsequent boundary distance regression and pixel wise classification networks. . doi:10.1016/j.media.2019.101602

Yinan, Y., Jiajin, L., Wenxue, Z., & Chao, L. (2016). Target classification and pattern recognition using micro-Doppler radar signatures. *Software Engineering, Artificial Intelligence, Networking, and Parallel/Distributed Computing, Seventh ACIS International Conference on*, 213-217.

Yin, Y., Zeng, Y., Chen, X., & Fan, Y. (2016). The Internet of Things in healthcare: An overview. *J. Ind. Inf. Integr.*, *1*, 3–13. doi:10.1016/j.jii.2016.03.004

Yogesh, H., Ngadiran, A., Yaacob, B., & Polat, K. (2017). A new hybrid PSO assisted biogeography-based optimization for emotion and stress recognition from speech signal. *Expert Systems with Applications*, *69*, 149–158. doi:10.1016/j.eswa.2016.10.035

Zeiler, M. D., & Fergus, R. (2013). *Visualizing and Understanding Convolutional Networks*. https://cs.nyu.edu/~fergus/papers/zeilerECCV2014.pdf

Zemouri, R., Zerhouni, N., & Racoceanu, D. (2019). Deep learning in the biomedical Applications: Recent and Future status. *Applied Sciences (Basel, Switzerland)*, *9*(8), 1526. doi:10.3390/app9081526

Zhang, R., Bernhart, S., & Amft, O. (2016). Diet eyeglasses: Recognising food chewing using emg and smart eyeglasses. *Proc. IEEE 13th Int. Conf. Wearable Implant. Body Sensor Netw.*, 7–12. 10.1109/BSN.2016.7516224

Zhang, K., Liang, X., Baura, M., Lu, R., & Shen, X. (2014). PHDA: A priority based health data aggregation with privacy preservation for cloud assisted WBANs. *Inf. Sci.*, *284*, 130–141. doi:10.1016/j.ins.2014.06.011

Zhang, Z. (2014). Big data and clinical research: Perspective from a clinician. *Journal of Thoracic Disease*, *6*(12), 1659–1664. PMID:25589956

Zhao, Z., Zheng, P., Xu, S., & Wu, X., (2017). Object Detection with Deep Learning: A Review. *IEEE Transactions on Neural Networks and Learning Systems for Publication*.

Zheng, Q., Furth, S. L., Tasian, G. E., & Fan, Y. (2019). Computer aided diagnosis of congenital abnormalities 28 of the kidney and urinary tract in children based on ultrasound imaging data by integrating texture image 29 features and deep transfer learning image features. *Journal of Pediatric Urology*, *15*(1), 75.e71–75.e77. doi:10.1016/j.jpurol.2018.10.020 PubMed

Zheng, Q., Warner, S., Tasian, G., & Fan, Y. (2018b). A Dynamic Graph Cuts Method with Integrated Multiple 2 Feature Maps for Segmenting Kidneys in 2D Ultrasound Images. *Academic Radiology*, *25*(9), 1136–1145. doi:10.1016/j.acra.2018.01.004 PubMed

Zhong, X., Bayer, S., Ravikumar, N., Strobel, N., Birkhold, A., & Kowarschik, M. (2018). Resolve intraoperative brain shift as imitation game. MIC- CAI Challenge 2018 for Correction of Brainshift with Intra-Operative Ultrasound (CuRIOUS 2018).

Zhou, L., Zhang, C., & Wu, M. (2018). D-linknet: Linknet with pretrained encoder and dilated convolution for high resolution satellite imagery road extraction. Proceedings of the IEEE Conference on Computer Vision and Pattern Recognition Workshops, 182–186. doi:10.1109/CVPRW.2018.00034

Zhou, S., Greenspan, H., & Shen, D. (2017). *Deep Learning for Medical Image Analysis*. Academic Press.

Zisserman, K. S. (2015). *Very deep convolutional networks for large-scale image recognition*. ICLR.

Zreik, M., van Hamersvelt, R. W., Wolterink, J. M., Leiner, T., Viergever, M. A., & Išgum, I. (2018). *Automatic detection and characterization of coronary artery plaque and stenosis using a recurrent convolutional neural network in coronary CT angiography*. Academic Press.

Zreik, M., Lessmann, N., van Hamersvelt, R. W., Wolterink, J. M., Voskuil, M., Viergever, M. A., Leiner, T., & Išgum, I. (2018). Deep learning analysis of the myocardium in coronary CT angiography for identification of patients with functionally significant coronary artery stenosis. *Medical Image Analysis*, *44*, 72–85. doi:10.1016/j.media.2017.11.008 PMID:29197253

About the Contributors

A. Suresh, B.E., M.Tech., Ph.D works as the Professor & Head, Department of the Computer Science and Engineering in Nehru Institute of Engineering & Technology, Coimbatore, Tamil Nadu, India. He has been nearly two decades of experience in teaching and his areas of specializations are Data Mining, Artificial Intelligence, Image Processing, Multimedia and System Software. He has published two patents and 90 papers in International journals. He has book authored "Industrial IoT Application Architectures and use cases" published in CRC press and edited book entitled "Deep Neural Networks for Multimodal Imaging and Biomedical Application" published in IGI Global. He has currently editing three books namely "Deep learning and Edge Computing solutions for High Performance Computing" in EAI/Springer Innovations in Communications and Computing, "Sensor Data Management and Analysis: The Role of Deep Learning" and "Bioinformatics and Medical Applications: Big Data using Deep Learning Algorithms" in Scrivener-Wiley publisher. He has published 15 chapters in the book title An Intelligent Grid Network Based on Cloud Computing Infrastructures in IGI Global Publisher and Internet of Things for Industry 4.0 in EAI/Springer Innovations in Communication and Computing. He has published more than 40 papers in National and International Conferences. He has served as editor / reviewer for Springer, Elsevier, Wiley, IGI Global, IoS Press, Inderscience journals etc... He is a member of ISTE, MCSI, IACSIT, IAENG, MCSTA and Global Member of Internet Society (ISOC). He has organized several National Workshop, Conferences and Technical Events. He is regularly invited to deliver lectures in various programmes for imparting skills in research methodology to students and research scholars. He has published four books, in the name of Data structures & Algorithms, Computer Programming, Problem Solving and Python Programming and Programming in "C" in DD Publications, Excel Publications and Sri Maruthi Publisher, Chennai, respectively. He has hosted two special sessions for IEEE sponsored conference in Osaka, Japan and Thailand.

R. Udendhran computer science researcher on Deep learning. He worked as a data scientist and presented research work in an international conference held at the University of Cambridge which is available in the ACM digital library. He has published research papers in international journals and his research area extensively focuses deep learning. He has completed M.Tech in computer science and engineering. His research work focuses on deep learning and cryptography.

S. Vimal is working in Department of Information Technology, National Engineering College, Kovilpatti, Tamilnadu, India. He has around Fourteen years of teaching experience and research experience. He is a EMC certified Data science Associate and CCNA certified professional too. He holds a Ph.D in Information and Communication Engineering from Anna University Chennai and he received

Masters Degree from Anna University Coimbatore. He is a member of various professional bodies and organized various funded workshops ,seminars from DST,DBT.AICTE. He has wide publications in the highly impact journals in the area of Data Analytics, Networking and Security issues and published 04 book chapters. He has hosted two special session for IEEE sponsored conference in Osaka, Japan and Thailand. He has acted as Session chair,organizing committee member, advisory committee and outreach committee member in various international conferences supported by IEEE. His areas of interest include Game Modelling, Artificial Intelligence, Cognitive radio networks, Network security, Machine Learning and Big data Analytics. He is a Senior member in IEEE and holds membership in various professional bodies. He has served as Guest Editor, reviewer for IEEE,Springer, Elsevier and Wiley journals.

* * *

Reyana A. is currently working as an Assistant Professor in the Department of Computer Science and Engineering at Nehru Institute of Engineering and Technology, Coimbatore, Tamilnadu, India. She is pursuing her doctoral degree in the field of Information and Communication Engineering at Anna University, Chennai. Has done her Masters in Computer Science & Engineering and research interests include Multi-sensor fusion, Internet of Things, Wireless Sensor Networks and Automata Theory. Has a decade of experience in academia with publications in Scopus indexed international, reviewer for Scopus international journals and a Member of IEEE and IAENG.

Sivakami A. is working as an Assistant Professor in Department of Physics, Bharat Institute of Engineering & Technology, India for the past two years. Her research interests include Nanotechnology, IoT and Deep learning.

M. S. Irfan Ahamed is currently working as Associate Professor, Department of Computer & Information Sciences, College of Science and Arts, Al Ula Branch, Taibah University, Saudi Arabia.

Hmidi Alaeddine received his Master's degree in embedded electronic systems from the University of Sousse, Tunisia, in 2017. He is currently pursuing a Ph.D. degree at the Faculty of Sciences, Monastir University, Monastir, Tunisia. His research interests are related to, deep neural network, Classification, reconfigurable architectures, network on chip, and image processing.

Ahmed Alenezi is currently working as Assistant Professor, Department of Computer & Information Sciences, College of Science and Arts, Al Ula Branch, Taibah University, Saudi Arabia.

Balamurugan is the professor and Head of Department of Computer Science and Engineering at Bharathidasan University. He has completed his PhD and guided 10 research scholars. His area of research includes cognitive science and data science. He has completed UGC funded research project and has a patent and published 30 papers in reputed journals. He is the board of member in different autonomous colleges.

Gopinath Ganapathy pursed Bachelor of Science and Master of computer applications from St. Joseph's College, Trichy and completed Ph.D from Madurai Kamaraj University. Currently working as a Registrar, Bhararhidasan University, Trichy. He has 30 years of total experience in academia, Industry, research &consultancy services. He has around 8 year International experience in the U.S and U.K. He served as a Consultant for a few fortune500 companies that include IBM, Lucent-Bell Labs, Merrill Lynch, Toyota, etc. He received the Young Scientist Fellow award for the year 1994 from the Govt. of Tamil Nadu with cash Rs 20000. His name is listed in "Who is Who in the World – 2009" by Marquis, USA for individual accomplishment He is a nominee for "Rashtriya Gouvrav" award. He is the recipient of "Best Citizens of India" award. He published and presented nearing100 research papers at international journals and conferences. He acted as academic advisor for mediating off-campus programs in University of Valley Creek, USA He is a member in several academic and technology councils in various Universities in India He convened many international conferences/workshops/seminars. He is a referee and editorial member in a few international journals. He is a Life Member in Indian Science Congress, Indian Society for Technical Education, and Computer Society of India.

Harini pursued bachelor of engineering from National Engineering College and currently pursuing Master of Engineering in Electronics and Instrumentation Department at National Engineering College. She has presented research work in international conferences and research publications included in Scopus indexed database and approved UGC journals and ongoing book chapters to be published by IGI Global and Wiley.

Muralikrishna Iyyanki, is Chief Advisor, UC Berkeley Andhra Smart Village Program. Former Raja Ramanna DRDO Distinguished Fellow, RCI, Ministry of Defence, Govt of India (2014-17). Professor of Excellence, Chiba University, Chiba, Japan. He is an Adjunct Professor, Asian Institute of Technology -Bangkok, Thailand. Member, GIS Academia Council of India. Professor and Head Spatial Information Tech. & Director [R&D] & Head, Centre for Atmospheric Sciences and Weather Modification Technologies, JNTUH, India. FIE, FIS, FAPAS, FISG, FIGU www.linkedin.com/in/iyyanki-v-muralikrishna.

Prisilla Jayanthi is Assistant Professor at K G Reddy College of Engineering and Technology, Hyderabad. She has done M.Tech in Computer Networks and Information Security. She has published International Conference Papers- 7 Springer and 2 IEEE and 1 ACM.

Balamurugan K. S. is working as an Associate professor in department of ECE, Bharat institute of Engineering and Technology, India for the past two years.

P. Karthika has received a BCA degree in 2008 from Trinity College for Women, Namakkal affiliate to Periyar University, Salem, India. She has received a MCA degree in 2011 from M. Kumarasamy College of Engineering, Karur affiliated from Anna University, Chennai. Currently, she is a full time research in the Department of Computer Applications in Kalasalingam Academy of Research and Education, Krishnankoil, TamilNadu. Research interests include Network Security and Steganography. Her research focus of Machine Learning IoT Image Security in Content based Video copy Detection for frame based fusion techniques using steganography. Her paper publication focus on mainly on peer-reviewed international journals indexed SCI, SCIE, and SCOPUS. She has membership of the International Association of Engineers (IAENG) has been received, and then The Society of Digital Information and Wireless Communications (SDIWC), and as well as the member of Universal Association of Arts And Management Professionals has been approved under Associate Member category (IRED) and also the member of International Computer Science and Engineering (ICSEC). She has worked for an organization member in "International Conference on Applied Soft Computing Technique (ICASCT'18)" during March 23-24, 2018 at Kalasalingam University, Krishnankoil, and TamilNadu, India. And also she has another one conference for organization member in "International Conference on Advances in Computer Science, Engineering and Technology (ICACSET' 18)" at Kalasalingam University, Krishnankoil, and TamilNadu, India.

Prathap M. had completed Ph.D from Bharathiar University and he is currently working as professor of Department of Information Technology, Wollo University, Ethiopia. He has more than 22 years of experience in Education. His specialization is Computer Networks, MANET, Network programming and simulations. Published a book title "Cracking IT Interview & Winning the Right Job "with ISBN number published at India. He had published more than 24 research papers in research journals like Springer, Scopus, etc. He offered many seminars in many universities and reputed colleges. His Project Costing USD 1, 31,600.00 was selected and sanctioned by the World Bank. He designed a latest curriculum for Computer Science and Engineering program under the ECBP Engineering Capacity Building Program and accredited by ACQUINN, Germany. He is a member of many professional organizations which include Membership in IEEE, Science Alert, CINA, INFLIBNET of UGC and others.

Jihene Malek was born in Tunis, on February 1 1973. She received the B.S. degree in Physics from Monastir University, in 1994; the M.S. and the Ph.D. degree in Electronics and Microelectronics from Faculte des Sciences de Monastir in 1996 and 2002 respectively. Since 2002 she has been maitre-assistant in Electronics and Microprocessors with the Electronics department, Institut Supérieur des Sciences Appliquées et de Technologie de Sousse. Her current research interests include biomedical engineering, medical image processing and visualization, code design, deep learning and big data.

V. R. S. Mani has completed his Ph.D. degree (2017) in Information and Communication Engineering and M.E. in Communication System (1994) from Anna University Chennai, in 1994. Between 1994-2017, worked as a Lecturer, Senior Lecturer, Assistant Professor, Associate Professor and currently working as Associate Professor (Senior Grade)in the Department of Electronics and Communication Engineering, National Engineering College Kovilpatti, Tamil Nadu. Specialized in, Machine Learning, Radar Signal and Image Processing. Published around 25 research articles in International and National level Journals and Conferences. Life member of IETE and ISTE.

Muthukumari pursued Master of Philosophy in Computer Science from Bharathidasan University, Trichy and currently pursuing Ph.D. in Department of Computer Science, Bharathidasan University. She has presented research work in international conferences and research publications included in Scopus indexed database and approved UGC journals.

Sudhagar Pitchaimuthu is a Sêr Cymru II-Rising Star Fellow awarded by European Union through Welsh Government. Currently, Dr. Sudhagar leads the "Multifunctional Photocatalyst & Coating" research group at SPECIFIC, College of Engineering, Swansea University, United Kingdom. His research background is on 'nanomaterial synthesis and coatings' towards constructing low-cost solar energy-driven photoelectrochemical cell. Mainly focuses on how transforming the abundance, sustainable solar energy into multifunctional applications such as energy conversion devices (solar cells and solar fuel cells), environmental clean-up (photocatalytic water treatment, air purification), and biosensors. His research group is currently developing a) Semiconductor thin film-based solar light photoabsorbers (metal oxide, metal chalcogenides), b) Quantum dot solids and films, c) Pt-free water oxidation/reduction catalyst and (d) Artificial Intelligence in Solar Catalysis. Dr. Sudhagar graduated from Physics discipline, Bharathiar University, India in 2009. He was served as Research Assistant Professor during 2009-2013 under Prof. Yong Soo Kang at Centre for Next Generation Dye-sensitized Solar Cells in Hanyang University, South Korea. Concurrently, he has been serving as a visiting scientist to the Department of Applied Physics, University of Jaume I, Spain from 2011. From 2013 November– 2015 October, he was working as JSPS Research Fellow under Prof. Akira Fujishima (original inventor of photoelectrocatalytic water splitting phenomena) at International Photocatalytic Research Center, Tokyo University of Science, Noda, Japan. Dr. Sudhagar has been published 93 articles in peer-reviewed journals (citations: 3208, h index: 32), and contributed 6 book chapters. Also, he has presented more than 50 articles in international conferences. He is the Founder and Chair for ANEH Conference Series and also he has been served as co-convenor for 5 international conferences and workshops. Currently, Dr. Sudhagar chairing the KSR Global Nano Research Network (GNRN) and India-UK Joint Research Centre.

J. Preethi is currently working as Professor in the Department of Computer Science and Engineering at Anna University Regional Campus, Coimbatore, Has more than a decade of experience in teaching & coordinating research activities. Has published twenty plus International Journals, is the Recognized supervisor for guiding Ph.D scholars of Anna University under the faculty of Information and Communication Engineering. Her research areas include Image Processing and Soft Computing. Acting as Director (i/c) for Centre for University Industry Collaboration, Anna University.

Dhinesh Kumar R. is currently working as Senior Consultant. He has more than ten years of industrial work experience. His is a senior consultant for projects in Capgemini.

K. M. Karthick Raghunath has received his B Tech. in Information Technology from Anna University in 2008 and M.E. in Pervasive Computing Technology from Anna University (BIT Campus) in 2011. In 2019, he has received his Ph.D. degree from Anna University, Chennai. His research interests include Wireless Sensor Networks, Pervasive/Ubiquitous Computing, and Embedded Systems. He is a lifetime member of ACM, Institute of Engineers (IE), IAENG and IACSIT.

George Dharma Prakash Raj is currently working as an Assistant Professor in the Department of Computer Science and Engineering and Applications, Khajamalai Campus, Bharathidasan University Tamilnadu, India. He has authored and co-authored several papers in various reputed journals and conference proceedings. He is a Senior Member in the Association of Computer Electronics and Electrical Engineers,1133 Broadway, Suite 706, New York, Ny 10010, USA. He was one of technical Program Committee in various National and International Conferences conducted in India, USA, United Kingdom, Czech Republic, Jordan, Tanzania, Libya, Thailand, Malaysia, Lithuania, Poland, Bahrain, Hong Kong, South Korea, Dubai, Turkey, Switzerland, Japan, Lebanon, Thailand, China, France and Bangkok.

Sebastin S. was born in Tirunelveli, India. He received the B.E. degree in civil engineering from Government College of Engineering, Tirunelveli, India in 2009, and the M.E degree in Structural Engineering from Regional Centre of Anna University, Tirunelveli, India, in 2012. In 2014, he joined the

Department of Civil Engineering, National Engineering College, Kovilpatti as an Assistant Professor. His current research interest is Advanced concrete Technology.

Murali Ram Kumar S. M. was born in Tuticorin, India. He received the B.E. degree in civil engineering from National Engineering College, Kovilpatti, India in 2017, and the M.E degree in Structural Engineering from Algappa Chettiar Government College of Engineering and Technology, Karaikudi, India, in 2019. In 2019, he joined the Department of Civil Engineering, National Engineering College, Kovilpatti as an Assistant Professor. His current research interests include No-fine concrete, High Performance Concrete and Sustainable Materials.

Bagyalakshmi Shanmugam is working as an Assistant Professor in Physics, Sri Ramakrishna Institute of Technology, India.

Hussain Sharif has professional experience of 13 Years. His specialization is Computer Networks. He published six research papers apart from presentations in various national and international conferences. He organized various workshops related to his field. He offered many seminars in many universities and reputed colleges. He is a member of many professional organizations. He guided many projects. Has been the ISTE coordinator and motivated the students to get registrations for opening the ISTE (Indian Society for Technical Education) Students Chapter in the college. He completed his Ph.D in 2018 and currently working as lecturer in Department of information Technology Kombolcha, Amhara, Ethiopia.

Carmel Sobia is working as Assistant Professor (Senior Grade) of Electronics and Instrumentation Department at National Engineering College with 10 years of experience. she has completed his Ph.D. She had guided several sponsored projects and also have several publications in reputed journals.

V. Suresh is working as the Associate professor of Electronics and Instrumentation Department at National Engineering College with 16 years experience. He has completed his Ph.D. and also completed his funded project under the Board of Research in Nuclear Science (BRNS), Mumbai. He has several International Journal publications and also a reviewer for De Gruyters Journal. He has published several papers in Measurement Science Review-Journal and Elsevier journals.

Krishnaprasath V. T. is currently working as an Assistant Professor in the Department of Computer Science and Engineering at Nehru Institute of Engineering and Technology, Coimbatore, Tamilnadu. He is pursuing doctoral degree in the field of Information and Communication Engineering at Anna University, Chennai. Has done her Masters in Computer Science & Engineering and research interests include Wireless sensor networks and automata theory. Has more than a decade of experience in academia with publications in Scopus indexed international, reviewer for Scopus international journals and a Member of IEEE and IAENG.

G. Yamini pursed Bachelor of Science, Master of Science, Master of Philosophy in Computer Science from Bharathidasan University, Trichy in1998, 2001 and 2004 respectively. Currently pursuing Ph.D. in Department of Computer Science, Bharathidasan University, Trichy, since 2011. She has presented many papers in conferences and participated in various seminars and workshops in various institutions in and around Trichy. Guided around 20 M.Phil graduates. Her main research work focuses on Software Metrics, Software Reliability, Prediction of reliability in software and has 12+ years of teaching experience and 2 years of research experience.

Index

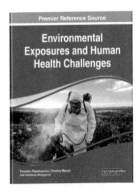

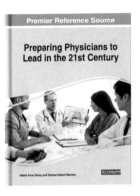

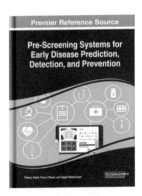
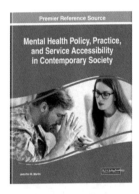

Ensure Quality Research is Introduced to the Academic Community

Become an IGI Global Reviewer for Authored Book Projects

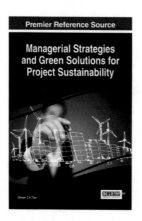

The overall success of an authored book project is dependent on quality and timely reviews.

In this competitive age of scholarly publishing, constructive and timely feedback significantly expedites the turnaround time of manuscripts from submission to acceptance, allowing the publication and discovery of forward-thinking research at a much more expeditious rate. Several IGI Global authored book projects are currently seeking highly-qualified experts in the field to fill vacancies on their respective editorial review boards:

Applications and Inquiries may be sent to:
development@igi-global.com

Applicants must have a doctorate (or an equivalent degree) as well as publishing and reviewing experience. Reviewers are asked to complete the open-ended evaluation questions with as much detail as possible in a timely, collegial, and constructive manner. All reviewers' tenures run for one-year terms on the editorial review boards and are expected to complete at least three reviews per term. Upon successful completion of this term, reviewers can be considered for an additional term.

If you have a colleague that may be interested in this opportunity,
we encourage you to share this information with them.

IGI Global Proudly Partners With eContent Pro International

Receive a 25% Discount on all Editorial Services

Editorial Services

IGI Global expects all final manuscripts submitted for publication to be in their final form. This means they must be reviewed, revised, and professionally copy edited prior to their final submission. Not only does this support with accelerating the publication process, but it also ensures that the highest quality scholarly work can be disseminated.

English Language Copy Editing

Let eContent Pro International's expert copy editors perform edits on your manuscript to resolve spelling, punctuaion, grammar, syntax, flow, formatting issues and more.

Scientific and Scholarly Editing

Allow colleagues in your research area to examine the content of your manuscript and provide you with valuable feedback and suggestions before submission.

Figure, Table, Chart & Equation Conversions

Do you have poor quality figures? Do you need visual elements in your manuscript created or converted? A design expert can help!

Translation

Need your documjent translated into English? eContent Pro International's expert translators are fluent in English and more than 40 different languages.

Hear What Your Colleagues are Saying About Editorial Services Supported by IGI Global

"The service was very fast, very thorough, and very helpful in ensuring our chapter meets the criteria and requirements of the book's editors. I was quite impressed and happy with your service."

– Prof. Tom Brinthaupt,
Middle Tennessee State University, USA

"I found the work actually spectacular. The editing, formatting, and other checks were very thorough. The turnaround time was great as well. I will definitely use eContent Pro in the future."

– Nickanor Amwata, Lecturer,
University of Kurdistan Hawler, Iraq

"I was impressed that it was done timely, and wherever the content was not clear for the reader, the paper was improved with better readability for the audience."

– Prof. James Chilembwe,
Mzuzu University, Malawi

Email: customerservice@econtentpro.com   **www.igi-global.com/editorial-service-partners**

Printed in the United States
By Bookmasters